## Advances in Karst Science

**Series Editor**

James LaMoreaux, Tuscaloosa, AL, USA

This book series covers advances in the field of karst from a variety of perspectives to facilitate knowledge and promote interaction thereby building stepping stones in this process. Methodologies, monitoring, data analysis and interpretation are addressed throughout the wide range of climatic, geological and hydrogeological contexts in which karst occurs. Case studies are presented to provide examples of advancement of the science.

Issues to be addressed include water supply, contamination, and land use management. These issues although occurring on a local basis share many of the same features on the global stage. This book series is a critical resource to the scientific community allowing them to compare problems, results, and solutions. The presented information can be utilized by decision makers in making decisions on development in karst regions. Contributions presented may be used in the classroom and to work with stakeholders, scientists, and engineers to determine practical applications of these advances to common problems worldwide.

The series aims at building a varied library of reference works, textbooks, proceedings, and monographs, by describing the current understanding of selected themes. The books in the series are prepared by leading experts actively working in the relevant field. The book series *Advances in Karst Science* includes single and multi-authored books as well as edited volumes. The Series Editor, Dr. James W. LaMoreaux, is currently accepting proposals and a proposal document can be obtained from the Publisher.

Bartolomé Andreo • Juan Antonio Barberá •
Juan José Durán-Valsero •
José Manuel Gil-Márquez •
Matías Mudarra
Editors

# EuroKarst 2022, Málaga

## Advances in the Hydrogeology of Karst and Carbonate Reservoirs

Springer

*Editors*
Bartolomé Andreo
Department of Geology
University of Málaga
Málaga, Spain

Juan Antonio Barberá
Department of Geology
University of Málaga
Málaga, Spain

Juan José Durán-Valsero
Geological Survey of Spain
Instituto Geológico y Minero de España-CSIC
Madrid, Spain

José Manuel Gil-Márquez
Department of Geology
University of Málaga
Málaga, Spain

Matías Mudarra
Department of Geology
University of Málaga
Málaga, Spain

ISSN 2511-2066 ISSN 2511-2082 (electronic)
Advances in Karst Science
ISBN 978-3-031-16878-9 ISBN 978-3-031-16879-6 (eBook)
https://doi.org/10.1007/978-3-031-16879-6



This Springer imprint is published by the registered company Springer Nature Switzerland AG
The registered company address is: Gewerbestrasse 11, 6330 Cham, Switzerland

# Preface

After a delay of two years due to the COVID, the *Eurokarst* conference, the largest event on karst hydrogeology and carbonate terrains in Europe, was held again. This third edition took place in Málaga, Spain, in June 2022, involving more than 150 researchers from 30 countries from all over 4 continents.

Organized by the Universities of Málaga, Neuchâtel, and Besançon, with the collaboration of the Spanish Geological Survey (IGME-CSIC), the Eurokarst conference series arose in 2014 with the aim to continue promoting advances in research in the field of karst and carbonate reservoirs, after more than 40 years of regular meetings. The first and second editions were organized in Neuchâtel (Switzerland) and Besançon (France) in 2016 and 2018, respectively. In 2022, for the third time, the Eurokarst offered the opportunity to share with the scientific community the most recent advances in karst hydrogeology and to increase the understanding of processes and impacts on this media.

Among the current topics addressed during the conference, and despite their recurrence, understanding groundwater flow systems within karst aquifers, its functioning, and the protection and management of water resources are still relevant issues, even more facing future scientifical and technical challenges posed by a changing world. In addition, there was a good number of research contributions dealing with developing new methods based on physico-chemical characteristics of karst groundwater, hydrodynamic studies, and hydrogeological parameter estimation.

It is clear that karst aquifers are complex hydrosystems, and their hydrological behaviour cannot be assessed without (1) geological knowledge, especially structural geology and epigenetic and hypogenic features and (2) long time series of fundamental parameters, such as rainfall, discharge flow, water electrical conductivity, chemistry, and many other parameters. Only with this information is it possible to interpret karst dynamics to an event or seasonal scale. Moreover, karst terrains require knowledge of catchments in natural areas, detailed information on mean residence time, and more studies about the natural attenuation of contaminants to enhance our knowledge on intrinsic and specific vulnerability assessment.

A continuous transfer of improvements from conceptual to numerical modelling approaches, and *vice versa*, is necessary to enhance knowledge of carbonate aquifer functioning and ultimately achieve better evaluation and management of water resources. The state of the art and the perspective of karst modelling were presented during the Eurokarst 2022. Various types of numerical models representing both innovative approaches and case studies using well-established modelling platforms were shown as examples on this topic.

Last but not least, some works contributed to advances in cave sciences, including new methods for caving research and monitoring dissolved gases, water, and even ice appearance in cavities. Also, geomorphological aspects of karst were discussed and some outstanding examples of geodiversity were shown during the Eurokarst 2022 conference.

Eurokarst 2022 demonstrated that further research, not only on groundwater resources assessment but also on the characterization of the hydrogeological functioning of the aquifers, is essential to manage the aquifer sustainably and solve the still uncertainties of this media.

This book includes 36 articles selected from the 131 communications presented to Eurokarst 2022, covering a wide variety of topics in many fields related to karst: karst hydrogeology and methods to study karst aquifers, karst caves, and geomorphology, among others.

The articles are organized around two main topics:

- Karst hydrogeology and methods to study karst aquifers
- Karst caves, geomorphology, landscape, and natural heritage.

Málaga, Spain Bartolomé Andreo
Málaga, Spain Juan Antonio Barberá
Madrid, Spain Juan José Durán-Valsero
Málaga, Spain José Manuel Gil-Márquez
Málaga, Spain Matías Mudarra

# Reviewers

Contributions published in this book have been reviewed by members of the Scientific Committee which includes the following professionals:

Bartolomé Andreo, University of Málaga, Spain
Augusto Auler, Instituto do Carste, Brazil
Alice Aureli, UNESCO, France
Juan Antonio Barberá, University of Málaga, Spain
Catherine Bertrand, University of Franche Comté, France
Lhoussaine Bouchaou, University of Ibn Zohr, Agadir, Morocco
Avi Burg, Geological Survey of Israel
Zhang Cheng, Karst Dynamics Laboratory, Guilin, China
Nathalie Doerfliger, BRGM, France
Joanna Doummar, American University of Beirut, Libanon
Juan José Durán-Valsero, IGME-CSIC, Spain
Dave Evans, FloSolutions, Lima, Perú
Francesco Fiorillo, University of Sannio, Italy
José Manuel Gil-Márquez, University of Málaga, Spain
Laurence Gill, Trinity College, Ireland
Nico Goldscheider, Karlsruhe Institute of Technology, Germany
Chris Groves, Western Kentucky University, USA
John Gunn, University of Birmingham, United Kingdom
Andreas Hartmann, University of Dresden, Germany
Hervé Jourde, University of Montpellier2, France
Martin Knez, Karst Research Institute, Postojna, Slovenia
Neven Kresic, Geosyntec, Washington, USA
Neno Kukuric, IGRAC, The Netherlands
James LaMoreaux, P.E. LaMoreaux and Associates, Inc., USA
Peter Malik, Geological Survey of Slovakia
Sergio Martos Rosillo, IGME-CSIC, Spain
Nicolas Massei, University of Rouen, France
Matías Mudarra, University of Málaga, Spain
Eulogio Pardo-Igúzquiza, IGME-CSIC, Spain
Marco Petitta, University of Sapienza, Roma, Italy
Nataša Ravbar, Karst Research Institute, Slovenia
Philippe Renard, University of Neuchâtel, Switzerland
Michael Sinreich, Swiss Federal Office for the Environment FOEN, Switzerland
Tadej Slabe, Karst Research Institute, Postojna, Slovenia
Zoran Stevanovic, University of Belgrade, Serbia
Ben Tobin, Kentucky Geological Survey, USA

Iñaki Vadillo, University of Málaga, Spain
George Veni, National Cave and Karst Research Institute NCKRI, USA
Steve Worthingthon, Worthington Groundwater, Canada
Pierre Yves Jeannin, SISKA, Switzerland

# Acknowledgments

The editors of this book are highly thankful to a good number of institutions, companies, and people who participated in the event preparation and without whom the book could not be published. First of all, we would like to thank the sponsors and collaborators who contributed financially to support the conference:

As sponsors:

- University of Málaga
- UNESCO. Intergovernmental Hydrological Programme
- Van Walt
- Valeport Water

As collaborators:

- The International Association of Hydrogeologist (IAH-AIH)
- Springer Verlàg AG
- Academia Malagueña de Ciencias
- Fundación Cueva de Nerja
- TRAQUA
- Sondeos Martínez
- Tetraedre
- Albillia
- FluoTechnik
- Picarro
- Distribuciones Industriales y Científicas (DICSA)
- EIC Group

The partner organizations were the following:

- The University of Málaga, Spain
- The Spanish Geological Survey (IGME-CSIC)
- The University of Neuchâtel, Switzerland
- The University of Franche-Comté, Besançon, France

Results of contributions published are related to selected projects supported by: EU-PRIMA research and innovation programme—KARMA project—(PCI2019-103675 in Spain), Spanish Research Agency (PID2019-111759RB-100), FEDER-Andalusia programme (UMA20-FEDERJA-009), and Research Group RNM 308 of the Andalusian Government.

We also want to thank the members of the Scientific Committee of the Eurokarst 2022 very warmly (see the list on the following page), who shared their expertise and knowledge with the authors in order to provide the best possible technical quality within the limited time frame available to edit the book. Finally, we want to thank Springer's staff in charge of the project: Jim LaMoreaux, Annet Buettner, and Ramamoorthy Rajangam.

Our special gratitude to the staff of the Centre of Hydrogeology at the University of Málaga, whose help was invaluable in organizing the conference. Thanks to José Francisco Martín, Jaime Fernández-Ortega, Lucía Ojeda, Alejandro Carrasco, Juan José Rovira, José María Ávila, Javier Buera, José Pablo González de Aguilar, Nuria Naranjo, Alejandro Millán, Lucía Sánchez, Antonio Linares, Sergio Durán, Christian Moreno, María Jesús Civantos and Iñaki Vadillo, for their help throughout this organization. Thanks also to the postgraduate students Evan Morel, Katheryn López-Lasserra, Guillermo Pérez, Rogério Tadeu Souza, and Víctor Maqueda.

The Organizing Committee

Bartolomé Andreo Navarro, Málaga, Spain

Juan José Durán Valsero, Madrid, Spain

José Manuel Gil-Márquez, Málaga, Spain

Juan Antonio Barberá Fornell, Málaga, Spain

Cristina Liñán Baena, Málaga, Spain

Matías Mudarra Martínez, Málaga, Spain

Beatriz de la Torre Martínez, Málaga, Spain

Catherine Bertrand, Besançon, France

Philippe Renard, Neuchâtel, Switzerland

Málaga (Spain), July 2022

# Contents

# Karst Hydrogeology and Methods to Study Karst Aquifers

# Initiative to Select, Label and Protect the World's Most Important Karst Springs

Zoran Stevanović

**Abstract**

Springs are important to the humanity because they provide potable water to many locations in the world, thus ensuring health, sanitary conditions, food production and economic development. Karst and mineral water springs are particularly important, but springs emerging from karst aquifers are by far the largest—some are even discharging entire underground rivers—whose flow sometimes exceeds 100 $m^3/s$. Having caused the establishment of permanent settlements and nearby cities, especially in the Roman times, many springs are also historically important. Although some of the best-known karst springs are actively used and very well protected from pollution, many others around the world have been contaminated, devastated by over-pumping or impounded by reservoirs. Taking advantage of the opportunity provided by the 50th anniversary of the IAH Karst Commission (KC) and the involvement of UNESCO in our karst research, this initiative (project MIKAS) aims to bring together both the KC members and many national experts to work, on a voluntary basis, to: 1. develop criteria for the selection of most important karst springs, which inter alia should include historic, aesthetic and scientific values; 2. establish the list of springs; 3. create the Code of Practice for these springs' utilisation and protection; and 4. promote these springs by their in situ labelling and Internet publicising. The idea to identify and protect selected springs does not imply prevention of their further use. To the contrary, this initiative intends to highlight their importance, defend them against possible devastation and ensure that any further intervention considers their protective status.

Z. Stevanović (✉)
Centre for Karst Hydrogeology, Department of Hydrogeology, Faculty of Mining and Geology, University of Belgrade, Djušina 7, 11000 Belgrade, Serbia
e-mail: zstev_2000@yahoo.co.uk

**Keywords**

Springs • Karst • Geo-heritage • Protection • Promotion

## 1 Introduction

Groundwater is a vital resource that provides almost half of all the drinking water that is available in the world. However, for most people, this invisible groundwater is out of sight and out of mind. In addition to many earlier campaigns to promote groundwater and its importance, the UN-Water has decided that the theme for the 2022 World Water Day (WWD) should be "Groundwater: Making the Invisible Visible". Not all groundwater is invisible, though. As aquifers' discharge points, natural springs provide specialists with an insight into underground secrets, and indirectly assess the complexity of groundwater origin and distribution (LaMoreaux and Tanner 2001; Bakalowicz 2005; Kresic and Stevanović 2010; White 2010). Karstic and mineral water springs are particularly important, but springs emerging from karst aquifers are by far the largest, some discharging more than 100 $m^3/s$ (Kresic 2013). They resemble true underground rivers.

Karst covers more than 15% of the continental ice-free land (Goldscheider et al. 2020) and karst aquifers supply approximately 9.2% of the world's population, or close to 678 million people, with potable water (Stevanović 2019). The intakes of springs are the most common tapping structures in karst environment, as channelling gravity springs and diverting water over long distances are still easier that drilling many wells in hard karstic rocks (Stevanović 2018). The latter is the second common way of tapping karst water. Karst springs are important because they provide precious water quality, also sustain ecosystems and maintain the baseflow of many rivers in the world (Cantonati et al. 2006; Bonacci et al. 2009; Stevens et al. 2011).


B. Andreo et al. (eds.), *EuroKarst 2022, Málaga*, Advances in Karst Science,
https://doi.org/10.1007/978-3-031-16879-6_1

The karst environment is so specific that it requires, almost as a rule, a multidisciplinary approach and engagement of specialists from various fields. In 1970, the Karst Commission (KC) was established under the umbrella of the International Association of Hydrogeology (IAH) as an answer to the demand to collectively integrate this highly specialised branch of hydrogeology. More than 50 years old today, the Commission plays the focus point for the exchange of ideas for further development of karst hydrogeology (Milanović and Stevanović 2021). The project of mapping karst aquifers all over the world was recently completed thanks to the support of the KC (Chen et al. 2017; Goldscheider et al. 2020). Through their published books and articles, as well as many activities, members of the KC also largely contributed to marking 2021 as the International Year of Caves and Karst, extended to also include the year 2022. This is why the author of this article is launching the initiative to bring these two together: karst springs and karst experts. The aim is to select, label, protect and promote the most important karst springs at the global and national level. Although the KC should lead this activity, it is expected to receive support from UNESCO (Gunn 2021) and its Intergovernmental Hydrological Programme (IHP) and many national experts who should work on a voluntary basis.

At the annual meeting of the KC IAH held on 23 June 2022 during the Eurokarst 2022 conference in Malaga, this initiative has been approved and following proposal of the author of this contribution the project titled most important karst aquifer's springs (MIKAS) launched. The initial project's advisory board has also been approved and consists of one representative from each continent, chairs of the KC IAH and UNESCO IHP and project's coordinator.

**Fig. 1** Some of candidates for global list of springs heritage—existing memorial plates at spring site are posted in left upper corner of each photograph. **a** Khanis spring in northern Iraq, cuneiform inscription carved in rocks; **b** Comal spring in Texas, USA, iron plate; **c** Almyros spring, Crete, Greece, plastic plate. All photos by Z. Stevanović

## 2 Brief History of Some Karst Springs Utilisation

Tapping spring water is an ancient art. Historically, in order to have easy access to water, people built their settlements near large springs. As a rule, cities with plentiful water drawn from successfully constructed intakes and reservoirs provided a base for prosperous development and a safe haven for their citizens. In contrast, cities that had no water supply from nearby springs were often destroyed or abandoned because they were unable to survive long sieges.

The history of capturing groundwater and diverting water to distant points is linked to ancient China, Mesopotamia and Egypt. However, the golden age was achieved in the Roman times.

The importance of precious spring water and its prevalence over nearby river water had been proven by the Assyrian emperor Sanherib, son of Sargon II (703–681 BC). He constructed the intake systems at the Khanis karstic gravity spring (near Atrush, northern Iraq, Fig. 1) to supply the historical city of Nineveh, located some 16 km away (near today's Mosul in Iraq on Tigris River). The Khanis water structure was one of the very first aqueducts ever constructed (Reade 1978; Stevanović 2010).

During that historical period, Hezakiah, the ruler of ancient Jerusalem, decided to dig a 500 m long tunnel under and through the city walls to ensure supply of water from the Gihon spring, which drains the Turonian limestone aquifer. Before the Assyrian siege of the city, the spring was connected to the Siloam water pool to prevent the spring from falling into the hands of the enemy (Frumkin and Shimron 2006).

The Romans' supremacy and dominance in the ancient world were demonstrated through their knowledge of water, including the art of tapping and delivering spring water. In the narrow historical centre of Rome, there were 23 springs that initially supplied small settlements around the city, while at the height of the Roman Empire, 11 long aqueducts delivered more than 13 $m^3/s$ of water to the city from distances ranging from 16 to 91 km (Lombardi and Corazza 2008). The Roman architect Vitruvius was the first to leave a written record showing that springs on mountain slopes can be recharged by atmospheric precipitation, finding their interaction in the rapid propagation of infiltrated water.

The Romans established many settlements across their empire. Thus, several major cities were constructed around major karst springs in the Adriatic part of the Mediterranean basin: Trieste—Timavo springs, Rijeka—Zvir group of springs, Split—Jadro spring, Dubrovnik—Šumet and Ombla springs, Kotor—Škurda and Gurdić springs (Milanović 1981, 2021; Stevanović 2010).

In the nineteenth century, water from karst springs continued to be an important source of water supply to cities that were developed nearby. The first 130 km long mountain pipeline was completed in 1873 to supply the city of Vienna from the Kaiserbrunn spring (Fig. 2). The quality of its water is excellent, as it generally requires only chlorination, primarily to clean the distribution pipes. Paris also obtained water from several captured springs 100 to 150 km away from the city, piped by aqueducts built between 1865 and 1893 (Margat et al. 2013).

## 3 Large Karst Springs Distribution

For the purpose of mapping large karst springs under the World Karst Aquifer Mapping (WOKAM) project, Chen et al. (2017) and Goldscheider et al. (2020) established the basic selection criteria that included the following:

- Permanent karst springs with a minimum discharge > 200 l/s (for print map), > 500 l/s (for database),
- Temporary or highly variable springs with a maximum discharge > 10 $m^3/s$ (50 $m^3/s$),
- Very important submarine springs (experts' opinion),
- Thermal water springs with discharge > 100 l/s that are > 4 °C warmer than the average air temperature (or 200 l/s and 10 °C warmer than air temperature),
- Karst springs > 100 l/s with peculiar gas composition, such as $CO_2$ or $H_2S$ (200 l/s).

Due to uneven distribution of karst springs and their most dense distribution in Dinaric karst (Stevanović et al. 2016), a more flexible approach has been applied to the final WOKAM map. Therefore, the number of springs per 1000 $km^2$ in the Alpine orogenic belt and the Dinaric system was reduced. For instance, in some small countries such as Bosnia and Herzegovina, there are all of 8 springs that regularly discharge more than 2000 l/s. Turkey has the same number of large springs, followed by Montenegro (5).

**Fig. 2** Some of candidates for global list of karst springs heritage—a) Bekhal spring, Taurids Mts. Iraq; b) Margoon spring, Zagros Mts. Iran; c) Vaucluse spring, Provence, France; d) Kaiserbrunn spring, Rax Mt. Austria; e) Inka's spring, Cusco, Peru. All photos by Z. Stevanović

Whereas, less restrictive criteria were applied for selecting springs in some other parts of the world.

In the Americas, the largest number of springs is located in the southern part of USA (Texas and Florida), Mexico (Yucatan) and Belize. There is a dense distribution of large springs in the Near and Middle East, China and Indonesia, although many regions with a wide distribution of karst rocks do not have very large springs. This is mainly due to the presence of platform type of karst, with a lesser karstification degree (e.g. Canada, South America, Great Britain, northern and eastern Africa).

The full-scale WOKAM map includes 201 selected karst water sources, i.e. 162 continental freshwater springs, 16 submarine springs, 8 thermal springs and 15 water abstraction structures (Goldscheider et al. 2020). The World Karst Spring Hydrograph Database (WOKAS) has also been prepared (Olarinoye et al. 2020) to complement data included in the WOKAM database.

## 4 Karst Springs Heritage—Selection Criteria and Application

The purpose of this initiative is to make a global karst springs heritage list. The list should be made based on common criteria but adapted to local conditions and recognising the specific circumstances of each country. Something that is important in one country does not have to be important in others. The above discussed karst springs' discharge can be viewed as an important but not critical criterion for selecting springs that should, in the end, be declared geo-heritage sites. The point of the project is not to make a long list, but to instead identify the most important springs in each of the karst countries, and propose that they be labelled, better protected and used in a sustainable fashion. Some large countries may have more than ten, while some others may have just one or two, maybe even no springs at all that would meet the listing criteria. However, some sense of balance between these two extremes is always recommendable.

The initiative proposes that the following steps be taken:

1. Developing criteria for the selection of most important karst springs;
2. Drafting the list of springs;
3. Creating the *Code of Practice* for these springs' utilisation and protection; and
4. Promoting the springs by their in situ labelling and Internet publicising.

The selection criteria should include, *inter alia*[1]:

- Historic,
- Aesthetic,
- Economic,
- Ecologic and
- Scientific values.

Some illustrations of great historic values were provided in the previous chapter. However, there are numerous springs in the world that have historic and/or cultural significance for local nations or community development (Figs. 1, 2).

The aesthetic criterion is always tricky, as *de gustibus non est disputandum*. However, something like a waterfall, a huge cliff or a cave behind a spring should commonly be judged as a nice and acceptable landscape for the list (Fig. 2).

The economic-management value should consider a spring's active use. Spring water can be used for potable water supply, irrigation or for supplying the local industry. In a few words, it can support the local economy by generating food and income to the local community or the country as a whole. Some springs are used for generating hydropower, providing geothermal energy or are applied in balneology and recreation.

Even if not tapped, water from karst springs can be essential from ecological point of view, to sustain ecosystems, maintain the baseflow of rivers or fill large reservoirs.

The scientific value may take into account specific discharge mechanisms of the springs such as large maximal yield, intermittent flowing, gas bubbling, changing water quality in coastal areas (fresh, brackish and saline) or some other properties that could be of research interest to the hydrogeological community.

Further discussion on selecting the criteria and drafting the *Guidelines* for inquiry should be the task of the project's Advisory Board representatives of different regions of the world. The *Code of Practice* for these springs should be prepared along with the *Guidelines* or as a part thereof. The idea of identifying and protecting selected springs does not imply prevention of their further use. On the contrary, the aim of the initiative is to highlight their importance, defend them from possible devastation and ensure that any further intervention takes into account their protected status.

Following in the footsteps of the WOKAM and WOKAS projects, members of the Advisory Board should call on regional and national experts to support the project by providing proposals and supporting the project's implementation in the field. The idea is to create unique panels with basic information (in local languages and English) about the springs, their history and importance, morphological characteristics, discharge mechanisms and other specific facts. The content and form of these unique panels should also be included in the *Guideline* prepared by the Advisory Board.

Information about the springs in question, presented on the Internet and in brochures, would help their promotion and could generate additional income for the local communities from geotourism. In case a spring is actively used for water supply, which would necessitate its greater protection, there would still be space for visits of organised groups during designated time slots.

## 5 Conclusions

Springs in karst are precious "eyes" that give insight into the often complex mechanism of groundwater creation and distribution. Many are properly utilised or flowing freely to ensure the existence of ecosystem, while others are over-pumped or improperly protected from pollution. The

[1] The criteria should be established through the joint work of the project's Editorial Board.

initiative to identify, label and promote the most important karst springs at the global level would spread knowledge about their presence and value, while the prepared and applied *Code of Practice* would help with their protection and more sustainable use.

By decision of the IAH Karst Commission, this initiative has been converted into a project MIKAS, to be led by its members and voluntarily engaged the Advisory Board. However, the MIKAS project would require the involvement of many regional and local experts, and the employment of their knowledge in the process of identification, description and further promotion of selected springs including in situ labelling by commonly agreed unique panel with emblems of the KC, respective national or local community authority and probably UNESCO IHP.

Eurokarst 2022 and the KC annual meeting were an excellent opportunity to present and discuss the project's concept, create the Advisory Board consisting of regional representatives and plan further steps.

The project further extension on selection of mineral and thermal springs for creation of their global list should be a task of the IAH Commission on Mineral and Thermal Waters.

## References


Bakalowicz M (2005) Karst groundwater: a challenge for new resources. Hydrogeol. J,13:148–160

Bonacci O, Pipan T, Culver D (2009) A framework for karst ecohydrology. Environ Geol 56(5):891–900

Cantonati M, Gerecke R, Bertuzzi E (2006) Springs of the Alps, sensitive ecosystems to environmental change: from biodiversity assessments to long-term studies. In: Lami A and Boggero A (eds) Ecology of high altitude aquatic systems in the Alps. Developments of hydrobiology. Hydrobiologia, p 59–96

Chen Z, Auler AS, Bakalowicz M, Drew D, Griger F, Hartmann J, Jiang G, Moosdorf N, Richts A, Stevanović Z, Veni G, Goldscheider N (2017) The world karst aquifer mapping project–concept, mapping procedure and map of Europe. Hydrogeol J 25(3):771–785

Frumkin A, Shimron A (2006) Tunnel engineering in the Iron age: geoarchaeology of the siloam tunnel, jerusalem. J Archaeol Sci 33:227–237

Goldscheider N, Zhao Ch, Auler A, Bakalowicz M, Broda S, Drew D, Hartmann J, Jiang G, Moosdorf N, Stevanović Z, Veni G (2020) Global distribution of carbonate rocks and karst water resources, Hydrogeol J 28(5):1661–1677

Gunn J (2021) Karst groundwater in UNESCO protected areas: a global overview. Hydrogeol J 29(1): 297–314

Kresic N, Stevanović Z (2010) (eds) Groundwater hydrology of springs: engineering, theory, management and sustainability, Elsevier Inc., BH, Burlington-Oxford

Kresic N (2013) Water in karst. Management, vulnerability and restoration. McGraw Hill

LaMoreaux PE, Tanner JT (2001) (eds) Springs and bottled waters of the world ancient history: Source, occurrence, quality and use. Springer-Verlag, Berlin, Heidelberg, p 315

Lombardi L, Corazza A (2008) L'acqua e la città in epoca antica. In: La Geologia di Roma, dal centro storico alla periferia, Part I, Memoire Service. Geol. d'Italia, S.E.L.C.A, Firenze, pp 189–219

Margat J, Pennequin D, Roux JC (2013) History of French hydrogeology. In: Howden N and Mather J (eds) History of Hydrogeology. Intern. Contrib. to Hydrogeol. 28. CRC Press/Balkema, Taylor & Francis Group, London, p 59–99

Milanović P (1981) Karst hydrogeology. Water Resources Publications, Littleton CO

Milanović P (2021) Karst of eastern Herzegovina and Dubrovnik littoral, HET, Trebinje, p 472

Milanović P, Stevanović Z (2021) Fifty years of history of the karst commission of the international association of hydrogeologists. Hydrogeol J 29:7–19

Olarinoye T, Gleeson T, Marx V, Seeger S, Adinehvand R, Hartmann A (2020) Global karst springs hydrograph dataset for research and management of the world's fastest-flowing groundwater. Sci Data 7:59. https://doi.org/10.1038/s41597-019-0346-5

Reade J (1978) Studies in assyrian geography, Part 1: Sennacherib and the waters of Ninveh. Revue D'assyriologie Orientale 72(47–72):157–175

Stevanović Z (2010) Utilization and regulation of springs. In: Kresic N and Stevanovic Z (eds) Groundwater hydrology of springs. Engineering, theory, management and sustainability. Elsevier Inc. BH, Amsterdam, p 339–388

Stevanović Z (2018) Global distribution and use of water from karst aquifers. In: Parise M, Gabrovsek F, Kaufmann G, Ravbar N (eds) Advances in karst research: theory, fieldwork and applications, geological society. vol 466, London, Special publications, p 217–236

Stevanović Z (2019) Karst waters in potable water supply: a global scale overview. Environ Earth Sci 78:662. https://doi.org/10.1007/s12665-019-8670-9

Stevanović Z, Goldscheider N, Chen Z, The WOKAM Team (2016) WOKAM—The world karst aquifer mapping project, examples from South East Europe, Near and Middle East and Eastern Africa, In: Stevanović Z, Krešić N, Kukurić N (eds) Karst without boundaries, CRC Press/Balkema, EH Leiden; Taylor & Francis Group, London, p 39–51

Stevens LE, Springer EA, Ledbetter DJ (2011) Inventory and monitoring protocols for springs ecosystems. Springs Stewardship Institute, Museum of North Arizona, Flagstaff. http://docs.springstewardship.org/PDF/Springs_Inventory_Protocols_110602.pdf. Accessed 16/12/2021

White W (2010) Springwater geochemistry, In: Krešić N and Stevanović Z (eds) Groundwater hydrology of springs. Engineering, theory, management and sustainability, Elsevier Inc. BH, Amsterdam, p 231–268

# Flood Hazard in the Classical Karst: The Case of Mucille Polje (NE Italy)

P. Turpaud, C. Calligaris, and L. Zini


## Abstract

In the north-western area of the Classical Karst (NE Italy), the Mucille depression, after abundant precipitations, is subject to frequent floodings, which become problematic since 2001 as they more frequently affect housing and recreational areas, leading the population to believe that the swallow holes draining the area stopped functioning. The climate changes as well as the increased frequency of intense rainfall events led to evaluate the draining capacity of the swallow holes in order to provide fundamentals for the mitigation measures. The depression is fed by a spring area and drained by two swallow holes one of which is permanently active, while the other functions only during floods. About 24 h after the onset of heavy rains, the whole depressed area is flooded. About 8 days later, the water level begins to decrease, coming back to its initial height in about fifteen days. During floods, while springs and swallow holes discharges measurements are impossible, the extension of the flooded areas has been mapped. The obtained flooded surface together with high resolution DTM coverage allows to calculate the volume of surface water. Consequently, the hydrologic balance can be estimated during the whole event. This study provides meaningful evidences for the design of measures to mitigate the risk. It estimates the discharge of the swallow holes, confirming their efficiency. Nonetheless, it also emphasizes the need to improve their draining capacity, especially considering the unsuspected high outflow of the springs at the onset of the flood.



## Keywords

Karst hydrogeology • Flood hazard management • DEM analysis • GIS • Groundwater monitoring


## 1 Introduction

In the last decades, invasive and destructive flood events are constantly increasing (Hirabayashi et al. 2013; Kvočka et al. 2016; Tabari 2020) and karst areas are no exception. Their response to floods and to the water outflowing is dependent on the intrinsic characteristic of the karstified hydrostructure (conduits and fractures). The study of a small but meaningful area (the Mucille karst depression) in the Friuli Venezia Giulia Region (NE Italy) provides insights concerning the flood dynamics in such environment.

After abundant precipitations, the Mucille karst depression is subject to frequent floodings which become problematic since 2001 as they, more frequently, affect housing and recreational areas, leading the population to believe that the swallow holes draining the area stopped functioning. This belief, together with the climate changes that will most probably cause an increase of the frequency of intense rainfall events, led to evaluate the draining capacity of the swallow holes and the aquifer dynamics both by direct discharge measurements and by water balance modeling. These assessments provide insights for flood risk reduction measures and data that can be helpful in evaluating the efficiency of the realized mitigation actions.

The Classical Karst hydrostructure extends for approximately 750 km$^2$ from the Soča/Isonzo River to the Adriatic Sea and up to the town of Postoina. The springs draining it, drain waters of three main contributions: the Isonzo groundwaters, the Reka and Raša rivers input through swallow holes and the effective infiltration. The groundwaters of the Isonzo Plain are partially drained along the contact between the karst and the alluvial plain (Timeus 1928;


P. Turpaud · C. Calligaris (✉) · L. Zini
Dipartimento di Matematica e Geoscienze, Università degli Studi di Trieste, Trieste, Italy
e-mail: calligar@units.it





B. Andreo et al. (eds.), *EuroKarst 2022, Málaga*, Advances in Karst Science,
https://doi.org/10.1007/978-3-031-16879-6_2

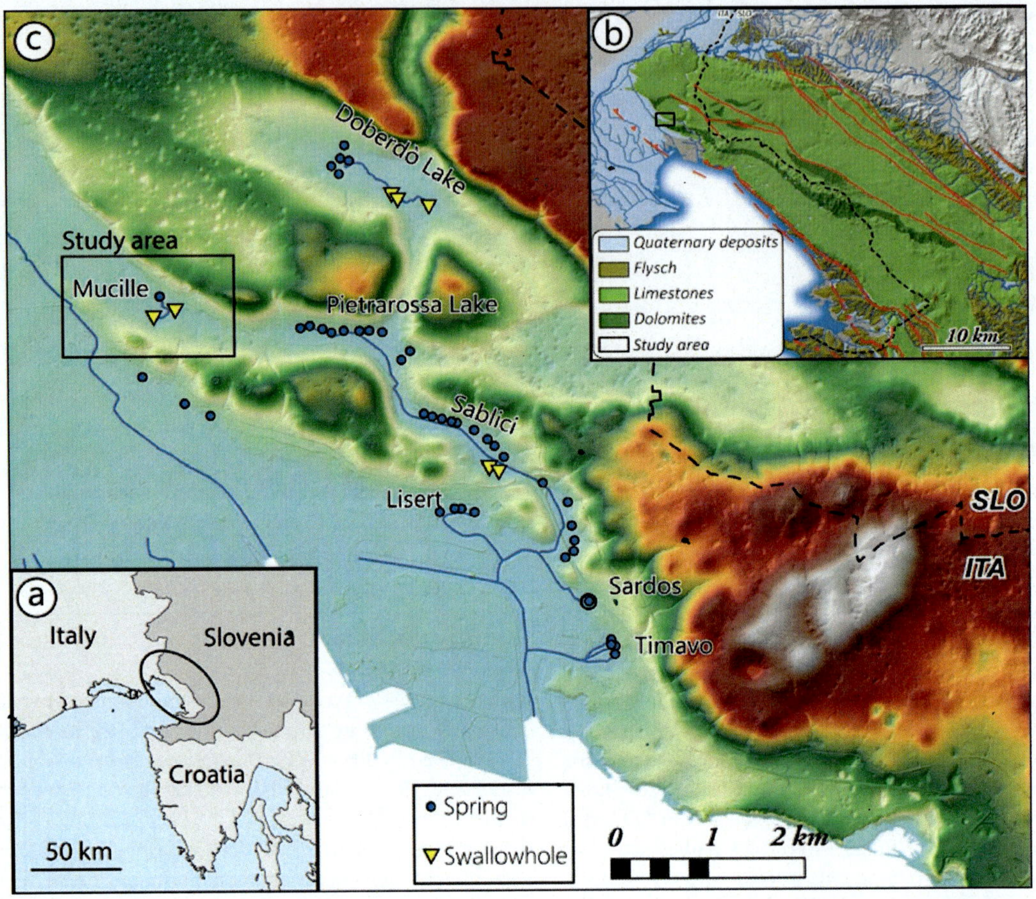


**Fig. 1** The study area is located along the south-western edge of the Classical Karst. The latter spans across the Italian-Slovenian border **a**, **b**. Simplified geological map of the Classical Karst after Jurkovšek et al. (2016). The Mucille depression belongs to a network of polje featuring numerous springs that drain the aquifer (**c**)

Mosetti and D'Ambrosi 1963; Gemiti and Licciardello 1977; Cancian 1987; Doctor et al. 2000; Samez et al. 2005; Urbanc et al. 2012; Zini et al. 2015). This contribution is calculated at approximately 10 $m^3/s$ (Zini et al. 2011, 2013; Calligaris et al. 2019). In the southeastern sector of the hydrostructure, in Slovenia, at the contact between flysch and carbonates, several swallow holes drain the Reka and Raša rivers. If the latter drains only small water quantities, Reka river input definitively influences the hypogean flows of the whole eastern sector of the Classical Karst as its discharge ranges between 0.18 $m^3/s$ and over 300 $m^3/s$ with an average of 8.26 $m^3/s$ (Gabrovšek and Peric 2006).

Effective infiltration represents the main contribution to the aquifer recharge with an average value of 20.6 $m^3/s$ (Civita et al. 1995). Rainfalls quickly infiltrate into the articulated paths of the karst structure thanks to an average high rainfall (between 1000 mm/year on the coast and 1800 mm/year in the innermost areas), a poor vegetation cover, the absence of thick soils and an intense and widespread karstification of the rock mass.

The study area, located at the contact between the south-western edge of the Classical Karst and the Isonzo Plain, is characterized by a series of polje NW–SE oriented which include, in addition to the Mucille plain, Doberdò and Pietrarossa lakes as well as the Sablici depression (Fig. 1).

The karst aquifer, which in the area can be found on average at elevations between 2 and 5 m a.s.l., is intercepted by these karst depressions giving rise to a complex system of springs characterized by extremely variable flows according to the rainfall regime. This hydrogeological context was heavily modified in the post-war period by the construction of a channel draining, toward the Adriatic Sea, Pietrarossa and Sablici lakes (Zini et al. 2022). The result is a lowering of their baseflow level of about 1 m. Mucille and Doberdò lakes were not directly involved by the reclamation activities and thus maintained their natural hydrography characterized by the presence of a spring area, a drainage channel and a series of sinkholes. Nonetheless, also Doberdò Lake evidenced a lowering of its baseflow level of about 1 m leading to longer drought periods. This lowering is not proven in the Mucille area due to the lack of data regarding the level monitoring but is can be expected.

The Mucille depression is laid on a gently SSW dipping cretaceous bedded limestone of the Povir formation (Jurkovšek et al. 2016).

It is fed by a spring area (S-02). During low-water conditions, the flow is channeled toward a swallow hole (Sw-01), while at high water table conditions, a second swallow hole (Sw-02) participates in draining the area (Fig. 2).

The minimum elevation within the Mucille depression of about 4.8/4.5 m a.s.l. is encountered along the channel connecting the spring area to the swallow holes. The rest of the area is almost flat with elevations ranging from 6.0 to 8.5 m a.s.l. The lower groups of houses can be found at 8.1 and 8.8 m a.s.l., respectively.

## 2 Methods

The hydrogeological characteristics of the area have been defined by field-survey and manual monitoring of water level, electrical conductivity and temperature. Alongside, water level in an adjacent piezometer (S20, Figs. 2 and 3) has been monitored continuously in between 2017 and 2019 using a CTD-Diver (Van Essen Instruments) multiparametric probe. Pressure was compensated using a Baro-Diver probe (Van Essen Instruments).

Hydrograph analysis of two flood events together with hydrologic balance and in situ discharge measurements helped to define a flood control strategy. Discharge measurements at various water level have been performed with an MF Pro (OTT HydroMet). While discharge measurements are possible in low-water conditions, during floods assessment of springs and swallow holes discharge is impossible due to channel overflow. Digital terrain model allows to calculate the surface water volume at incremental water table level (Fig. 4). Using phreatic levels registered in

**Fig. 2** Composite aerial photograph of the Mucille area. Spring area, swallow holes and monitoring piezometer are highlighted. The flooded surface on 18/12/2017 is mapped

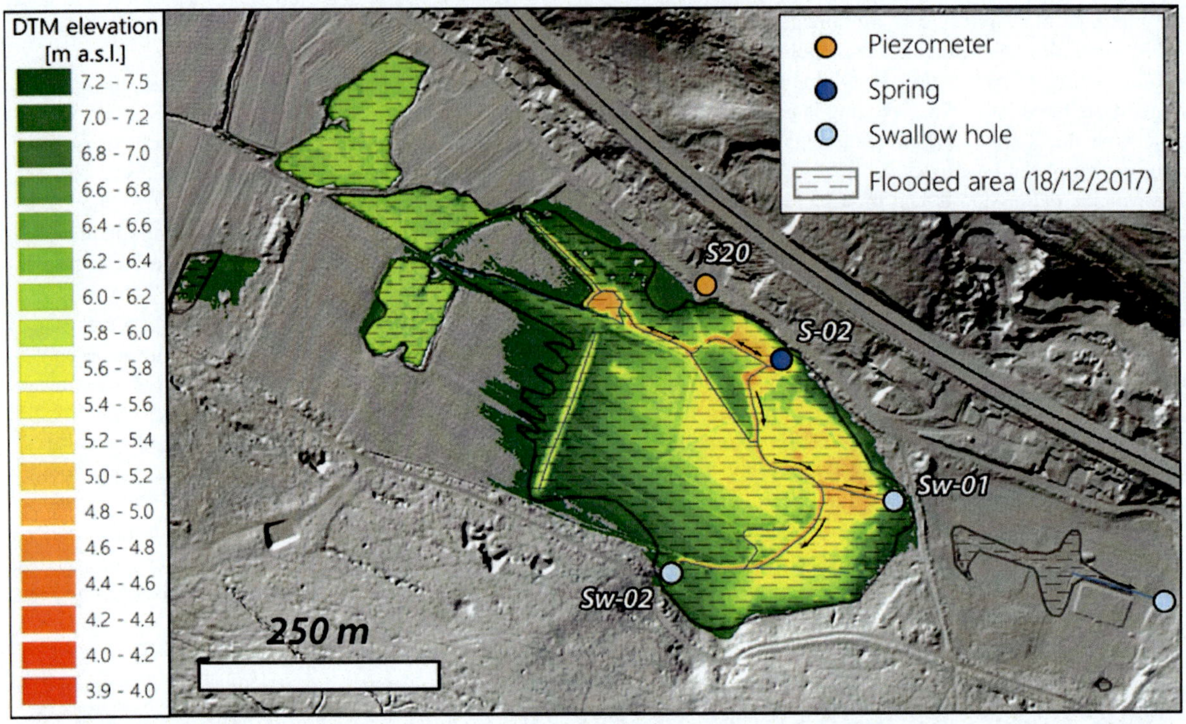


**Fig. 3** DTM of the Mucille depression. The flooded area at the peak of event 1 (18/12/2017) coincides with the 7.5 m a.s.l. elevation. The surface water volumes calculated with DTM allow the modeling of the water table elevation as function of the surface water volume

an adjacent piezometer (S20), this volume can be computed over time. Consequently, the hydrologic balance (inflow minus outflow) has been estimated during two flood events.

## 3 Results

The water balance and draining capacity of the swallow holes were analyzed for two different flood events. The results of in situ discharge measurements acquired during the second one and at various stages during other floods (not presented here) are also discussed.

### Event 1

The flood was caused by two rainfall events: a first one between 25th November and 1st December (76 mm) and a second one between 8 and 16th December (153 mm) 2017 (Fig. 5). The flood recorded a maximum water table at 7.5 m a.s.l. The fast water rise required a mean balance of 880 l/s (inflow minus outflow) for one day (to be compared with the low-water conditions springs discharge of about 50 l/s). The following day it decreased to 280 l/s. The next 5 days, the water balance oscillated between 0 and about 200 l/s. From the eighth day on, while water level began to lower, the balance became negative for 8 days with values around −200 l/s. As springs were still active, what obtained demonstrates the high drainage capacity of the swallow

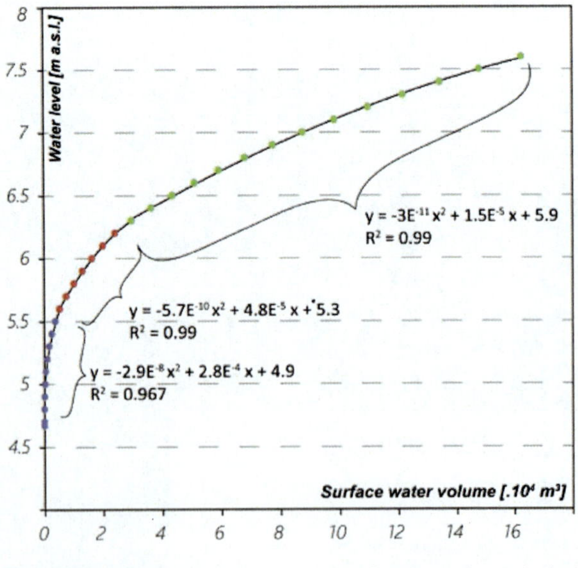


**Fig. 4** Relationship between water level and surface volume obtained by DTM elaboration. Successively this model was used for water balance calculation

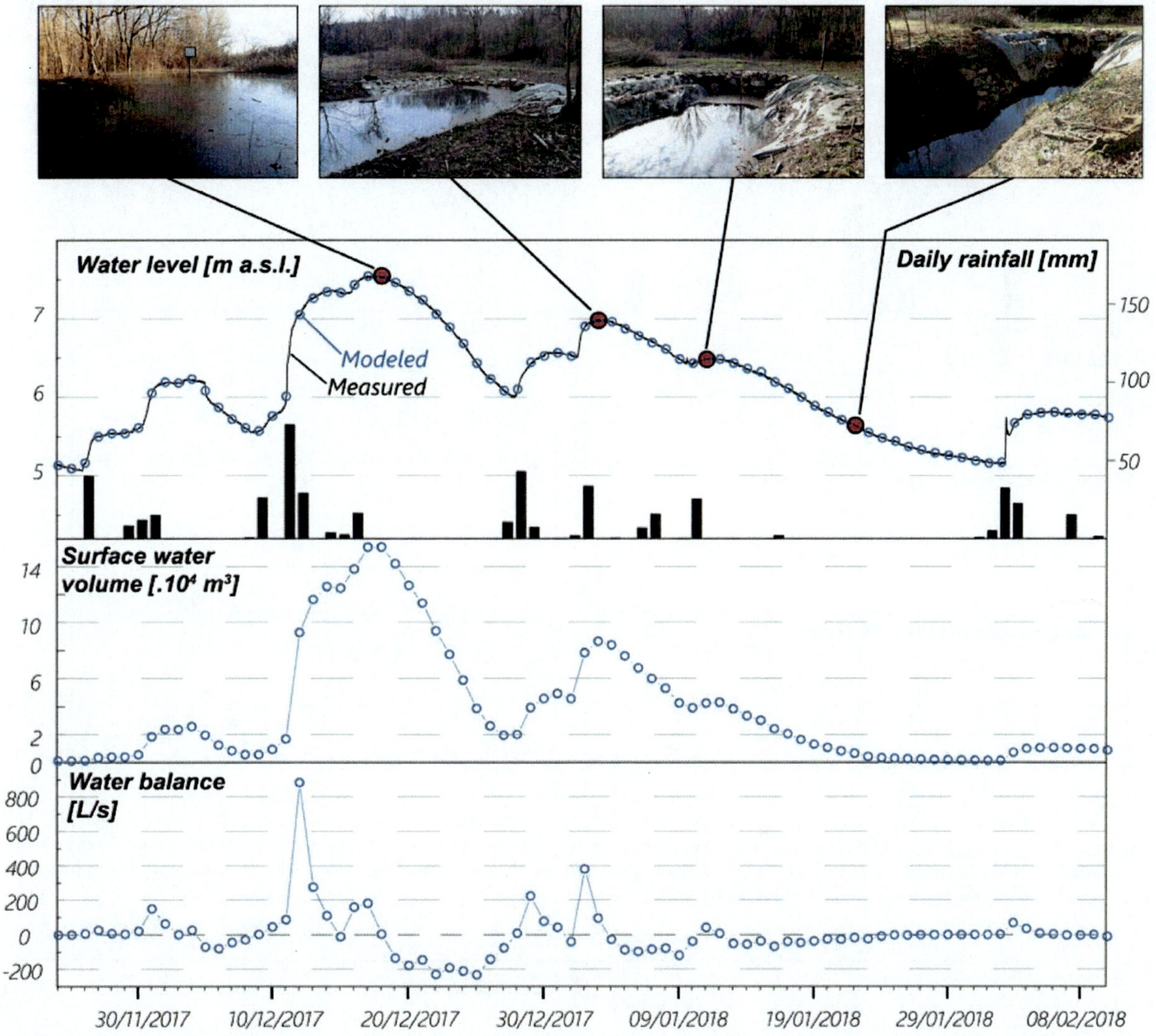


**Fig. 5** Water level measured in the S20 piezometer and daily rainfall data from Gorizia Italy (provided by ARPA-OSMER) during event 1. Photographs illustrate the situation in the Sw-02 at different levels. The water level allows the calculation of the surface water volume (middle graphic) and consequently the daily mean water balance (lower graphic)

holes at high-water conditions when direct discharge measurements are impossible.

### Event 2

The flood was caused by two precipitation pulses which took place on 28th February (15 mm) and between 1st and 4th of February (88 mm) 2018 (Fig. 6). This event illustrated a more common and less intense rise of the water table with a maximum of about 6.5 m a.s.l. The excess input of about 300 l/s for one day became negative after 5 days. It remained negative (at ca. −50 l/s) for the next 7 days. The last 7 days of the modeled period were characterized by a slightly negative balance (between −7 and −2 l/s).

After the peak of event 2, in situ discharge measurements have been performed to assess directly the swallow holes drainage capacity. Before becoming inactive due to water lowering, Sw-02 absorbs about 80 l/s. Sw-01 absorbs ca. 70–80 l/s before the return to low flow condition.

### Discharge Rating Curves

Further in situ discharge measurements have been performed at different stages. The discharge rating curves of both

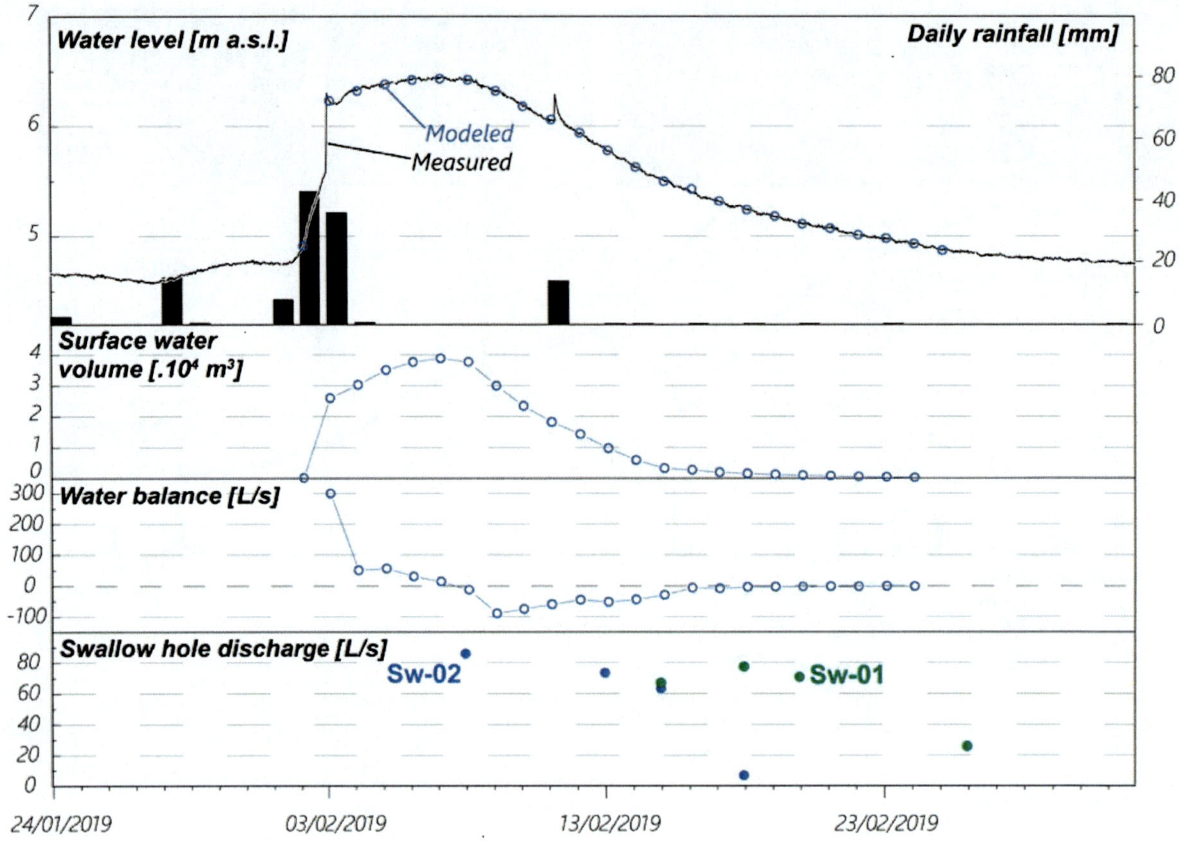


**Fig. 6** Water level measured in the S20 piezometer and daily rainfall data from Gorizia Italy (provided by ARPA-OSMER) during event 2. The water level allows the calculation of the surface water volume and consequently the daily mean water balance. Discharge values (in situ measurements) of both swallow holes are plotted in the lower graphic

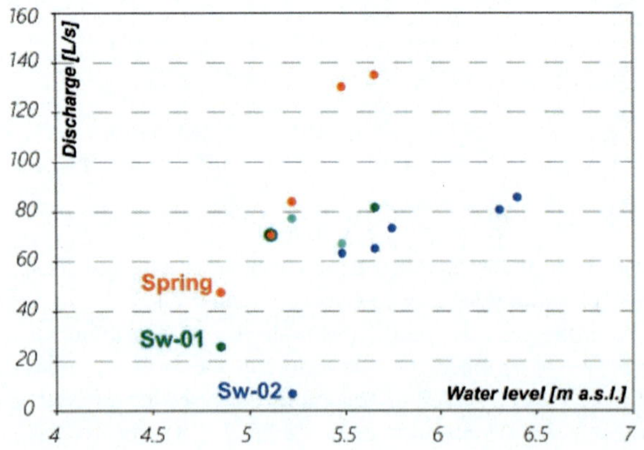


**Fig. 7** Discharge rating curves of the two swallow holes and of the spring area

swallow holes and spring area are plotted in Fig. 7. Data confirm the swallow hole draining capacity of at least 80 l/s each, up to a water level of 6.5 m a.s.l.

## 4 Conclusions

This study provides insights concerning the flood dynamics in karst environment through direct and indirect assessments.

In detail, the direct discharge measurements confirm the functioning of the swallow holes even at relatively high water levels (6.5 m a.s.l.). Hydraulic balance model allows to indirectly assess their draining capacity during floods when in situ measurements are impossible to be done (up to about 7.5 m a.s.l.).

The 880 l/s difference between input and output at the incipient of large floods was not expected. It implies that the maximum discharge at the spring should be about 1 $m^3/s$. Consequently, the conventional pumping to lower water level would be insufficient to mitigate damages.

The proposed solution involves first of all the regular maintenance and cleaning of both swallow holes. On the other hand, a drainage by channel would be prohibitive due to costs. Housing and recreational areas could be protected with the construction of an embankment and the mechanical enlargement of the southern swallow hole (active only during floods) in order to improve its drainage capacity. The comparison of the discharge rating curves and of the water balance before and after the intervention will permit to assess the efficiency of the intervention.

**Acknowledgements** This study has been funded by the municipality of Ronchi dei Legionari (Italy) in the framework of the project between the municipality itself and the Department of Mathematics and Geosciences of the Trieste University "Elaborazione del modello idrogeologico del sottosuolo dell'area del Parco delle Mucille di Selz" and its extensions (POS. 2175 del 30/06/2017) of which Prof. Luca Zini was the Scientific Coordinator.

## References

Calligaris C, Casagrande G, Iervolino D, Lippi F, Olivo P, Ramani M, Treu F, Zini L (2019) Water–budget as a tool to evaluate the sustainable use of groundwater resources (Isonzo Plain, NE Italy). ROL 47:7–12. https://doi.org/10.3301/ROL.2019.02

Cancian G (1987) L'idrologia del Carso goriziano-triestino tra l'Isonzo e le risorgive del Timavo. Studi Trentini Di Scienze Naturali 64:77–98

Civita M, Cucchi F, Eusebio A, Garavoglia S, Maranzana F, Vigna B (1995) The river Timavo: an important supplementary water resource which needs to be protected and regained. Acta Carsologica 24:169–186

Doctor DH, Lojen S, Horvat M (2000) A stable isotope investigation of the classical Karst aquifer: evaluating karst groundwater components for water quality preservation. Acta Carsologica 29(1):79–82

Gabrovšek F, Peric B (2006) Monitoring the flood pulses in the epiphreatic zone of karst aquifer: the case of Reka river system, Karst plateau, SW Slovenia. Acta Carsologica 35(1):35–45

Gemiti F, Licciardello M (1977) Indagini sui rapporti di alimentazione delle acque del Carso triestino e goriziano mediante l'utilizzo di alcuni traccianti naturali. Annali Gruppo Grotte Ass. 30 Ott. 6:43–61

Hirabayashi Y, Mahendran R, Koirala S, Konoshima L, Yamazaki D, Watanabe S, Hyungjun K, Shinjiro K (2013) Global flood risk under climate change. Nat Clim Change Lett 3:816–821. https://doi.org/10.1038/NCLIMATE1911

Jurkovšek B, Biolchi S, Furlani S, Kolar-Jurkovšek T, Zini L, Jež J, Tunis G, Bavec M, Cucchi F (2016) Geology of the classical Karst region (SW Slovenia–NE Italy). J Maps 12(1):352–362. https://doi.org/10.1080/17445647.2016.1215941

Kvočka D, Falconer RA, Bray M (2016) Flood hazard assessment for extreme flood events. Nat Hazards 84:1569–1599. https://doi.org/10.1007/s11069-016-2501-z

Mosetti F, D'Ambrosi C (1963) Alcune ricerche preliminari in merito a supposti legami di alimentazione fra il Timavo e l'Isonzo. Boll Geofis Teor Appl 5:69–83

Samez D, Casagrande G, Cucchi F, Zini L (2005) Idrodinamica dei laghi di Doberdò e di Pietrarossa (Carso Classico, Italia). Relazioni con le piene dei fiumi Isonzo, Vipacco e Timavo. Atti e Memorie Commissione Grotte "E. Boegan" 40:133–152

Tabari H (2020) Climate change impact on flood and extreme precipitation increases with water availability. Sci Rep 10:13768. https://doi.org/10.1038/s41598-020-70816-2

Timeus G (1928) Nei misteri del mondo sotterraneo: risultati delle ricerche idrogeologiche sul Timavo 1895–1914, 1918–1927. Atti e Memorie Commissione Grotte "E. Boegan" 22:117–133

Urbanc J, Mezga K, Zini L (2012) An assessment of capacity of Brestovica–Klariči karst water supply (Slovenia). Acta Carsologica 41(1):89–100

Zini L, Calligaris C, Zavagno E (2013) Classical Karst hydrodynamics: a sheared aquifer within Italy and Slovenia. In: Evolving water resources systems: understanding, predicting and managing water-society interactions. IAHS Publication, vol 364, pp 499–504. ISSN 0144-7815

Zini L, Calligaris C, Cucchi F (2015) The challenge of tunneling through Mediterranean karst aquifers: the case study of Trieste (Italy). Environ Earth Sci 74:281–295. https://doi.org/10.1007/s12665-015-4165-5

Zini L, Calligaris C, Cucchi F (2022) Along the hidden Timavo. Geol Field Trip Maps 14/1.3:1–69. https://doi.org/10.3301/GFT.2022.03

Zini L, Visintin L, Cucchi F, Boschin W (2011) Potential impact of a proposed railway tunnel on the karst environment: the example of Rosandra valley, Classical Karst Region, Italy-Slovenia. Acta Carsologica 40(1): 207–218

# Impacts of Recharge and Discharge on Sustainability of the Trinity Aquifers of Central Texas

B. A. Smith, B. B. Hunt, J. Watson, and J. Camp

**Abstract**

The karstic Trinity Aquifers of central Texas provide baseflow to streams and are used extensively as water supplies for domestic, municipal, agricultural, and industrial purposes. Rapid increases in population in the area are placing significant demands on the aquifers in an area that has very limited surface water supplies and is prone to significant droughts. Droughts occur frequently in this area and current levels of pumping in the Trinity Aquifer have resulted in both the capture of springflow in karstic areas, resulting in a major spring ceasing flow during drought and groundwater mining in other less karstic areas of the Trinity. The most significant historical drought in central Texas occurred in the 1950s and lasted for up to 10 years. Tree-ring data show that even more significant droughts have occurred over the past thousand years. Predictions for changes to precipitation due to climate change are for more extensive flooding and more severe drought. Long-term trends in water levels are downward with only limited recovery during very wet periods. An understanding of how these aquifer systems function is key to proper aquifer management which is likely to involve limits to pumping during average conditions and more severe reductions during periods of drought and incorporating alternative water supplies such as rainwater harvesting, aquifer storage and recovery, and perhaps importation of water from distant sources.

**Keywords**

Karst • Trinity Aquifer • Drought • Sustainability

B. A. Smith (✉) · J. Watson · J. Camp
Barton Springs/Edwards Aquifer Conservation District, 1124 Regal Row, Austin, TX, USA
e-mail: brians@bseacd.org

B. B. Hunt
Bureau of Economic Geology, University of Texas, Austin, TX 78704, USA

## 1 Introduction

The Trinity Aquifers of central Texas have been used as a source of water for many years, and water discharging from springs provides base flow to streams in the area. Base flows in streams that are fed by the Trinity Aquifers provide recharge to downgradient portions of the Trinity Aquifers and to the Edwards Aquifer. Rapid population growth in recent years has significantly increased the demand for groundwater in the area. Hays County, which is at the center of the study area, has been the fastest growing county in the USA from 2010 to 2020 (U.S. Census Bureau 2021). The certain increase in demand for groundwater coupled with severe drought will likely cause significant reductions in springflow and decreases in water levels. This will cause many water-supply wells to cease the ability to yield water and negative ecological impacts along spring-fed streams, both of which will lead to serious economic consequences for the area. Furthermore, reduction of Trinity Aquifer springflow will reduce recharge to the Edwards Aquifer that hosts a variety of endangered aquatic species at Barton and San Marcos Springs.

Studies are being conducted to better understand the aquifers of central Texas so that proper management of these aquifers will allow for sufficient groundwater of good quality to be available for human and ecological purposes.

## 2 Hydrogeology

The geologic units that make up the Trinity Aquifers in the study area (Fig. 1) are largely limestones and dolomites of Early Cretaceous age (Fig. 2). These sediments were deposited on a broad shelf that separated the deeper water of the Gulf of Mexico to the southeast and land of the Llano Uplift to the northwest. Some of the depositional environments in which these sediments were deposited are coastal plains, tidal flats, shorelines, and a variety of shallow-water


B. Andreo et al. (eds.), *EuroKarst 2022, Málaga*, Advances in Karst Science,
https://doi.org/10.1007/978-3-031-16879-6_3

environments (Hunt et al. 2020). These units are nearly flat lying with a gentle dip to the east and southeast (Wierman et al. 2010). Major faulting occurred during the early Miocene along the Balcones Fault Zone. The strikes of these faults are generally north-northeast, and most of the faults occur in central and eastern Hays County and in adjacent counties to the north and south (Fig. 1). The Balcones Fault Zone consists of a series of en-echelon normal faults, down-to-the southeast. Fracturing associated with the faulting shattered the brittle carbonate rocks, providing pathways for groundwater movement and the development of solution conduits.

The eastern portion of the study area contains the Edwards Group at and near the surface. Stratigraphically beneath the Edwards Group is the Trinity Group that is exposed in the western portion of the study area. In the western side of the study area, the outcrops are dominated by the Lower Glen Rose Member within the river valleys, with the Upper Glen Rose Member making up the areas of higher elevation.

The Upper Glen Rose Member (Fig. 2) is 108 m (355 ft) thick in the upper reaches of the Onion Creek watershed and thickens to about 137 m (450 ft) in the eastern portion of the study area. In outcrop, the Upper Glen Rose Member is subdivided into eight informal lithologic units, which correlate to the classic work of Stricklin et al. (1971). These units generally consist of stacked and alternating limestones, dolomites, mudstones, and marls.

The Lower Glen Rose Member (Fig. 2), about 76 m (250 ft) thick, is characterized by fossiliferous limestone units with well-developed rudistid reef mounds and biostromes often found near the top and base of the unit. The shaley and dolomitic Hensel, about 10 m (33 ft) thick, is also exposed in the incised river valleys and locally provides semi-confining aquifer properties. The Cow Creek Limestone is about 23 m (75 ft) thick. The upper portion of the

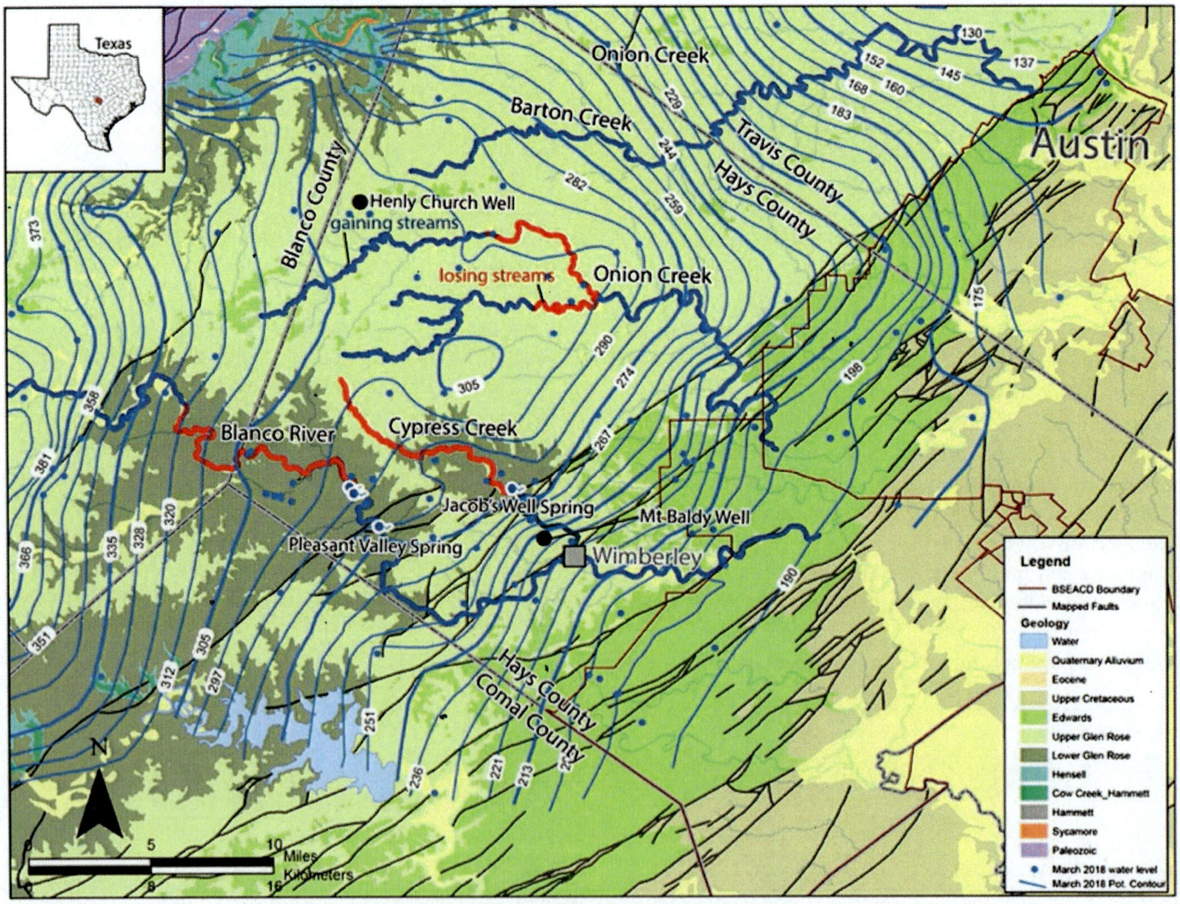


**Fig. 1** Geologic map of study area including county boundaries, major surface streams, locations of wells for water-level measurements, and potentiometric contours. The Barton Springs/Edwards Aquifer Conservation District (BSEACD) boundary is represented by dark red lines in eastern Hays County. The Hays Trinity Groundwater Conservation District (HTGCD) makes up the remainder of Hays County west of BSEACD. Potentiometric contours are in meters

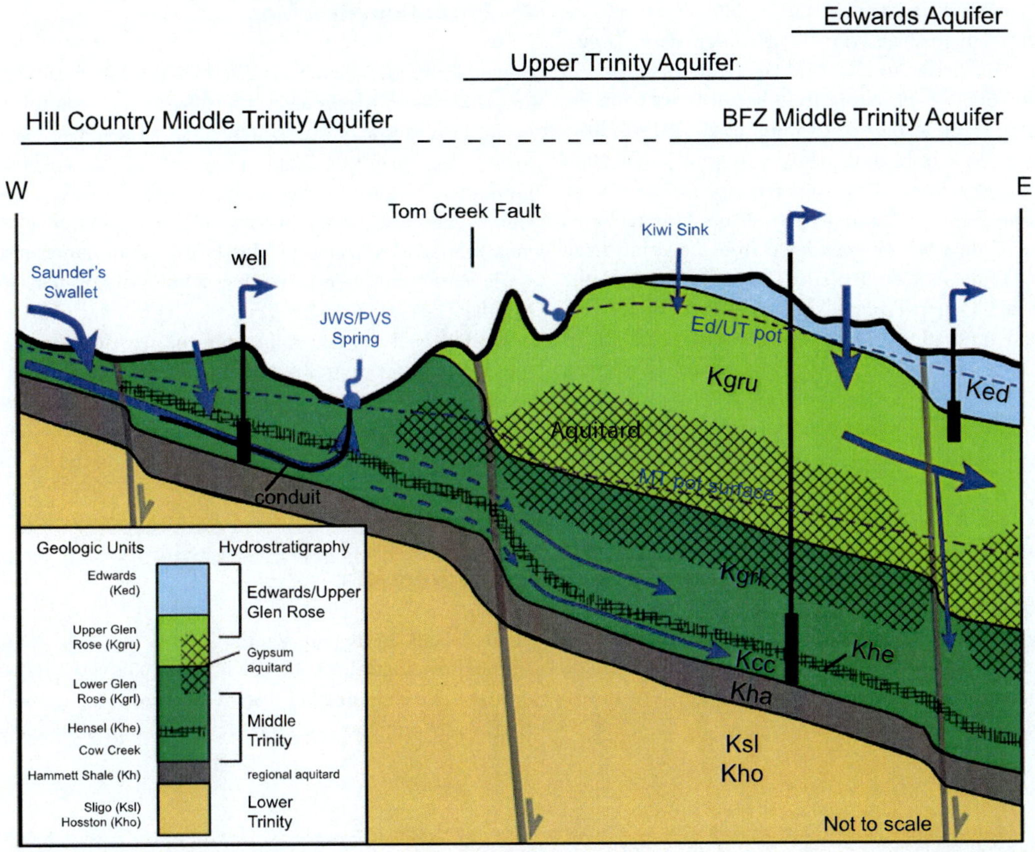


**Fig. 2** Schematic cross-section of the aquifers and formations that are the subjects of this paper and arrows showing generalized directions of flow (Smith et al. 2018)

Cow Creek Limestone is a cross-bedded grainstone unit that is often limestone but can also be dolomite. The lower portion of the Cow Creek Limestone becomes more dolomitic and silty with depth grading into the underlying Hammett Shale. The Hammett Shale serves as a confining unit between the Middle and Lower Trinity Aquifers.

These carbonate units have undergone varying degrees of karstification by hypogenetic processes and by later eogenetic processes. Groundwater flow in these aquifers is a combination of flow through the primary porosity matrix of the rocks, flow through localized pores that have developed by dissolution of the rock, and through conduits that have developed by dissolution along faults, fractures, and bedding planes. Little is known about the distribution of these porosity types within the aquifers because of the lack of accessibility to the deeper portions of these aquifers. Observations of structural and karst features at the surface provide some clues to the nature of the deeper karst. Solution voids, such as caves and sinkholes, are common throughout the study area where these limestones and dolomites are exposed at the surface.

## 3 Recharge

Recharge to the Trinity Aquifers has not been well quantified and the distribution of recharge across geographic locations has not been determined in detail. However, numerical modeling efforts that are underway are evaluating rainfall, streamflow, and water-level data to gain a better understanding of recharge in the study area. Part of these efforts will be to determine how much recharge is taking place along streams and how much is taking place in the uplands.

Recent studies (Smith et al. 2015; Hunt et al. 2017) indicate significant recharge to the Upper and Middle Trinity Aquifers occurs from losing streams, such as the Blanco River, Cypress Creek, Onion Creek, and Barton Creek (Fig. 1). The losing reaches of the Blanco River and Cypress Creek sustain Pleasant Valley Spring and Jacob's Well Spring, respectively. Any water in the Cow Creek that does not exit the aquifer at the springs provides flow to the deeper part of the Middle Trinity Aquifer to the east.

Another source of recharge to the Middle Trinity is vertical leakage from the overlying Upper Glen Rose (Jones et al. 2011). Recharge to the Middle Trinity Aquifer in central Hays County is primarily from lateral flows from the updip recharge areas to the west (Smith et al. 2015). However, there is some indication of a hydrologic connection (vertical leakage) from the Upper Glen Rose into the Middle Trinity either due to significant head gradients from recharge in the Upper Glen Rose or drawdown from pumping from the Middle Trinity Aquifer (BSEACD 2017). Much of the recharge to the Upper Trinity Aquifer is from direct precipitation and infiltration in upland areas where the Upper Trinity geologic units are exposed at the surface (Wierman et al. 2010).

Recharge from Barton Creek to the Trinity Aquifers has not been delineated in detail, but flow measurements and dye-trace studies have shown that recharge to the Trinity Aquifers is taking place along some segments of Barton Creek. At one location on Barton Creek, about 7 km (4.3 miles) above the Hays–Travis County boundary, a dye-trace study was conducted in which dye was injected into swallets in the bed of Onion Creek where the Upper Glen Rose Member outcrops. The dyes were detected in several Middle Trinity water-supply wells as much as 5.3 km (3.3 miles) downgradient of the injection points.

## 4 Potentiometric Map

Some of the tools being used to characterize these aquifers are a network of pressure transducers in monitor wells, synoptic water-level measurements, streamflow measurements for gain/loss studies, geochemical analyses of groundwater and surface water, aquifer testing, dye-trace studies, and analytical and numerical modeling. Water-level measurements collected in March 2018 from more than 110 wells were used to construct the potentiometric map for the Middle Trinity Aquifer (Fig. 1). The map shows that groundwater flows generally from the west to the east over the western and central portions of the study area. Near the center of Hays County, groundwater flow diverges with flow to the northeast and to the southeast. Further east in the study area, the geologic units are faulted deeper into the subsurface where there are no wells.

## 5 Hydrographs

Hydrographs from two Middle Trinity monitor wells and from a flow gage at Jacob's Well Spring show the impacts of drought and pumping on water levels and spring discharge (Fig. 3). Water levels in the Henly Church well in the far

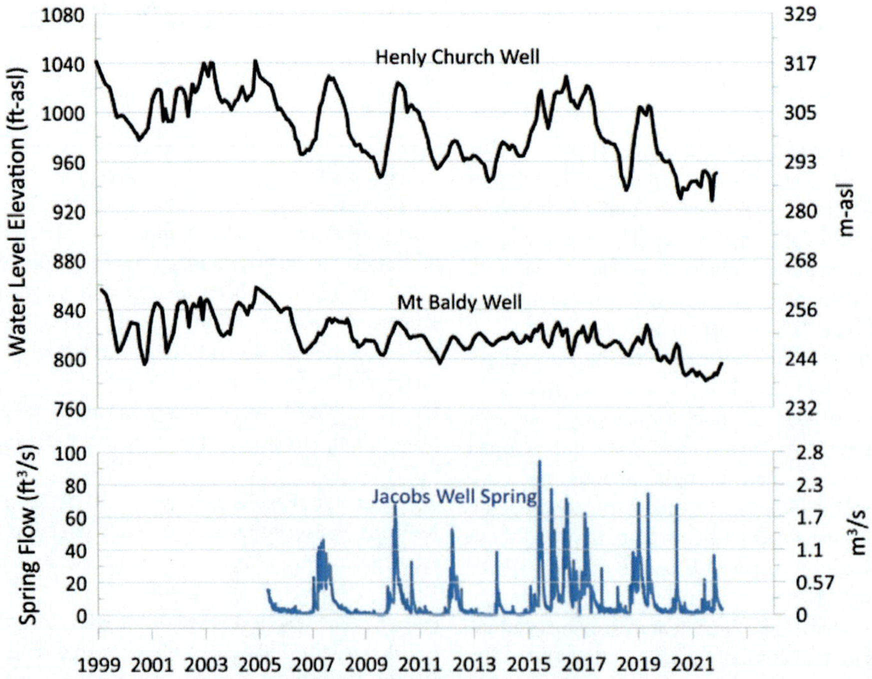


**Fig. 3** Hydrographs of water levels in Henly Church and Mt. Baldy wells and flow from Jacob's Well Spring

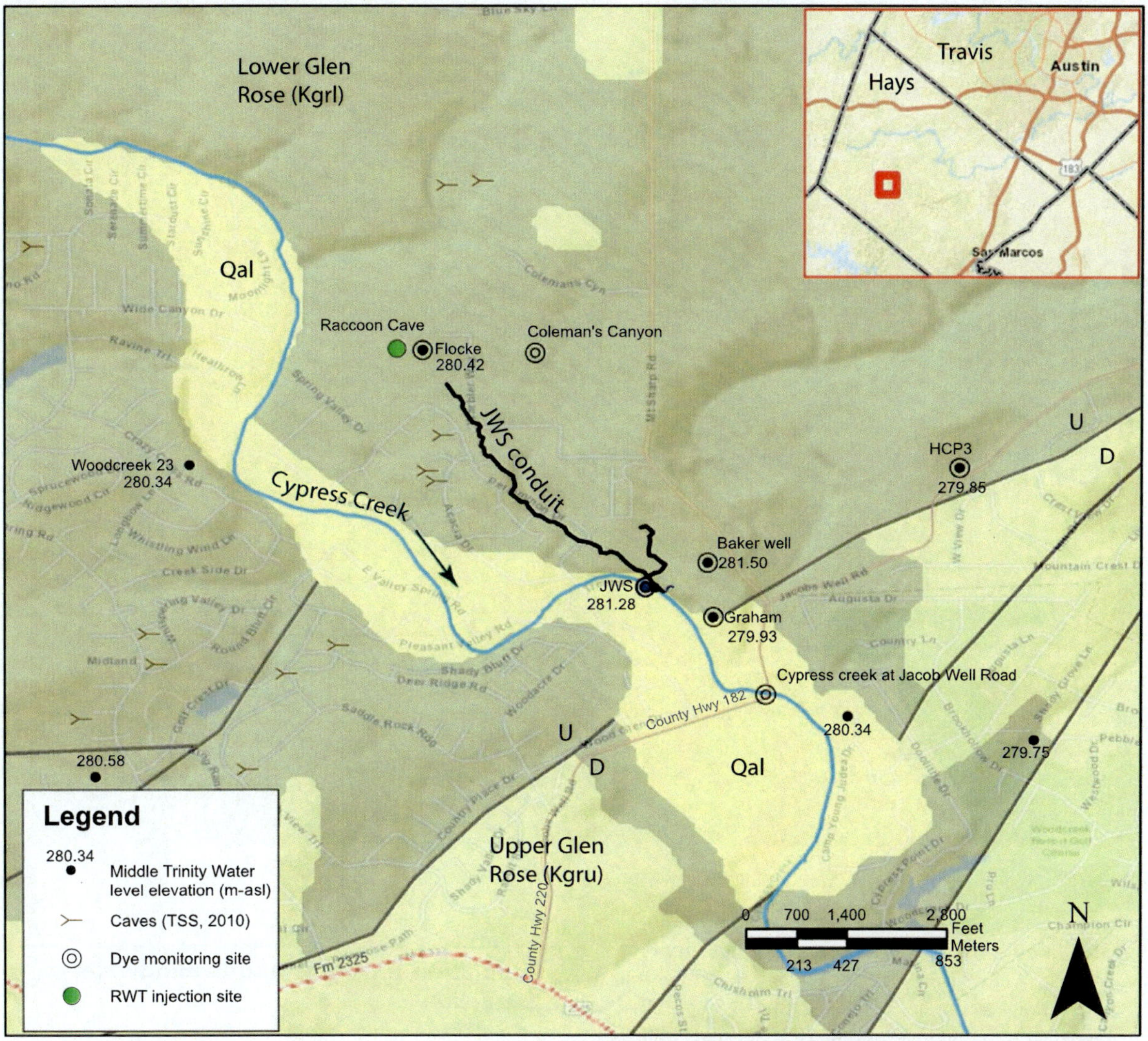


Sources: Esri, HERE, DeLorme, USGS, Intermap, increment P Corp., NRCAN, Esri Japan, METI, Esri China (Hong Kong), Esri (Thailand), MapmyIndia, © OpenStreetMap contributors, and the GIS User Community, Sources: Esri, DeLorme, USGS, NPS, Sources: Esri, USGS, NOAA

**Fig. 4** Geologic map of the area around Cypress Creek including Jacob's Well Spring (JWS) and the underwater cave passage shown with heavy black line

western portion of Hays County (Fig. 1) show an overall decline from 1999 with distinct highs and lows representing wet and dry conditions, respectively. Water levels in the Mt. Baldy well northwest of Wimberley (Fig. 1) show more moderate highs and lows, but the overall trend shows a more moderate decrease from 1999 until 2019 when water levels make a steeper drop.

Flow from Jacob's Well Spring is the primary source of flow in Cypress Creek which feeds into the Blanco River in the town of Wimberley (Figs. 1, 4, and 5). A large conduit, accessible by cave divers, is the main pathway for water to discharge from Jacob's Well Spring. This cave has been mapped to a length of 1723 m (5655 ft) with the main conduit having a length of 1323 m (4341 ft) and a northwest-southeast strike. The main conduit is situated in the Cow Creek and Hensel formations.

Continuous flow measurements from Jacob's Well Spring began in 2005 (Fig. 3). Long-term average discharge from Jacob's Well Spring is about 0.5 $m^3/s$ (17 $ft^3/s$-cfs). Reports from residents indicate that during the drought-of-record of

**Fig. 5** Jacob's Well Spring on Cypress Creek, central Hays County

the 1950s, Jacob's Well Spring never ceased flowing. Flow from Jacob's Well Spring ceased altogether during four periods of drought between 2005 and 2014. Since 2014, there have only been moderate droughts in the study area as reflected in the hydrograph for Jacob's Well Spring (Fig. 3). However, with above average rainfall, including record-setting rains in 2015, the Henly Church and Mt. Baldy monitor wells show water levels in 2020 and 2021 were lower than any time since 1999 when water-level measurements were first made with continuously recording instruments. And the water-level highs from 2015 to 2017 were lower than the highs in 2002 and 2004 when rainfall amounts were considerably less. The most likely explanation for this discrepancy is the increases in groundwater extraction that have been taking place in recent years.

## 6 Population Increases and Groundwater Extraction

There is very little firm data for the number of water-supply wells in the area and even less for the amount of extraction from these wells. Most of the wells that extract groundwater from the Trinity Aquifers are low-production domestic wells for which there are no meters to record production. Wells that supply water to housing developments or to commercial entities are permitted by the groundwater districts and in recent years, there are records for this extraction. Groundwater districts in the study area are collecting more information about new wells being installed and learning more about wells that were installed in the past. But any estimates for groundwater extraction is very approximate. Better estimates of increases in groundwater extraction can be made from population records. Considering that there is very little surface water available in the study area, increases in groundwater extraction should coincide with changes in population. According to the U.S. Census Bureau (2021), Hays County is the fastest growing county in the United States for counties with populations greater than 100,000 for the census period 2011–2020. The population of Hays County increased by 53% since 2010 to 241,067 people. These trends in population growth and groundwater extraction suggest that unreasonable impacts to the Trinity Aquifers are likely to happen in the not-too-distant future. Recent studies have shown that in portions of southwest Travis County, the Middle Trinity Aquifer is depleted of groundwater such that new wells are being completed into the Lower Trinity Aquifer (Hunt et al. 2020).

## 7 Groundwater Management

Groundwater management in Texas is conducted largely by groundwater conservation districts that are authorized by the State of Texas to regulate the production of groundwater. Groundwater conservation districts cover less than two thirds of the state, and the districts have varying authority to regulate groundwater. Groundwater production in other parts of the state is not regulated. Within the study area (Fig. 1), which consists of the western two thirds of Hays County, two groundwater districts regulate the use of groundwater, the Barton Springs/Edwards Aquifer Conservation District (BSEACD), and the Hays Trinity Groundwater Conservation District. BSEACD manages the aquifers within the district by assessing the sustainable yield of the aquifers. This involves regulating the aquifers in a manner to avoid unreasonable impacts to the aquifers and related water resources. Unreasonable impacts are considered by BSEACD to be loss of yield of groundwater to existing wells, degradation of water quality, depletion of groundwater over a long-term basis, significant decreases in springflow, and ecological impacts to surface waters. Avoidance of unreasonable impacts to the aquifers is achieved by conducting scientific studies and promulgating rules in a process that involves various stakeholders in the development of groundwater management policy.

## 8 Conclusions

Increases in population in the study area are bringing about a significant increase in groundwater extraction. Multiple sources of data indicate that water levels and springflow are decreasing even as rainfall is increasing. Recent data suggest that groundwater in the study area is being depleted to varying degrees. Some areas have already lost the ability to extract groundwater from some portion of the aquifers and must now drill deeper or look for other sources of water such as rainwater harvesting, aquifer storage and recovery, desalination of saline groundwater, and importing water by pipelines from distant sources.

Because of the lack of groundwater data from the most intense droughts in central Texas, collecting data from numerous monitoring points is critical to this process. This involves developing more monitoring wells with a broad geographic distribution. Measuring flow in the streams that contribute water to the aquifers and that receive flow from springs must be conducted under various flow conditions. Only limited tracer tests have been conducted in the study area, so more tracer testing needs to be conducted to better understand groundwater flow paths and velocities of flow. Numerical modeling of these aquifers will be a critical tool in understanding these aquifers and making predictions about future groundwater availability under various pumping and drought scenarios. A better understanding of population growth and groundwater extraction is needed. Proper management of the aquifers will provide long-term use of the aquifers for groundwater users and protection for the various ecologic resources associated with these aquifers.

The keys to effective management of the Trinity Aquifers will be conducting a high level of scientific studies and involving an advisory team of groundwater scientists with involvement of a diverse group of stakeholders to advise the process and promulgation of rules that meet the legal and societal requirements for groundwater management.

## References

BSEACD (Barton Springs/Edwards Aquifer Conservation District) (2017) Hydrogeologic setting and data evaluation: 2016 electro purification aquifer test, cow creek well field: Hays County, Texas. Technical Memo 2017–1010. 39p. + Appendices

Hunt BB, Smith BA, Gary MO, Broun AS, Wierman DA, Watson J, Johns DA (2017) Surface-water and groundwater interactions in the Blanco River and Onion Creek watersheds: implications for the Trinity and Edwards Aquifers of Central Texas. South Texas Geol Soc Bull 57(5):33–53

Hunt BB, Cockrell LP, Gary RH, Vay JM, Kennedy V, Smith BA, Camp JP (2020) Hydrogeologic atlas of southwest Travis County, central Texas: Barton Springs/Edwards Aquifer conservation district report of investigations 2020–0429, April 2020, 80p. + digital datasets

Jones IC, Anaya R, Wade SC (2011), Groundwater availability model: Hill Country portion of the Trinity Aquifer of Texas. Texas Water Development Board Report 377, 165p

Smith BA, Hunt BB, Andrews AG, Watson JA, Gary MO, Wierman DA, Broun AS (2015), Hydrologic influences of the Blanco River on the Trinity and Edwards Aquifers, Central Texas, USA. In: Andreo B (ed) Hydrogeological and environmental investigations in karst systems. Springer, Berlin, pp 153–161

Smith BA, Hunt BB, Wierman DA, Gary, MO (2018) Groundwater flow systems in multiple karst aquifers of central Texas: NCKRI symposium 6, 15 Sinkhole conference, West Virginia

Stricklin FL Jr, Smith CI, Lozo FE (1971) Stratigraphy of lower cretaceous trinity deposits of Central Texas. The University of Texas at Austin, Bureau of Economic Geology Report of Investigations No. 71. 63p

U.S. Census Bureau (2021) https://data.census.gov/cedsci/profile?g=0500000US48209

Wierman DA, Broun AS, Hunt BB (eds) (2010) Hydrogeologic atlas of the Hill Country Trinity Aquifer, Blanco, Hays, and Travis Counties, Central Texas. Barton Springs/Edwards Aquifer conservation district 17 (11 × 17 inch) plates

# Exploration of Karst Groundwater and Surface Water in Central Middle Atlas, Morocco

Abdelghani Qadem, Sébastien Lebaut, and Zohair Qadem

**Abstract**

The Middle Atlas, water tower of Morocco, is largely composed of a limestone substratum, which implies groundwater redistribution towards karstic aquifers. In its central part, the Middle Atlas is drained by Sebou River upstream. Many authors have studied this redistribution essentially from the group of springs flows downstream of the Upper Sebou catchment and from the upstream of the Oum Er Rbia basin which drains the southern Middle Atlas. We refine and discuss these proposals on the basis of new results concerning hydrological balances calculated from flow data available at 5 hydrometric gauges and the spatialization of climatic variables $P$ and ET in this area. It seems that 30% of the rainfall infiltrated into the Upper Sebou basin is transferred outside the basin, 25% to the Oum Er Rbia and 5% to the Saïs plain. We hypothesize that the group of springs located at the extreme downstream end of the basin and making up a third of the Sebou flow at its outlet from the Middle Atlas is fed by infiltration into the upstream sub-basins that are right bank tributaries of the main course of the Sebou.

**Keywords**

Middle Atlas • Karst • Water budget • Spring

A. Qadem (✉)
Laboratory DLRH, Sultan Moulay Slimane University, Beni Mellal, Morocco
e-mail: qademabdo@gmail.com

S. Lebaut
UR LOTERR, University of Lorraine, Nancy, France

Z. Qadem
GeoEA, Sidi Mohamed Ben Abdelah University, Fes, Morocco

## 1 Introduction

The central Middle Atlas has already been the subject of numerous karst hydrology studies. These researches, punctual in space or in time, have demonstrated the complexity of the hydrological and hydrogeological functioning due to its essentially karstic substratum. Indeed, net precipitation infiltrates widely and is subject to underground redistribution due to tectonic accidents or karst landforms. The studies, carried out mainly on springs data, show that there is no concordance between topographical and functional catchment areas. While all authors agree on the role played by the Liassic carbonates in the infiltration of rainwater which they drain to recharge underground reservoirs, there is no consensus on the general modalities of transfer/circulation of this water within the sector and towards the surrounding areas. The purpose of this paper is to hypothesize inter-basin transfers and the volumes involved using hydrological balances at hydrometric stations. This is a classical hydrological approach to establishing and performing hydrological balances from recorded hydroclimatic data. These assumptions are made by means of a spatialized quantification of climatic variables, precipitation and ET from ground-based and spatialized data and from earth observation satellites data. These input values of the hydrological system compared to the observed flows allow hypotheses to be made regarding the underground redistribution of the infiltrated water. These hypotheses are discussed on the basis of the conclusions already reached by authors who have taken an interest in the subject.

## 2 Study Area

In geological terms, the Middle Atlas belongs to the intracontinental Atlas system which crosses North Africa from west to east, from the Atlantic to the central Mediterranean along the 34° north latitude parallel. It constitutes a branch



B. Andreo et al. (eds.), *EuroKarst 2022, Málaga*, Advances in Karst Science,
https://doi.org/10.1007/978-3-031-16879-6_4

which crosses the Mesetas domain in an oblique way (Fig. 1) with altitudes exceeding 3000 m asl (Michard et al. 2008).

The central Middle Atlas is part of two morphostructural units. To the west, dominating the western Meseta, is the tabular Middle Atlas; it corresponds to a karst (Causse in French) established in carbonated sedimentary sequences of the Lower Jurassic (Trias, Lias) whose altitudes are around 1200 m asl. This karst is marked by volcanic units and basaltic flows dating from the Cenozoic (Frizon de Lamotte et al. 2008). The eastern part corresponds to the folded Middle Atlas, which is bordered by two major NE-SW faults. It corresponds to a succession of synclinal basins in which marls outcrop and anticlinal ridges in dolomites corresponding to a mountain relief whose highest point is 3081 m asl (Jebel Bou Iblane). But the main characteristic of the central Middle Atlas is the influence of limestone and karstic phenomena (De Waele 2009) (Fig. 2) due to the erosion and dissolution processes that affected the Eocene palaeo-surface (Beaudet and Martin 1967). The hydrological functioning of the hydrosystem results from this geological/hydrogeological situation. On the Causses, which are crossed by valleys with no flow, drainage is almost absent over large areas. The generalized infiltration of water in the very permeable substratum feeds the resurgences at the base of the hillslopes in contact with impermeable formations (Amraoui et al. 2003). In this case, underground water circulation is the dominant feature of the hydrosystem. The other feature of the landscape is related to the surface forms of the karst. Indeed, there are a large number of lakes in the poljes, dry or wet valleys and dolines resulting from the intense Mio-Pliocene karstification phase, which is now almost inactive (Martin 1981).

From a climatic perspective, the Haut Sebou basin is in a transitional zone. In the west of the basin, the climate is rather humid, with significant rainfall on the Middle-Atlasic Causse (Ifrane, Azrou). The high reliefs can receive more than 1000 mm of rainfall. The basins such as the Guigou plain or the Maâsser basin are considered to be under a semi-arid influence with rainfall of around 500–600 mm due to the protection of the reliefs (Ait Khabbach, Sefrou), although there are distinctions to be identified. These climatic nuances within the catchment area result from the

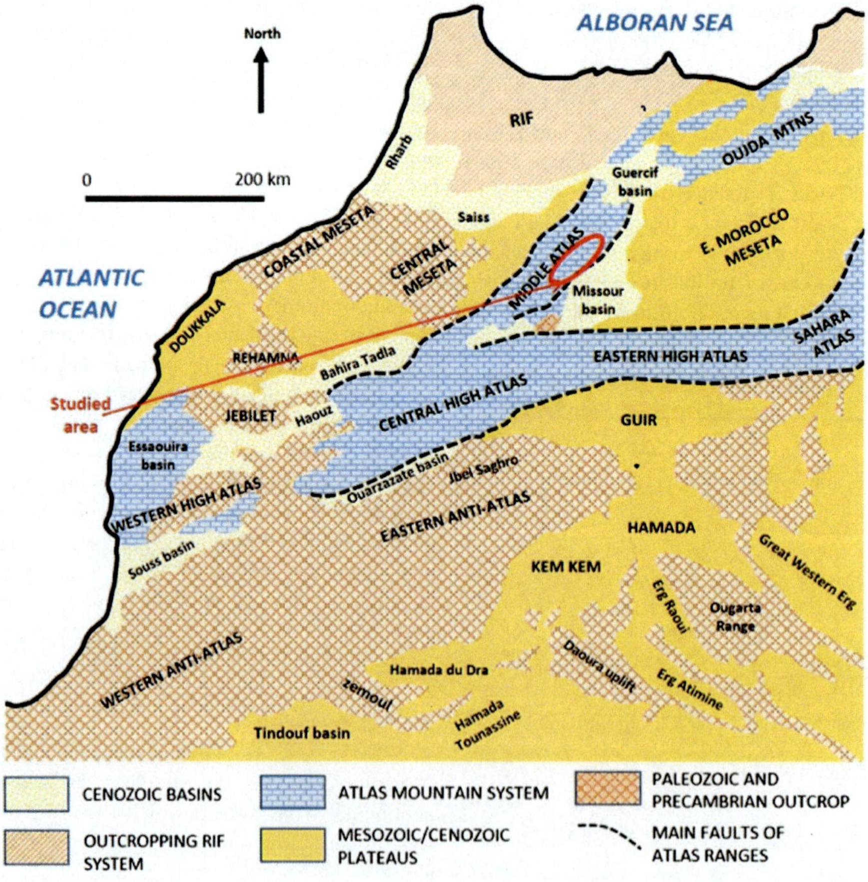


**Fig. 1** Geological context of the central Middle Atlas at the scale of Morocco (adapted from Michard et al. 2008)

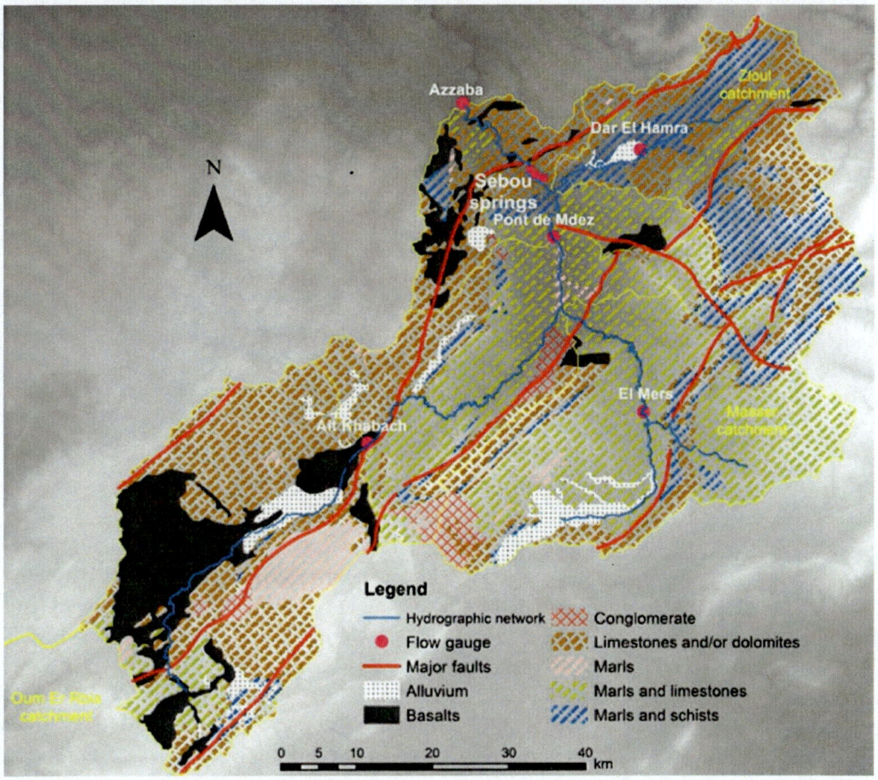


**Fig. 2** Geology of the Sebou basin (according to Terminer 1936)

different morphologies of the relief as well as the phenomenon of continentalization.

## 3 Material and Method

### 3.1 Water Balance

In a catchment area, water balance can be represented by conservation equations, and the simplest model is characterized by its water reserve $S$ and the quantities of incoming water $P$ and outgoing water $Q$ and ET between two-time references $t1$ and $t2$, most often corresponding to the hydrological year (Fig. 3):

$$S|t2 - S|t1 = \Delta S = P - Q - \text{ET}$$

Term $P$ represents precipitation. Water reserves are formed by groundwater, soil and surface water. Loss terms $Q$ and ET are, respectively, the runoff and the total evapotranspiration of the catchment. When the change in water supply between $t1$ and $t2$ can be considered negligible a balance equation is:

$$\text{ET} = P - Q = \text{FD} \quad \text{with FD: flow deficit}$$

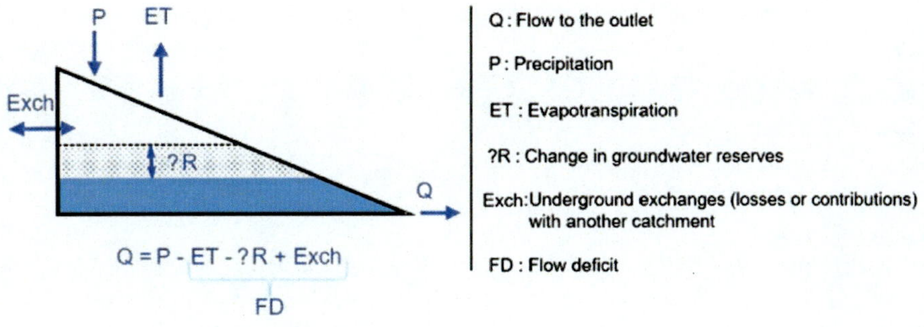


**Fig. 3** Representation of the components of the water balance

Characteristic of a bioclimatic context (Réméniéras 1986), ET is more or less stable in a homogeneous climatic zone. By comparing FD with ET of a particular catchment, it is possible to detect whether FD represents only ET or whether it is composed of another variable. In the latter case, one can assume open catchments and calculate the volumes involved in these inter-catchment transfers (Fig. 3).

In this case, the runoff deficit, in addition to evapotranspiration, includes exchanges between basins as well as possible variations in the hydrogeological reserve on either side of the hydrological year considered. Applied to karst basins, these volumes allow the calculation of a real hydrogeological catchment area and the inference of water volume transfers between catchments.

## 3.2 Data

- **Runoff data**

Hydrological analysis of the Upper Sebou basin can be based on the flow values observed from 5 hydrometric gauges managed by the Sebou Basin Hydraulic Agency (Table 1). Three stations are set up on the main drain and two on the most important tributaries (Mâasser and Zloul). The analysis is also based on the flows observed at the three springs Aïn Sebou, Timedrine and Ouemander.

Our analysis is based on two hydrological years for which we have a maximum of data in all compartments: surface flows, spring flows and precipitation. As the Aït Khabbach station ceased to operate in 1982, the flows we use are the average flows for the period 1970/1982.

- **Rainfall**

Topography conditions the spatial distribution of rainfall through the orographic effect due to the ascent of air masses, with adiabatic cooling inducing condensation of moist air on windward slopes (Chow et al. 1988). However, the relationship between precipitation and elevation is complex and depends on the way the region is exposed to the prevailing winds and on the synoptic conditions. To overcome this problem, we adopted a precipitation modelling method that aims to predict the values taken by a group of variables *Y* (variables to be predicted, target variables and explained variables) from a series of variables *X* (predictor variables, the descriptors and explanatory variables).

This method, presented extensively in the thesis of Qadem (2015), gives similar results to those obtained by Nejjari (2002) by kriging. For the hydrological years 1984/1985 and 1985/1986, the rainfall calculated by the model on the Haut Sebou catchment area in Azzaba is, respectively, 338 and 462 mm. They are very close to those of Nejjari calculated by kriging which are, respectively, 359 and 451 mm, that is to say differences of the order of 5% for the two methods.

- **Evapotranspiration**

Remote sensing data that are spatially distributed can be relevant control variables in the calculation of ET, and there is a large number of publications showing that these satellite data can be used to determine ET from regional to global scales (Bastiaansen et al. 1998). We rely on the MOD 16 product values available since 2000 at a spatial resolution of 1000 m. Estimates based on MODIS products although unable to account for soil surface heterogeneity at the plot scale are able to reproduce the average catchment response (Velpuri et al. 2013). For arid and semi-arid environments, several authors have shown that results are acceptable at regional and watershed scales (Aguilar et al. 2018).

# 4 Results and Discussion

We do not have MODIS ET values for the years 1984/86 but a regression between P and ET over the period 2000/2010 shows an acceptable $R^2$ (0.75), so it was possible to calculate ET values from rainfall values. The ET values for the Upper Sebou basins are between 189 and 226 mm/year (Table 2). These values are close to those found by Chaponnière (2005) for the Rerhaya basin located slightly further south for which about 225 mm is proposed.

**Table 1** Flow gauges data

| River | Flow gauge | Putting in to operation (year) | Catchment area (km$^2$) | Elevation outlet (m asl) | *X* | *Y* |
|---|---|---|---|---|---|---|
| Guigou | Ait khabach | 1970 | 1200 | 1478 | 557.00 | 314.80 |
| Maasser | El Mers | 1981 | 985 | 818 | 593.00 | 318.20 |
| Guigou | Pont Mdez | 1955 | 3460 | 725 | 581.40 | 341.90 |
| Zloul | Dar El Hamra | 1984 | 640 | 830 | 593.00 | 318.20 |
| Sebou | Azzaba | 1957 | 4677 | 478 | 596.65 | 359.57 |

**Table 2** Rainfall and ET values for the five basins

| | Azzaba | | Pont de Mdez | | El Mers | | Ait Khabbach | | Dar El Hamra | |
|---|---|---|---|---|---|---|---|---|---|---|
| mm | Rainfall | ETR | Rainfall | ETR | Rainfall | ETR | Rainfall | ETR | Rainfall | ETR |
| 1984/85 | 338 | 201 | 295 | 186 | 320 | 192 | 350 | 182 | 378 | 216 |
| 1985/86 | 462 | 218 | 364 | 196 | 484 | 210 | 457 | 195 | 496 | 236 |
| Mean | 400 | 210 | 330 | 191 | 402 | 201 | 404 | 189 | 437 | 226 |

**Table 3** Runoff means at flow gauges

| | Azzaba | Pont de Mdez | El Mers | Ait Khabbach | Dar El Hamra |
|---|---|---|---|---|---|
| Area ($km^2$) | 4677 | 3460 | 985 | 1200 | 640 |
| $m^3/s$ | 16.3 | 3.9 | 2.3 | 1.2 | 2.7 |
| $hm^3$ | 514.4 | 122.3 | 71.6 | 36.9 | 84.3 |
| mm | 110 | 35 | 73 | 31 | 132 |

**Table 4** Flow data at springs for 1984 to 1986

| | Aïn Oumender | | Aïn Timedrine | | Aïn Sebou | | 3 springs | |
|---|---|---|---|---|---|---|---|---|
| | $m^3/s$ | $hm^3$ | $m^3/s$ | $hm^3$ | $m^3/s$ | $hm^3$ | $m^3/s$ | $hm^3$ |
| 1984/85 | 0.23 | 7.29 | 0.70 | 22.15 | 2.06 | 65.07 | 6.49 | 94.51 |
| 1985/86 | 0.27 | 8.43 | 1.46 | 46.01 | 5.76 | 181.77 | 12.38 | 236.21 |
| Means | 0.25 | 7.86 | 1.08 | 34.08 | 3.91 | 123.42 | 9.44 | 165.36 |

**Table 5** Comparison of observed flows with theoretical flows from the water balance

| | | Azzaba | Pont Mdez | El Mers | Ait Khabbach | Dar El Hamra |
|---|---|---|---|---|---|---|
| P-ETR (mm) | | 190 | 139 | 201 | 215 | 211 |
| Losses | mm | 80 | 104 | 128 | 184 | 79 |
| | $hm^3$ | 374.2 | 358.6 | 126.4 | 221.1 | 50.8 |

On average for the two years considered, flow at the Azzaba station, the outlet of the Upper Sebou basin, is 16.3 $m^3/s$, which represents an average annual volume of 514.4 $hm^3$ (Table 3). At the same time, the flows measured on the three springs represent an average flow of 9.4 $m^3/s$ for an annual volume of 165.4 $hm^3$ (Table 4).

The first observation is that on the whole catchment area, in Azzaba, there is a loss of 80 mm of flow, which represents an average of 374 $hm^3$ over the two years in question (Table 5). At the same time, the upstream station of Aït Khabach is missing 221 $hm^3$. As many authors have shown, this upstream part of the tabular Middle Atlas belongs to the hydrogeological basin of the Oum Er Rbia located to the south. This volume of water therefore feeds the basin via the sources of the Oum Er Rbia. Considering the values of rainfall and evapotranspiration calculated on this part of the Upper Sebou basin, this transfer of 221 $hm^3$ corresponds to a drained impluvium surface of 871 $km^2$ for an average flow of 7 $m^3/s$ and a water level of 184 mm. The surface value is very close to the 884 $km^2$ proposed by Bentayeb et al. (1977) and 850 $km^2$ by Moniod (1973). From the considered transfer flow perspective, our values are also close to those proposed: Moniod (1973) proposes 196 mm and Loup (1962) proposes a value of 9 $m^3/s$ for the sources of the Oum Er Rbia fed by the upstream part of the Sebou. In the light of the literature, we can already say that our values confirm what other authors have proposed on the transfers at the level of the Upper Sebou/Oum Er Rbia interfluve.

Further downstream on the Upper Sebou, there is a loss of 358 $hm^3$ at the Pont de Mdez station. This missing volume is explained on the one hand by the 221 $hm^3$ transferred to the Oum Er Rbia basin and on the other hand by the 126 $hm^3$ missing at the El Mers station on the right bank tributary Maasser (Fig. 4). The missing 11 $hm^3$ can be explained by underground transfers towards the west in the direction of the edge of the tabular Middle Atlas from which many karstic springs emerge (Latati 1985; Belhassan et al. 2010). Further downstream, the Zloul, another right bank tributary of the

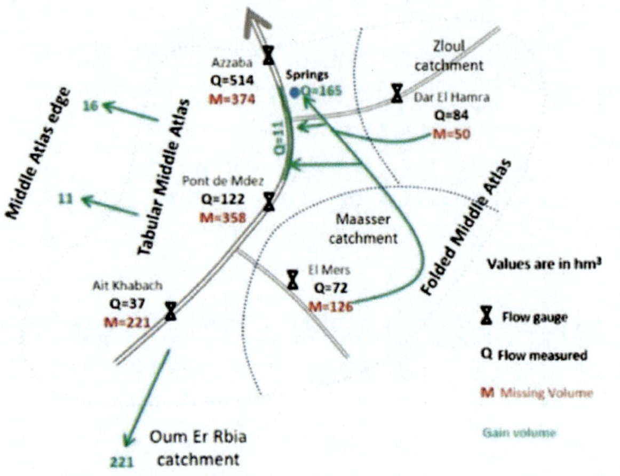


**Fig. 4** Groundwater circulation in the Upper Sebou basin

Upper Sebou, has a shortfall of 50 hm$^3$ at the Dar El Hamra station. From a geological point of view, it is located in the same formations as the Maasser. At the outcrop, the formations are essentially marl and limestone of the Toarcian and Middle Jurassic, but which can be largely fractured, whereas the culminating massifs correspond to the limestones and dolomites of the Lower and Middle Lias (Fig. 2). The 176 hm$^3$ missing from the El Mers and Dar El Hamra stations is therefore infiltrated into the deep karstic aquifers. This value is very close to the 165 hm$^3$ of the 3 Sebou springs. It is therefore quite possible to attribute the supply of the springs to losses on the left bank formations of the Sebou. This is in line with the hypothesis of El Khalki (1990) or Akdim et al. (2011) who attribute the supply of the group of springs to the limestone formations at the outcrop in the massifs located east of Haut Sebou catchment. Indeed, given this redistribution, it seems doubtful that it is the Causse of the Middle Atlas, on the right bank, which feeds the springs as proposed by Nicod (1993) or Nejjari (2002), or it would be necessary to admit that the volumes lost are drained outside the basin towards the east, which is not possible given the geological structure. This hypothesis is never mentioned in the literature and is ultimately not very probable as the geological structure as a whole suggests a convergence of underground flows towards the Mdez basin and the downstream part of the Upper Sebou basin. On the assumption of the quantities precipitated and evapotranspiration on the Maaser and Zloul basins, the 176 hm$^3$ corresponds to an infiltration on a surface of approximately 850 km$^2$. These two basins contain a total of 670 km$^2$ of "limestone and dolomite" formations which correspond to the mountainous massifs that dominate the Upper Sebou basin to the east. On these massifs the rainfall values are higher than average, so this surface is in line with the necessary impluvium.

We suggest that the 11 hm$^3$ difference between the loss on the Zloul/Maaser and the spring flows feed the Sebou wadi via the bottom of the bed between the Pont de Mdez station and the dyke complex, as we have highlighted (Qadem 2015), following other authors, by using punctual flow measurement campaigns. On the base of all the assumptions, the remaining difference between the volumes at the Pont de Mdez and Azzaba stations, 16 hm$^3$, would be transferred west towards the edges of the Tabular Middle Atlas in the same configuration as mentioned for 11 hm$^3$ further upstream. Although the distance is significant, several tens of kilometres, this water would feed the large karstic springs at the contact of the Tabular Middle Atlas and the Saïs plain as highlighted by Dauteuil et al. (2016).

## 5 Conclusion

The hydrological balances made from the flows measured at 5 hydrometric gauge on the hydrographic network of the Upper Sebou and on 3 springs have made it possible to infer the underground transfers within the well-developed karstic network in the Middle Atlas. In this respect, the great interest of the ET data from remote sensing should be underlined, as they allow the balance to be completed without unknowns. Before this type of data, the ET value was unknown, and it was therefore difficult to distinguish between the part of the runoff deficit attributable to ET and that attributable to underground transfers. It appears that almost half of the useful precipitation falling on the Upper Sebou basin does not contribute to the flow of the Upper Sebou but is transferred either to the Oum Er Rbia basin or to the Saïs plain. For the remaining water, a large part also escapes the flow within the upstream basin of the Upper Sebou to be found from the karstic emergences in its most downstream part. The populations of the Middle Atlas and their activities are therefore far from benefiting from all the water potential of the area. It would be interesting, in order to clarify matters, to be able to take into account the flows of the smallest springs that are found in the Middle Atlas, particularly in the basalts.

## References

Aguilar AL, Flores H, Crespo G, Marin M, Campos I, Calera A (2018) Performance assessment of MOD16 in evapotranspiration evaluation in Northwestern Mexico. Water 10(7):901. https://doi.org/10.3390/w10070901

Akdim B, Sabaoui A, Amyay M, Laaouane M, Gille E, Obda K (2011) Influences hydro karstiques du système sourcier Aïn Sebou-Timedrine-Ouamender Sur l'hydrologie de l'oued Sebou (Moyen Atlas, Maroc). Z Geomorphol 56(2):165–181

Amraoui F, Razack M, Bouchaou L (2003) Turbidity dynamics in karstic systems. Example of Ribaa and Bittit springs in the Middle Atlas (Morocco). Hydrol Sci J 48(6):971–984

Bastiaanssen WGM, Menentia M, Feddes RA, Holtslag AM (1998) A remote sensing surface energy balance algorithm for land (SEBAL). 1 Formulation. J Hydrol Vol 212–213:198–212

Beaudet G, Martin J (1967) Observations morphologiques sur les bordures nord-ouest et oust du Moyen-Atlas. Rev Géogr Maroc 12:113–142

Belhassan K, Hessane MA, Essahlaoui A (2010) Interactions eaux de surface–eaux souterraines: bassin versant de l'Oued Mikkes (Maroc). Hydrol Sci J 55(8):1371–1384. https://doi.org/10.1080/02626667.2010.528763

Bentayeb A, Leclerc C (1977) Les ressources en eau du Maroc, tome 3, Domaines atlasique et sud-atlasique. Edition du service géologique du Maroc, pp 37–84

Chaponnière A (2005) Fonctionnement hydrologique d'un bassin versant montagneux semi-aride: cas du bassin versant du Rehraya (Haut Atlas marocain). Thèse de doctorat de l'université de Toulouse, 268p

Chow VT, Maidment DR, Mays LW (1988) Applied hydrology. McGraw-Hill Education, 572p

Dauteuil O, Moreau F, Qarqori K (2016) Structural pattern of the Saïss basin and Tabular Middle Atlas in northern Morocco: Hydrological implications. J Afr Earth Sci 119:150–159

De Waele J, Melis MT (2009) Geomorphology and geomorphological heritage of the Ifrane-Azrou region (Middle Atlas, Morocco). Environ Geol 58:587–599

Frizon de Lamotte D, Zizi M, Missenard Y, Hafid M, El Azzouri M, Maury RC, Charrière A, Taki Z, Benammi M, Michard A (2008) The Atlas system—continental evolution: the geology of Morocco. Lecture notes in earth sciences, vol 116. Springer, Berlin

El Khalki Y (1990) Etudes hydro-géomorphologiques au Haut Sebou. Thèse de l'Université d'Aix-Marseille II

Latati A (1985) Thoughts on karst water resources in the Middel Atlas Mountains, Morocco. In: Karst water resources, proceedings of the Ankara-Antalya symposium, IAHS, vol 161, pp 633–642

Loup J (1962) L'Oum er Rebia. Etudes sur une grande rivière des montagnes marocaines. Revue de géographie alpine, tome 50 (4):519–555

Martin J (1981) Le Moyen Atlas central: Etude géomorphologique. Notes et mémoires du service géologique N°258 bis, 107p

Michard A, Frizon de Lamotte D, Saddiqi O, Chalouan A (2008): An outline of the geology of Morocco—continental evolution: the geology of Morocco. Lecture notes in earth sciences, vol 116. Springer, Berlin

Moniod F (1973) Etude hydrologique de l'Oum Er R'bia (Maroc). Cahier o.r.s.t.o.m. Série Hydrologie X(2):153–170

Nejjari A (2002) La sécheresse, l'eau et l'homme dans le bassin versant du Haut Sebou (Moyen-Atlas septentrional – Maroc). Thèse de doctorat, 304p

Nicod J (1993) Impacts des déboisements et défrichements récents sur les plateaux et moyennes montagnes calcaires du domaine méditerranéen. Actes du 116ème Congrès national des sociétés savantes. Sci Nat Et Montagnes Paris, pp 53–66

Qadem A (2015) Quantification, modélisation et gestion de la ressource en eau dans le bassin versant du Haut Sebou (Maroc). Thèse de doctorat, Université de Lorraine et Fès, 272p

Rémérniéras G (1986) L'hydrologie de l'ingénieur. Edition Eyrolles, 456p

Terminer H (1936) Etudes géologiques sur le le Moyen Atlas central, et le Moyen Atlas septentrional, 1566p

Velpuri NM, Senay GB, Singh RK, Bohms S, Verdin JP (2013) A comprehensive evaluation of two MODIS evapotranspiration products over the conterminous United States: using point and gridded FLUXNET and water balance ET. Remote Sens Environ 139:35–49

# Updating the Water Budget of the Gran Sasso Carbonate Fractured/Karstified Aquifer (Central Italy) for a Sustainable Management of Groundwater Resources

V. Lorenzi, C. Sbarbati, F. Banzato, M. Manetta, and M. Petitta

**Abstract**

Karst aquifers are fundamental in the water supply of European countries, where outcrops of carbonate rocks are very common, providing abundant groundwater resources. The Gran Sasso karst aquifer, selected as representative study area for Italy in the EU-funded KARMA project, is characterized by both high percentage of withdrawals for drinking purposes and significant interaction between groundwater and underground works. The recharge evaluation of the aquifer has been carried out considering the 2001–2020 monitoring period, comparing three different methods: the Turc and APLIS methods, on annual scale, and the Thornthwaite method, on monthly scale, territorially distributed by 100 × 100 m cells on GIS basis. The total recharge considers not only rainfall but also the contribution of snow melting on infiltration. The results show similar mean recharge values in 2001–2020 for all methods, corresponding to 19.9, 18.5, and 19.4 $m^3/s$, respectively, from Turc, Thornthwaite, and APLIS methods. A significant contribution to recharge from snowmelt has been confirmed (3.2 $m^3/s$ included in the above-mentioned values). These values can be considered reliable with respect to real discharge of the regional aquifer. The obtained results can be used to provide updated information to the drinking water companies for a suitable management of the available resource.

**Keywords**

Recharge assessment • Karst aquifers • Central Italy • Groundwater budget

V. Lorenzi · C. Sbarbati · F. Banzato · M. Manetta · M. Petitta (✉)
Earth Sciences Department, Sapienza University of Rome, Rome, Italy
e-mail: marco.petitta@uniroma1.it

## 1 Introduction

Karst aquifers are complex systems of heterogeneous nature and are fundamental in the water supply of European countries and those of the Mediterranean area, where outcrops of carbonate rocks are very common. About a quarter of the world's population depends, in whole or in part, on the drinking water of this type of groundwater (Goldscheider et al. 2008). In detail, the karst rocks are widespread mainly in the Mediterranean area and constitute 21.6% of the European land surface (Goldscheider et al. 2008). The importance of monitoring available water resources and updating the aquifer recharge is mainly due to the ongoing impact of climate change on these resources and to the increase in human abstraction. These two factors can modify the recharge/discharge ratio of groundwater resources. Although groundwater has always shown excellent resilience in adapting to climate change, it is important to create solutions to avoid future water supply issues. This is relevant for karst aquifers that are particularly susceptible to climate change impacts (Taylor and Greene 2008). These aquifers need to be studied in detail to understand their characteristics and assess their degree of vulnerability, to achieve sustainable management of karst groundwater resources in the face of reduced recharge and increased groundwater use. In this framework, the KARMA project has the aim to improve the hydrogeological knowledge and the sustainable management of karst water resources to obtain valuable information on recharge and groundwater vulnerability. The Gran Sasso aquifer has been chosen as the area of study for Italy, and it has been studied in detail during the last 25 years (Adinolfi Falcone et al. 2008; Amoruso et al. 2011; Barbieri et al. 2005; Petitta and Tallini 2002; Petitta et al. 2015; Tallini et al. 2013). The infiltration values of the system range between 500 mm/y (Scozzafava and Tallini 2001) and about 800 mm/y (Petitta and Tallini 2002). Scozzafava and Tallini (2001) provided a partially modified application of the Thornthwaite method, while Boni et al. (1986) proposed a


B. Andreo et al. (eds.), *EuroKarst 2022, Málaga*, Advances in Karst Science,
https://doi.org/10.1007/978-3-031-16879-6_6

"direct" method to assess effective infiltration. The aim of this work is to carry out a detailed analysis of the distribution of recharge over the considered area. The 20 year period from 2001 to 2020 was examined, and three different methods were applied to assess the recharge. For a detailed analysis of the budget, in addition to rainfall and temperature data, the contribution of snowmelt was also considered.

## 2 Hydrogeological Setting

The Gran Sasso aquifer is a calcareous-karstic ridge about 1034 km$^2$ wide. It is one of the most representative karst aquifers in Central-Southern Italy, for the amount of water resources, for the interaction between the aquifer and underground works and in relation to the valorization of the protected areas. The characteristic of this area is linked both to the huge availability of groundwater resources, and also to the impacts of human activities, such as the drilling of the Gran Sasso motorway tunnel. Moreover, its location within a National Park shows the need to satisfy human requirements but at the same time not to compromise the environment and its ecosystem services (Fig. 1). The Gran Sasso hydrogeological system is characterized by Meso-Cenozoic carbonate units (main aquifer). It is bounded by terrigenous units represented by Miocene flysch (regional aquiclude) along its northern and eastern sides and by Quaternary continental deposits (regional aquitard) along its southern side. It includes a uniform regional-wide groundwater table with a hydraulic gradient of 5–20% (Tallini et al. 2013). Its considerable importance comes from the high permeability due to fracturing and karst, and the high effective infiltration that feeds important springs located at its boundary, characterized by a total discharge that ranges between 18 and 25 m$^3$/s (Amoruso et al. 2011; Petitta and Tallini 2002). The massif's core, an endorheic basin having tectonic-karst origin, called Campo Imperatore basin (mean elevation 1650 m a.s.l.), acts as preferential recharge area.

The tectonic and structural discontinuities act as permeability limits that cause variations in the water level of the regional aquifer (Celico et al. 2005). Permeability limits are located at the northern and eastern boundaries, with E–W and N–S directions. The preferential directions of groundwater flow are locally conditioned by the main tectonic discontinuities and are guided by the altitude of the hydrogeological limits. In fact, most of groundwater is directed towards the most depressed sectors, where the permeability limits are placed at lower altitudes. Groundwater moves vertically, in the unsaturated zone, which has a thickness of about 1000 m. Its groundwater velocity and the discharge inside the karst conduits/main fractures depend on their width and on the outcrop lithology. Then, below the water table, it flows horizontally in the saturated zone to reach the main springs located at the aquifer boundaries (Petitta and Tallini 2002).

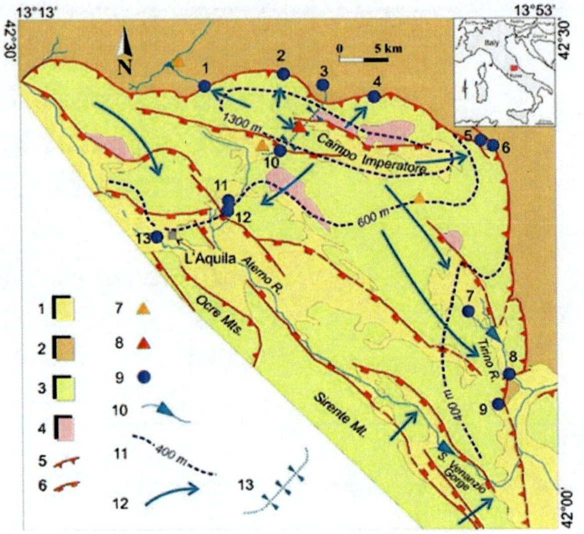


**Fig. 1** Simplified hydrogeological setting of Gran Sasso aquifer. 1 Aquitard (continental detrital units of intramontane basins, Quaternary), 2 aquiclude (terrigenous turbidites, Mio-Pliocene), 3 aquifer (calcareous sequences, Meso-Cenozoic), 4 low permeability bedrock (dolomite, upper Triassic), 5 main thrust, 6 main extensional fault, 7 thermo-pluviometric rain gauge, 8 thermo-snow gauge, 9 considered main springs, 10 streambed spring, 11 presumed water table elevation (m a.s.l.), 12 regional groundwater flowpath, and 13 highway tunnel drainage (modified from Petitta et al. 2015)

## 3 Recharge Evaluation Methodology

The water budget of the Gran Sasso aquifer has been evaluated to improve the knowledge about its recharge rate with respect to the previously collected data. In the Gran Sasso aquifer, the existing water budgets, applying indirect method, represent a relevant basis of knowledge. The methodology applied by Scozzafava and Tallini (2001) to determine the effective infiltration of the Gran Sasso aquifer adopted the modified Thornthwaite method (Thornthwaite and Mather 1957), according to local hydrogeological characteristics. In our research, the evaluation of the aquifer recharge has been obtained based on the temperature and precipitation gradients among the ridges. Following the authors methodology (Scozzafava and Tallini 2001) and based on 20 years of climate data monitoring (2001–2020), two annual rainfall gradients have been calculated. Specifically, the gradients correlate altitude and rainfall values of the 21 different thermo-rain gauges. The gauge of Campotosto (1344 m a.s.l.) for the northern side and the gauge of

Assergi (992 m a.s.l.) for the southern side were selected as representative gauges. They have been used as a reference point for calculating the rainfall value over the entire aquifer for each year using a gradient of 30/100 m for the southern side and of 42/100 m for the northern side. An isohyet map has been created using the ArcGis software system with 100 × 100 m pixel raster (reference system WGS 1984 UTM 33 N) (Fig. 2a).

By the same methodology applied to rainfall, an estimation of the temperature gradient and consequently isotherm map has been computed, selecting as representative site the Castel del Monte thermo-pluviometric gauge (1346 m a.s.l.).

A relevant parameter for recharge evaluation in karst mountainous areas is the snowmelt. To consider the snow contribution to the aquifer recharge, a monthly gradient has been calculated. This has been obtained by creating a correlation line between altitude and snow values in the period 1960–1990, referring to Fazzini and Bisci (1999). This gradient has been applied to the snow data recorded at the Campo Imperatore gauge (2152 m a.s.l.) to derive the recharge fraction due to snow melt (Fig. 2b).

From the distributed data and related maps, the hydrological budget for the Gran Sasso aquifer has been assessed using three different methods: Thornthwaite (Thornthwaite and Mather 1957), Turc (Turc 1954), and APLIS (Andreo et al. 2008), for the 2001–2020 period.

The Turc method (Turc 1954) has been used for annual water budget. The modified Thornthwaite method, described by Scozzafava and Tallini (2001), has been adopted to calculate the effective infiltration of the Gran Sasso aquifer at monthly scale. The Thornthwaite's method has been modified mainly to distinguish between the contributions of run-off and the net infiltration, using the curve number (CN), associating to each curve number a specific ST value. The ST is the maximum water storage in the soil that we adopted as field capacity values.

On the other hand, the APLIS method allows to estimate the mean annual recharge in carbonate aquifers, expressed as a percentage of precipitation, based on the variables of altitude, slope, lithology, infiltration landform, and soil type (Andreo et al. 2008). With appropriately combined and applied scores, using the GIS software, the rate of recharge and its spatial distribution have been estimated. The superimposition of each map for single variables allows to determine the known recharge value and the spatial distribution of recharge rate within the Gran Sasso karst aquifer. The mean value of annual recharge rate has been grouped into five regular recharge class. Finally, the overlay of the APLIS layer with isohyet map (Fig. 2a) estimates the aquifer recharge through direct infiltration.

## 4 Results and Discussion

The initial recharge values have been indicated as the sole contribution of rainfall. The run-off value is about 0.3% of the total rainfall, according to Scozzafava and Tallini (2001). As for the Turc method, the average recharge value for the considered twenty years is 16.7 $m^3/s$. Year 2007 and year 2013 represent the driest and the rainiest year, respectively. In 2007, the recharge is 10 $m^3/s$, while in 2013 the value of recharge is 26 $m^3/s$. The mean evapotranspiration is equal to

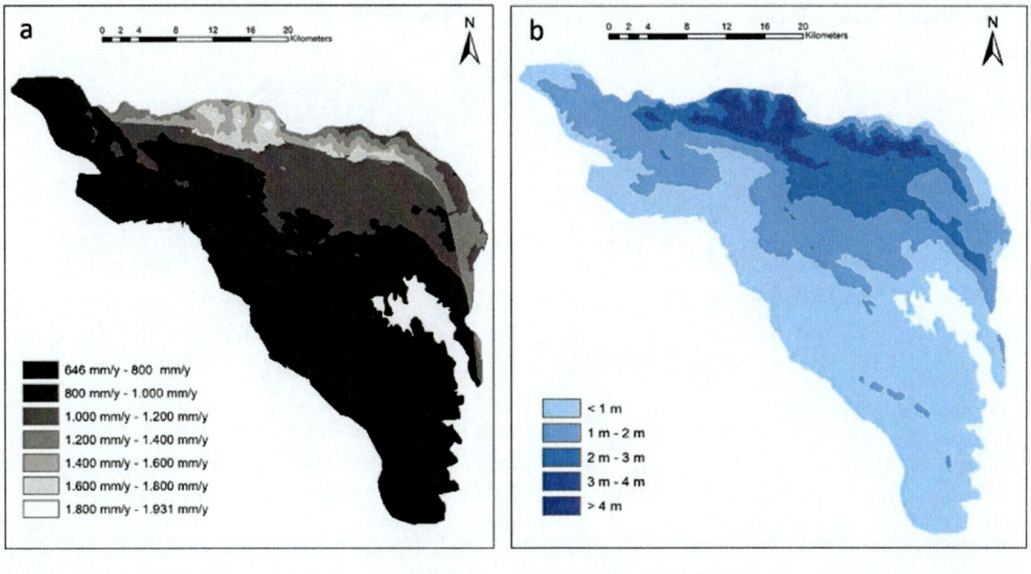


**Fig. 2** Rainfall distribution map **(a)** long-term average values, period 2001–2020 and **(b)** snowfall recharge map, period 2010–2020

444 mm/y, and most of the aquifer is characterized by values between 400 and 500 mm/y (Fig. 3a). The infiltration average value is 508 mm/y over a long period, without considering the infiltration snow component, which is 98 mm/y. The infiltration values are variables from below 200 mm/y in the lower elevation areas to over 1400 mm/y, as shown in Fig. 3b.

Using the Thornthwaite method, the average recharge of the long period is about 15.3 $m^3/s$ and the average infiltration is 462 mm/y. The average evapotranspiration is about 491 mm/y, slightly higher than those of the Turc method, ranging from a minimum of 400 mm/y at the high-altitude areas to a maximum of over 600 mm/y observed at low altitudes (Fig. 4a). The infiltration values are generally higher than 200 mm/y, up to more than 1400 mm/y at the peak zones (Fig. 4b).

Through the APLIS method, the Gran Sasso aquifer recharge rate results with a percentage of effective infiltration of 51.6% with respect to total rainfall. Overlaying the obtained recharge rate map to the raster rainfall map

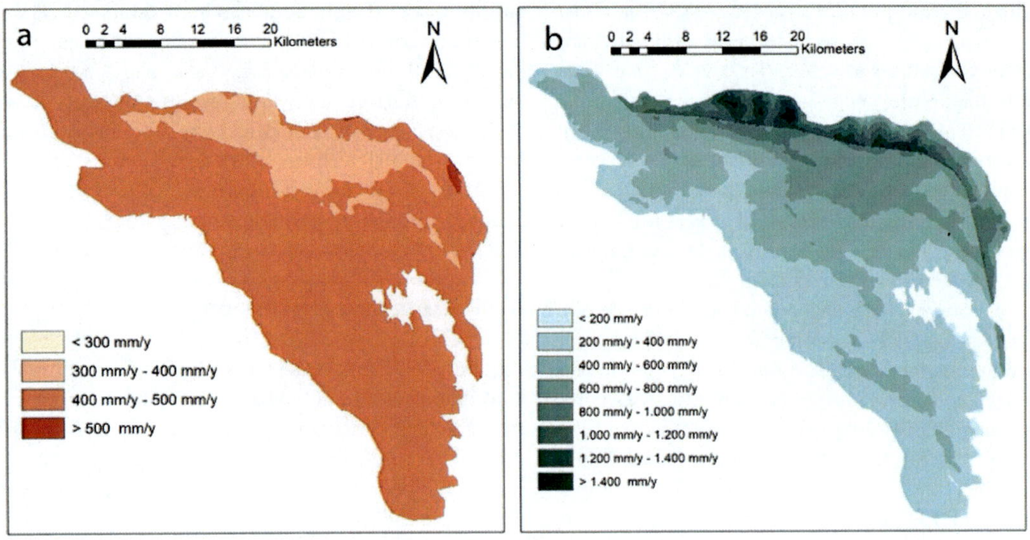


**Fig. 3** Maps of real evapotranspiration **a** and infiltration, **b** calculated by the Turc method

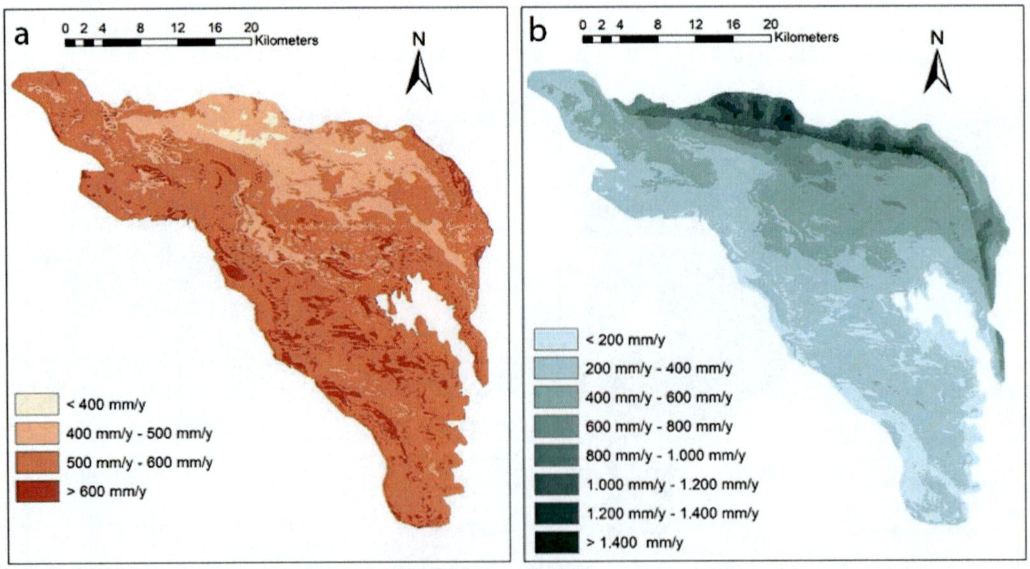


**Fig. 4** Maps of real evapotranspiration **a** and infiltration, **b** calculated by the Thornthwaite method

obtained for each year, the aquifer recharge has been inferred. The average recharge rate is 19.4 $m^3/s$, and the average infiltration value is 594 mm/y. 2007 represents the year with minimum recharge rate of 13.7 $m^3/s$, while 2013 is determined by the highest recharge rate of 21.3 $m^3/s$ (Fig. 5).

Rainfall recharge obtained through the different methods over the entire monitoring period, has been also evaluated in terms of distribution over time and space, and it has been distributed with elevation basing on five different altitude belts. Specifically, rainfall recharge distribution for the average (2001–2020), the driest and the rainiest years have been analysed calculating, for each altitude belts, recharge values and the corresponding percentage on the total balance budget (Table 1).

A large variability of percentage of recharge over the entire area has been assessed. As expected, the minor contribution comes from the lowest altitude belt (< 600 m) with values less or equal to 3% of total recharge. This altitude belt is, in fact, characterized by the smaller extension where the rainfall vertical gradient, as function of increasing altitude elevation, has negligible impact.

Ascending to 600–1000 m altitude belt (the widest one), the contribution to the aquifer recharge obviously increases, showing the wider variation among the different recharge conditions. Percentage values are comparable for the average and the rainiest years by both Turc and Thornthwaite, ranging between 19% and 23.5%. Differently, during the driest year, recharge decreases to minimum values of 14% and 16%, respectively, for Thornthwaite and Turc, highlighting the strong impact of dry periods at limited altitudes. For both the altitude ranges of 1000–1400 and 1400–1800 m, there are no significant differences among the three scenarios in terms of recharge, which vary from a minimum of 27% to a maximum of 30%. The highest altitude belt (> 1800 m) differently contributes to aquifer recharge, showing the relatively higher percentage values in the driest year. This evidence points out the fundamental role of the high-elevation areas in aquifer recharge, especially in drought periods.

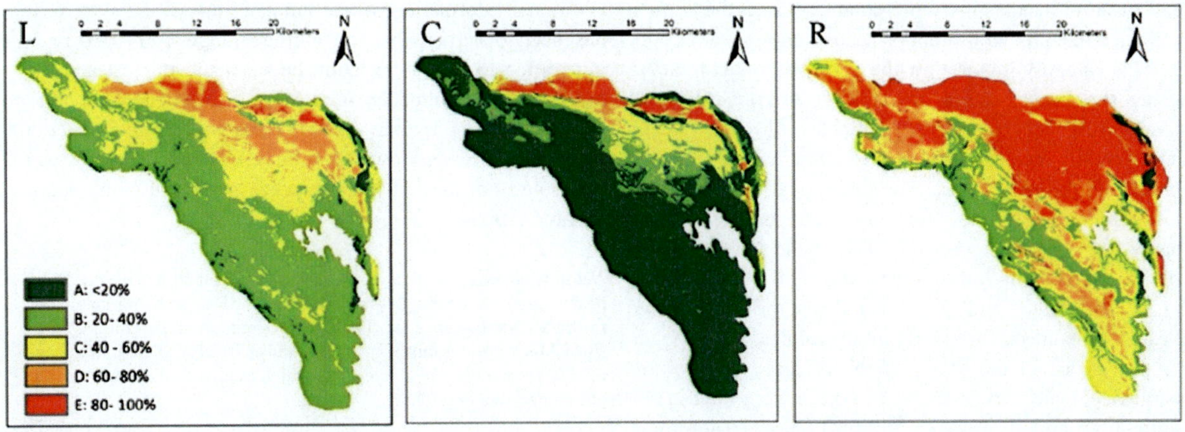


**Fig. 5** Recharge rate distribution calculated by the APLIS method for the long monitoring period (left panel, *L*), for the driest year (central panel, *C*), and for the rainiest year (right panel, *R*)

**Table 1** Recharge values calculated with the Turc (TU) and Thornthwaite (TH) methods for each altitude belt for the average, the driest and the rainiest years

| | | Average 2001–2020 | | | | Driest year | | | | Rainiest year | | | |
|---|---|---|---|---|---|---|---|---|---|---|---|---|---|
| Altitude range (m a.s.l.) | Area ($km^2$) | TU ($m^3/s$) | % | TH ($m^3/s$) | % | TU ($m^3/s$) | % | TH ($m^3/s$) | % | TU ($m^3/s$) | % | TH ($m^3/s$) | % |
| < 600 | 55 | 0.4 | 2 | 0.3 | 2 | 0.1 | 1 | 0.1 | 1 | 0.7 | 3 | 0.7 | 3 |
| 600–1000 | 340 | 3.6 | 22 | 2.9 | 19 | 1.6 | 16 | 1.2 | 14 | 6.1 | 23 | 5.1 | 23 |
| 1000–1400 | 310 | 4.9 | 29 | 4.4 | 29 | 3.0 | 30 | 2.3 | 27 | 7.8 | 30 | 6.5 | 29 |
| 1400–1800 | 227 | 4.7 | 28 | 4.4 | 29 | 3.3 | 33 | 2.7 | 32 | 7.0 | 27 | 6.2 | 27 |
| > 1800 | 102 | 3.1 | 19 | 3.3 | 21 | 2.0 | 20 | 2.2 | 26 | 4.4 | 17 | 4.1 | 18 |
| TOTAL | 1034 | 16.7 | 100 | 15,3 | 100 | 10 | 100 | 8,5 | 100 | 26.0 | 100 | 22.6 | 100 |

Additional insights about recharge distribution can be assessed by comparing the recharge values in relation to the area of each altitude belts. In fact, the medium–low (600–1000 and 1000–1400 m) and medium–high elevations (1400–1800 and > 1800 m), which cover, respectively, 650 and 329 km$^2$, contribute almost equally, in terms of percentage, to the recharge for the average year. During the rainiest year, the medium–low altitude belt fed the aquifer with a slightly higher contribution (+9%) with respect to the medium–high belt; conversely in the driest year, the higher elevation mainly contributes to the total balance (+17% in Thornthwaite). For the Gran Sasso aquifer, the main contribution to the total budget comes from the higher elevation areas, especially in the driest years as stated above. The crucial role played by the high-elevation recharge area of the massif has been also highlighted in previous studies (Amoruso et al. 2014; Petitta and Tallini 2002), according to which groundwater follows a radial flowpath from the aquifer core, characterized by the highest belts, towards the boundary of the massif where the main springs groups are located.

Results from the APLIS method for the average year (Fig. 5 *L*) show that the most representative class is the "Low" one characterized by an infiltration value ranging from 20 to 40%. This category impacts for the 58% of the whole area, covering the low-elevation areas (< 1000 m), while the remaining 42% is characterized by variable infiltration values, which generally increase according to altitude and reach the maximum at the peak areas. The driest year (Fig. 5 *C*) is mainly represented by the "Very Low" category (< 20% of recharge rate), which covers the 68% of the recharge area involving the medium–low altitude belts (< 1400 m). Even in this case, the "High" and "Very High" categories cover the highest altitude areas. Differently, in the rainiest year (Fig. 5 *R*), a prevalence of the "Very High" category (> 80% of recharge rate) which covers the 32% of the recharge area at the medium–high altitude belts (> 1400 m) has been inferred; otherwise, the medium–low altitude belts are characterized by a strong heterogeneity in recharge rate distribution.

Using the available literature gradient (Fazzini and Bisci 1999), the average recharge value (2010–2020) due to the snow component and the related distribution map (Fig. 2b) has been inferred. The snow contribution was calculated as a cumulative average, considering the snowy months since November to April. The average recharge value for the period 2010–2020 is 3.9 m$^3$/s, ranging from 5.4 m$^3$/s (2018) to 1 m$^3$/s (2020).

## 5 Conclusion

An analysis of the water balance of Gran Sasso has been carried out on annual and monthly scale, considering the period 2001–2020 and identifying the different contributions of rain and snow melt. The purpose of using three methods is to find out if they are efficient and if they can offer perspectives for the future management of groundwater resources. The improved knowledge will lead to a withdrawal optimization coupled with sustainability goals, aiming to an active management able to balance human needs and environmental issues. The similarity of the results obtained by the different methods confirms the reliability of the water budget.

The need to evaluate the annual water balance is confirmed by the great variability and oscillation of the annual recharge values. The importance of monitoring recharge is quite useful for improving water resource management and promoting the adaptation of abstractions to the natural oscillation of groundwater recharge.

Another important factor for the knowledge of the system and for the optimal management of the resources is the contribution of the recharge in the different altitude ranges. In fact, by analysing the long-term average recharge values referred to the different altitude ranges and comparing them with the driest and rainiest year, it is possible to have additional information on the vulnerability of the system. It has been observed that the altitude range 600–1000 m is most affected by the variation between the rainiest and the driest year. The influence on recharge of high-elevation areas increases during dry periods, while they have little impact in rainy years, when low-elevation carbonate outcrops can contribute to the infiltration process as much as high-elevation areas.

**Acknowledgements** This research has been supported by the European Commission through the Partnership for Research and Innovation in the Mediterranean Area (PRIMA) programme under Horizon 2020 (KARMA project, grant agreement number 01DH19022A). We thank the reviewer for useful suggestions which improved the content of the preliminary manuscript.

## References

Adinolfi Falcone R, Falgiani A, Parisse B, Petitt M, Spizzico M, Tallini M (2008) Chemical and isotopic (δ18O‰, δ2H‰, δ13C‰, 222Rn) multi-tracing for groundwater conceptual model of carbonate aquifer (Gran Sasso INFN underground laboratory—central Italy). J Hydrol 357:368–388

Amoruso A, Crescentini L, Petitta M, Rusi S, Tallini M (2011) Impact of the 6 April 2009 L'Aquila earthquake on groundwater flow in the Gran Sasso carbonate aquifer, Central Italy. Hydrol Process 25:1754–1764

Amoruso L, Crescentini S, Martino M, Petitta M, Tallini M (2014) Correlation between groundwater flow and deformation in the fractured carbonate Gran Sasso aquifer (INFN underground laboratories, central Italy). Water Resour Res 50:4858–4876

Andreo B, Vias J, Duran J, Jimenez P, Lopez-Geta JA, Carrasco F (2008) Methodology for groundwater recharge assessment in carbonate aquifers: application to pilot sites in southern Spain. Hydrogeol J 16:911–925

Barbieri M, Boschetti T, Petitta M, Tallini M (2005) Stable isotope (2H, 18O and 87Sr/ 86Sr) and hydrochemistry monitoring for groundwater hydrodynamics analysis in a karst aquifer (Gran Sasso, Central Italy). Appl Geochem 20:2063–2081

Boni C, Bono P, Capelli G (1986) Schema idrogeologico dell'Italia Centrale, [Hydrogeological scheme of central Italy]. Mem Soc Geol It 35(2):991–1012

Celico P, Fabbrocino S, Petitta M, Tallini M (2005) Hydrogeological impact of the Gran Sasso motor-way tunnels (Central Italy). Giorn Geol Appl 1:157–165

Fazzini M, Bisci C (1999) Clima e neve sul massiccio del Gran Sasso. Neve e Valanghe 36

Goldscheider N, Chen Z, Auler AS, Bakalowicz M, Broda S, Drew D, Hartmann J, Jiang G, Moosdorf N, Stevanovic Z, Veni G (2008) Global distribution of carbonate rocks and karst water resources. Hydrogeol J 28:1661–1677

Petitta M, Tallini M (2002) Idrodinamica sotterranea del massiccio del Gran Sasso (Abruzzo): nuove indagini idrologiche, idrogeologiche e idrochimiche (1994–2001). Boll Soc Geol It 121(3):343–363

Petitta M, Caschetto M, Galassi DMP, Aravena R (2015) Dual-flow in karst aquifers toward a steady discharge spring (Presciano, Central Italy): influences on a subsurface groundwater dependent ecosystem and on changes related to post-earthquake hydrodynamics. Environ Earth Sci 73:2609–2625

Scozzafava M, Tallini M (2001) Net infiltration in the Gran Sasso Massif of central Italy using the Thornthwaite water budget and curve-number method. Hydrogeol J 9:461–475

Tallini M, Parisse B, Petitta M, Spizzico M (2013) Long-term spatio-temporal hydrochemical and 222Rn tracing to investigate groundwater flow and water–rock interaction in the Gran Sasso (central Italy) carbonate aquifer. Hydrogeol J 21:1447–1467

Taylor CJ, Greene EA, 2008. Hydrogeologic characterization and methods used in the investigation of karst hydrology. Field techniques for estimating water fluxes between surface water and ground water. In Rosenberry DO, LaBaugh JW (eds) US geological survey. Reston, Virginia (EUA), pp 71–114

Thornthwaite CW, Mather JR (1957) Instructions and tables for computing potential evapotranspiration and the water balance. Publ, Climatol., p 10

Turc L (1954) Le bilan d'eau des sols: Relations entre les precipitations, l'evaporation et l'ecoulement. Annales Agronomiques 5:491–595; 6:5–131

# Combining Quantitative Analysis Tools (Cross-Correlation Analysis and Dye Tracer Tests) to Assess Response Times in Karst Aquifers. The Ubrique Karst System (Southern Spain)

J. F. Martín-Rodríguez, M. Mudarra, B. de la Torre, and B. Andreo

**Abstract**

In this work, a preliminary conceptual model of the hydrogeological functioning of a binary karst aquifer located in southern Spain is presented, based on the natural responses of the springs and the dye tracer tests. Results derived from the long-term analysis of cross-correlation function (CCF) (global average time of recharge input) have been validated with those obtained from a dye tracer test, performed during a specific recharge event. The overall response times (CCF results) obtained for the three springs studied were 58 (SP-1), 72 (SP-3), and 81 h (SP-2). This pattern in response times was contrasted with the first detection times of the BTC-Tracer Breakthrough Curve—at the same three springs (30.9, 35.3, and 47 h) and their respective recovery rates (2.03, 21.9, and 78.4% for S1, S2, and S3, respectively). The distribution of response times, both those determined by long-term statistical analysis and the ones derived from the use of dye tracer test, is directly related to the karst network development in each of the drainage routes. Additionally, the rate of tracer recovery at each of the springs is linked to the hierarchy of flow paths from the selected injection point toward each spring, within the overall drainage network scheme. An accurate knowledge of all factors affecting the hydrogeological behavior as well as groundwater flow paths (from swallow holes to the springs) is crucial for suitable management of the resource and to mitigate the pollution risks derived from fast infiltration of surface water.

**Keywords**

Cross-correlation • Dye tracer test • Karst aquifer • Time lag • Aquifer functioning

## 1 Introduction

Carbonate massifs commonly constitute complex aquifer systems located in mountainous areas in which groundwater flow occurs through the fracture and conduits network enlarged by dissolution surfaces within the rock matrix; in many cases conditioned by the geological structure. These aquifers have a high degree of heterogeneity and spatial anisotropy in their hydrogeological properties, an aspect that distinguishes them from other types of aquifers (Bakalowicz 2005; Ford and Williams 2007). Moreover, karst aquifers are much more vulnerable to contamination than other hydrogeological systems (Doerfliger and Zwahlen 1998; Hartman et al. 2021), so any research to better understand their functioning will result in an improvement in the management and protection of the water resources stored in them (Marín et al. 2021).

To quantify the response times of the system supposes a suitable approximation to understand the behavior of karst aquifers. Cross-correlation analysis has been traditionally used to measure the relationship between the input (rainfall) and output (spring discharge) signals (Mangin 1984; Kovačič 2010). The time lag elapsed between precipitation and its reflection in the hydrodynamic load of the springs can be considered an indicator of the development of the karst drainage network. On the other hand, dye tracer tests are a powerful tool used to obtain quantitative information on groundwater flow in complex hydrogeological environments, where groundwater drainage pathways often show intricate circulation patterns, besides the time-lag values between tracer injection and detection (Käss 1998).

J. F. Martín-Rodríguez (✉) · M. Mudarra · B. de la Torre · B. Andreo
Department of Geology, Center of Hydrogeology, University of Malaga (CEHIUMA), E-29071 Malaga, Spain
e-mail: josefranciscomr@uma.es


B. Andreo et al. (eds.), *EuroKarst 2022, Málaga*, Advances in Karst Science,
https://doi.org/10.1007/978-3-031-16879-6_7

Despite the potential and the utility of both types of approaches, they are not commonly used in conjunction to investigate the hydrogeological functioning of an aquifer (Mudarra et al. 2014). This contribution aims to advance the knowledge of the hydrogeological functioning of a geologically complex binary karst aquifer (the Ubrique karst system, southern Spain), through the joint use of tracer tests and the long-term analysis of the cross-correlation function of the hydrodynamic response of the three main springs. This has permitted to development of a preliminary conceptual model, which can be very useful for urban supply planning and the establishment of protection perimeters.

## 2 Site Description

The Ubrique karst aquifer extends over an area of 27.4 km$^2$ in the NE part of Cádiz province, southern Spain (Fig. 1). It is defined by two main mountain ranges, Caíllo and Ubrique (mean altitude of permeable outcrops of 1186 and 863 m a.s.l., respectively), with a ragged orography dominated by outcrops of carbonate rocks and an absolute altitude range from 317 to 1395 m a.s.l. The predominant climate has a remarkable seasonal pattern in the annual distribution of precipitation and air temperature. Precipitation occurs mainly from autumn to spring, associated with wet winds from the Atlantic Ocean. The mean annual precipitation and temperature values recorded in this area were, respectively, 1305 mm and 14.8 °C (Sánchez et al. 2017).

The aquifer (Penibetic Unit) consists of around 500 m thick Jurassic dolostones and limestones, which present low permeability rocks (Triassic impervious bed) at the bottom and Early Cretaceous–Tertiary marly limestones and marls at the top (Martín-Algarra 1987). Surrounding and overlaying the Penibetic Unit, outcrops of Tertiary flysch—type clays and sandstones appear (Fig. 1).

From the geological structure perspective of the aquifer, two sectors can be distinguished. On the one hand, the southern carbonate outcrop of the Sierra de Ubrique is characterized by the presence of a flat-hinged anticline fold and affected by recent tectonic features (block collapse and gravity-driven faults) on its western edge, where the springs are located. On the other hand, the Sierra del Caíllo area shows an intricate structure, where low permeability rocks are imbricated between the aquifer formations, as a result of reverse faults and accretionary overthrust (Fig. 2).

Groundwater discharge occurs along the southwestern edge of the carbonate outcrops, through two perennial springs located within the Ubrique urban area (S1 349 m a.s.l. and S2 317 m a.s.l.), and an overflow spring sited on the southern edge of the Sierra de Ubrique (S3, at 422 m a.s.l.). Recharge is produced by two main mechanisms (binary karst system): by direct infiltration of rainwater on the bare carbonate outcrops (autogenic component) and by the concentrated infiltration of runoff water from the small catchment area, formed by low permeability outcrops (allogenic component). When the precipitation is sufficiently intense, the resultant surface runoff infiltrates through the SH sinkhole and ends up incorporating into the underground water flow. Particularly relevant at the test site are the pollution processes of poorly treated wastewater (from Villaluenga del Rosario town) and the washing of livestock waste originated within the SH caption area (Fig. 1), which constitutes a focal point of groundwater contamination. These processes imply an undesirable risk for the healthiness of the Ubrique city water supply, entirely supported by the aquifer.

## 3 Methods

The installation and calibration of water-level data loggers (ODYSSEY™ Capacitance Water Level Logger) in each of the springs studied permit obtaining continuous series (with hourly measurement frequency) of discharge variations, by applying the corresponding rating equations. From that information, cross-correlation analyses have been performed (Fig. 3) using also rainfall data series (hourly records registered at Grazalema meteorological station (Fig. 1), from October 2013 to September 2018 (5 hydrological years).

For the dye tracer test, 3 kg of amidorhodamine G (acid red 50, CAS: 5873–16-5) was injected into the Villaluenga del Rosario shallow hole (SH in Fig. 1). The injection coincided with a recharge event during which a total rainfall amount of 225 mm was recorded in 2 days. Totally 124 water samples with detectable values of tracer were collected. A Perkin-Elmer LS55 spectrofluorometer was used for the detection and quantitative determination of the dye in spring water. Breakthrough curves and hydrographs of the spring have been jointly used to quantify the total mass of tracer recovered (Fig. 4).

## 4 Results

The *k* values obtained for each spring in the CCF analysis have been considered as the mean response time or time lag (hours). A high value in the time lag between the input signal and the hydrodynamic variations observed in the springs is indicative of a limited development of the conduit system within the global scheme of the drainage network, which leads to a certain potential to attenuate and modulate the hydrodynamic load variations induced by rainfall (Lorette et al. 2018; Fiorillo and Doglioni 2010). Instead, short time lags illustrate a high capacity of conduits to transfer rainwater from recharge zones to outlets (high degree of karstification development).

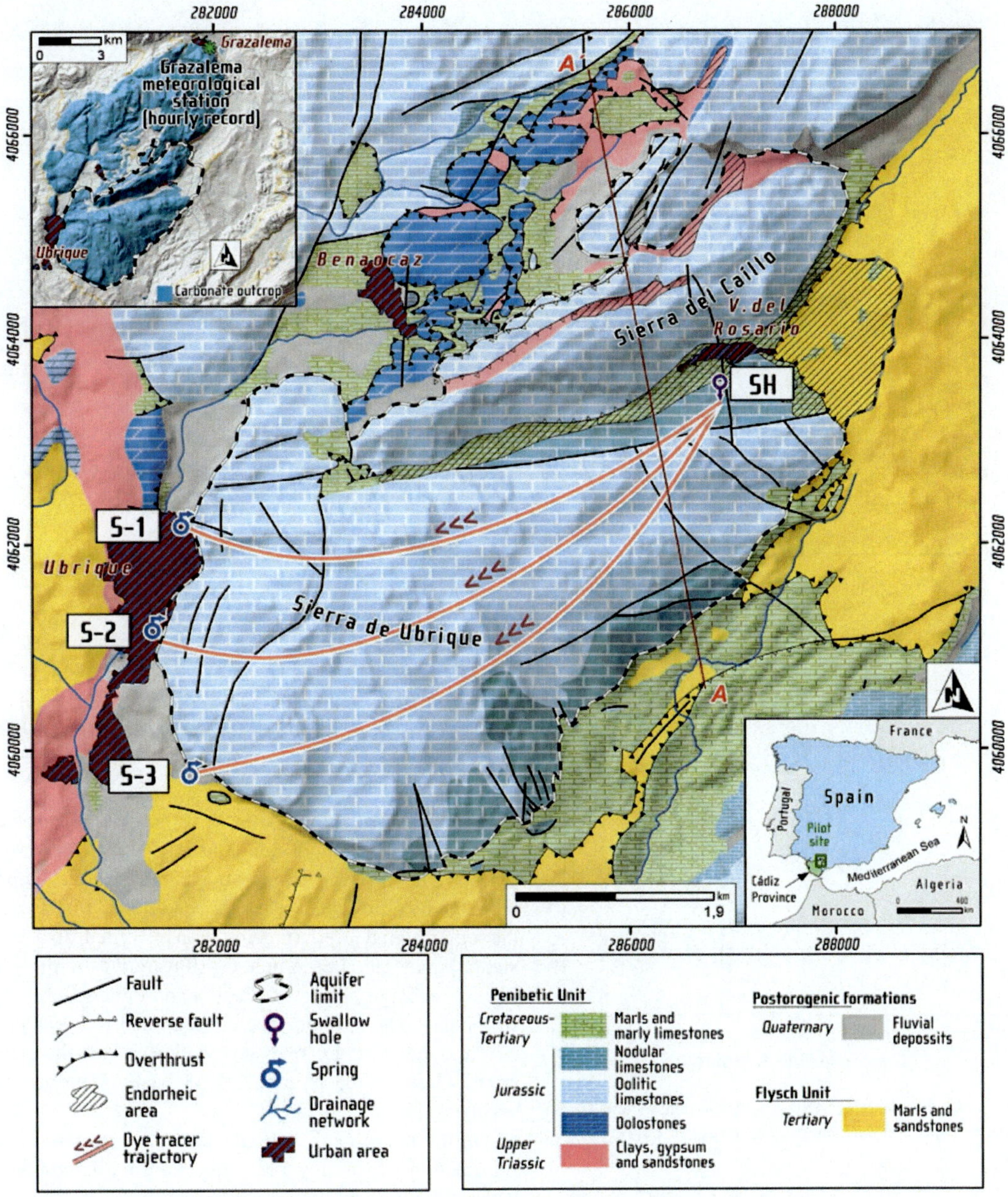


**Fig. 1** Location of the study site and geological and hydrogeological features of Ubrique karst aquifer. The figure also shows the location of the injection point during the dye tracer test (Villaluenga swallow hole, SH), the drainage system springs (S1, S2, and S3), and the cross-section A-A (Fig. 2)

The fastest response times have been detected in S1 spring (30.9 and 38.5 h., respectively, $-k$ value: 58.0 h-). In opposite, the largest travel times were always observed in S2 spring (47.0 and 64.3 h. versus 81 h. of $k$ value in CCF analysis). S3 spring shows intermediate time-lag values, both for $k$ parameter and for dye detection times (Table 1). Effective velocity considering the maximum concentration peak as reference was determined, which showed results according to the time-lag distribution: 159.8, 149.6, and 111.8 m/h for springs S1, S3, and S2, respectively).

The dye tracer concentration curves in the spring water (Fig. 4) show, in general terms, distributions typical of aquifers with well-developed karst drainage, with sharp rises and falls in their concentration rates, evidencing a relatively

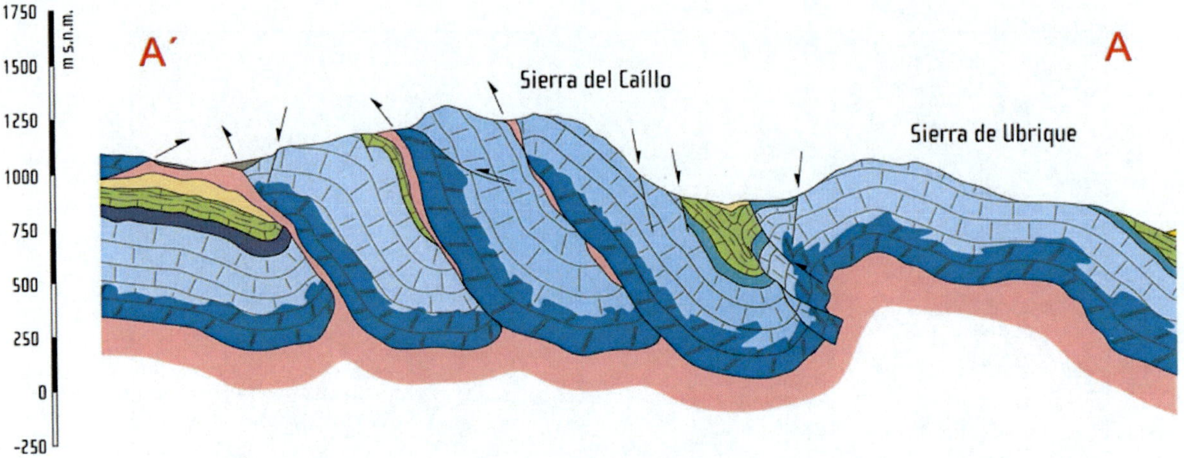


**Fig. 2** Geological—hydrogeological cross sections of Sierra de Ubrique, showing the geometry of the aquifer. The vertical and horizontal scales are the same. See location and key in Fig. 1

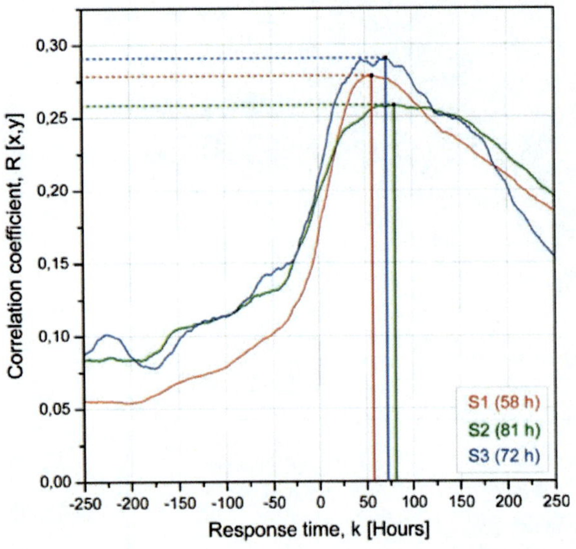


**Fig. 3** Cross-correlation function (CCF) between rainfall and spring discharge (hourly data) and calculated mean response times for each spring

rapid movement of the water flow from the injection points (recharge areas). Springs S2 and S3 show a secondary increase in tracer concentration after the main peak, resulting in a distortion of the decreasing concentration phase of the BTC curve, which contrasts with the morphology of the curve observed in *S*1, which has a single concentration peak and a gradual decrease in tracer concentration (Fig. 4).

In the study area, *S*3 spring showed the highest recovery value (68.4%), as well as the highest average flow rate during the test (4.9 $m^3/s$). Similar average flow rates were measured in springs *S*1 and *S*2 (1.7 and 2.0 $m^3/s$, respectively), although the mass recovered was noteworthy lower in S1 (2.03%) than in S2 (21.9%).

## 5 Discussion

### 5.1 Representativeness and Methodological Limitations

The analysis of global mean response times obtained through statistical approximations (CCF) provides information on the underground drainage network development, although its representativeness is conditioned by the fact that the input signal used (precipitation recorded at a specific station) may not be completely illustrative of the current aquifer recharge processes (Padilla and Pulido-Bosch 1995). In a binary karst system, the concentrated recharge component can be of major relevance and represent the principal contribution of the hydraulic load peaks identified in the springs.

On the other hand, artificial tracer tests provide detailed information concerning the connection and hydraulic characteristics along the flow path located between the injection point (commonly sinkholes or swallow holes connected to endorheic catchment areas, where the concentrated infiltration of runoff water occurs) and the detection points. In this situation, the representativeness stays restricted to the drainage path between the two points. As the tracer test has been performed starting from a single injection point, the mass recovered of dye tracer in each of the springs is proportional to the flow route hierarchy within the general scheme of the drainage network and provides information on the network drainage development. A high recovery rate commonly denotes a preferential flow path between injection and

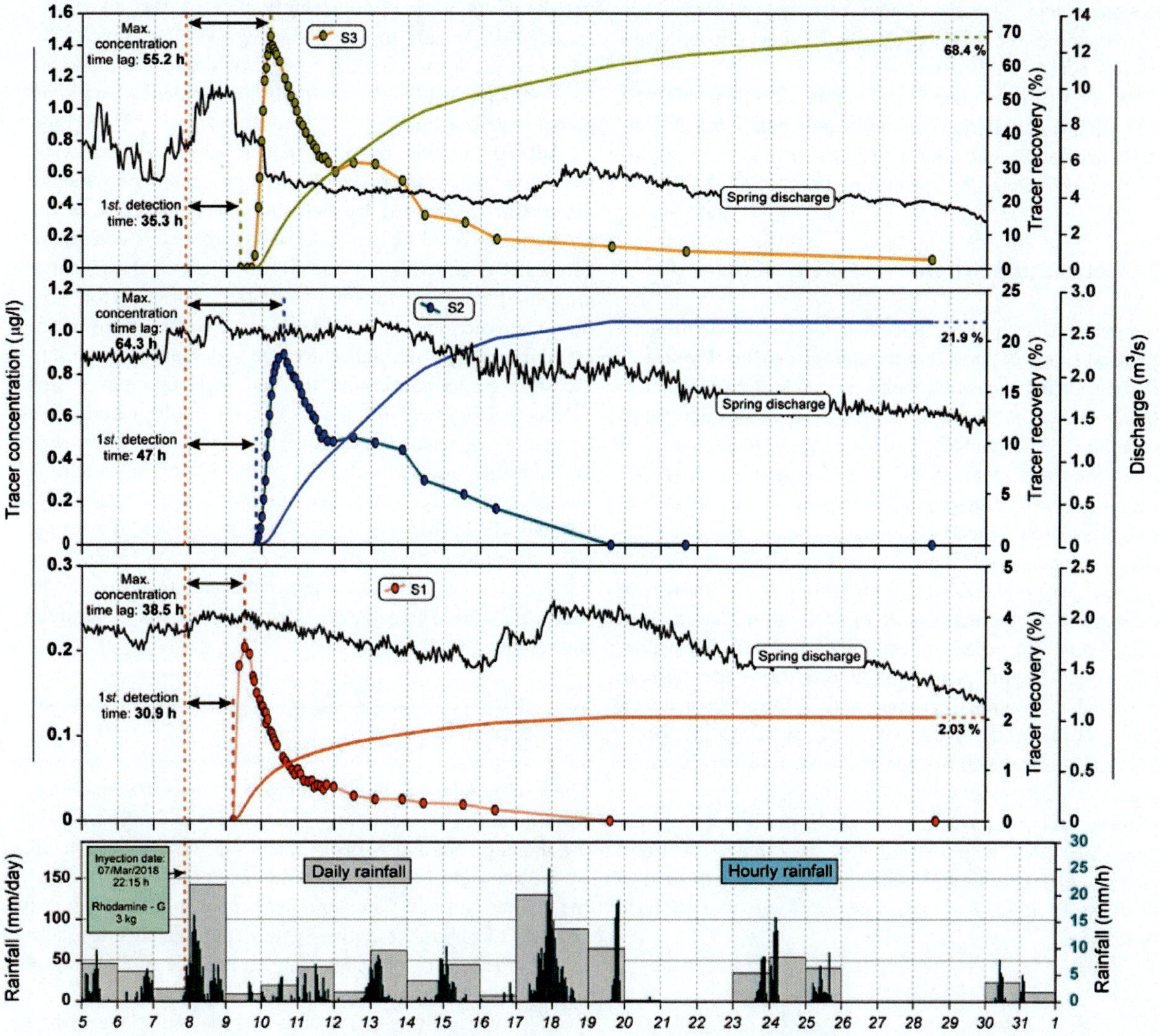


**Fig. 4** Dye tracer test results (concentration, recovery, and time parameters) carried on for the common injection point SH (see location in Fig. 1) from 5 March to 1 April 2018

**Table 1** Calculated time lags of the considered springs derived of CCF analysis and parameters inferred through dye tracer test

| Spring (m s.n.m.) | CCF analysis | | Dye tracer test | | | | |
|---|---|---|---|---|---|---|---|
| | Global time lag (h) | Mean discharge rate (l/s) | 1st. detection time lag (h) | Max. concentration time lag (h) | Recovery tracer rate (%) | Average speed flow (m/h) | Mean discharge ($m^3/s$)[a] |
| S1 (349) | 58 | 332 | 30.9 | 38.5 | 2.03 | 159.8 | 1.7 |
| S2 (317) | 81 | 133 | 47.0 | 64.3 | 21.9 | 111.8 | 2.0 |
| S3 (422) | 72 | 171 | 35.3 | 55.2 | 68.4 | 149.6 | 4.9 |

[a] Mean discharge during the dye tracer test

detection points, whereas a low recovery percentage is related to the existence of a marginal connection between them (Barberá et al. 2018).

The joint use of these techniques complements the methodological limitations of both approaches when considered separately and allows the establishment of a conceptual model of aquifer functioning (Koit et al. 2022).

## 5.2 Conceptual Scheme of Functioning

In general terms, a conceptual scheme for the hydrogeological behavior of the Ubrique aquifer can be developed, including clearly two subsystems (Fig. 5). A first sector is defined by the preferential drainage pathways addressed toward S2 (133 l/s) and S3 (171 l/s) springs, which represent most of the tracer recovery (92.3%) and, therefore, a greater contribution of the allogeneic component occurring in Villaluenga swallow hole (SH) to the recharge. In both BTC curves, secondary unsharpened concentration peaks have been detected with intensity in proportion to the maximum concentration of the main peak in each curve, and a comparable time lag since injection (5–6 days). This phenomenon has been related to the diversion of tracer through the secondary conduits systems, with a lesser karst development, or due to the hydrodynamic interactions that occur within the aquifer, such as conduits—matrix water exchange (Brick et al. 2006; Field and Leji 2012).

The second sector basically would be drained through S1 spring (332 L/s), with a very low percentage of tracer recovery (2.01%), although with the highest values of flow velocities (Table 1). The small tracer recovery rate obtained in S1 points to a minor contribution of the allogeneic recharge originating in the Villaluenga swallow hole (SH); however, the karstic drainage network shows a remarkable development, justified by the high flow velocities and short global response time.

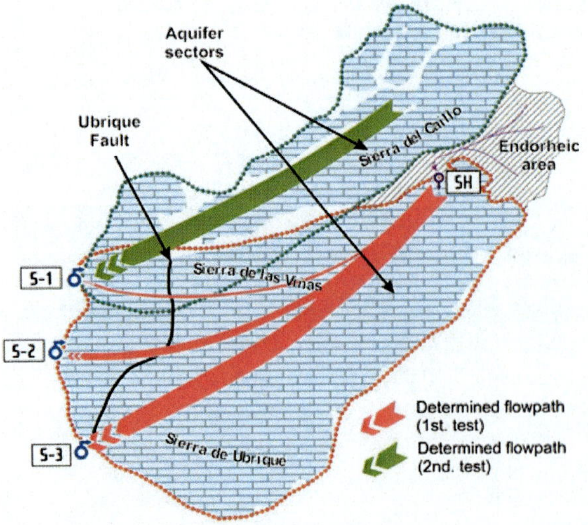


**Fig. 5** Schematic sketch of the Sierra de Ubrique aquifer

Furthermore, the recharge area related to S1 must be larger than that of the rest of the springs, considering that the mean flow discharged by this spring is the largest of the discharge edge (133 l/s). The near carbonate outcrops of Sierra del Caíllo (Figs. 1 and 5) are a part of the recharge area involved in the functioning of this aquifer sector. This flow path was revealed by Marín et al. (2021), using artificial dye tracers. The higher average elevation of this sector (1,186 m a.s.l.) concerning the carbonate outcrops of the Sierra de Ubrique (863 m a.s.l.) could imply a noticeable hydraulic gradient, causing a higher velocity of water flow and a sharper peak morphology of the BTC. This condition has been described by other authors in karst aquifers of northern Spain (Morales et al. 2010). As a result of a further dye tracer test, (Marín et al. 2021) validated the connection between several preferential infiltration points located in the Sierra del Caíllo and the S1 spring, by means of artificial tracers (Fig. 5).

# 6 Conclusions

The hydrogeological functioning of karst aquifers is strongly conditioned by the anisotropy inherent to the development of the drainage conduit network itself and, on the other hand, by the geological structure and tectonic processes that affect the aquifer bedrock. The combined application of long-term analysis of natural responses with experimental tools, such as dye tracer tests, provides insights into the functioning of the Ubrique karst system.

Thus, the analysis of the natural responses of the springs reveals that the aquifer has a generally high degree of karstification, with quick draining and low capacity of natural regulation (conduit flow system).

The obtained results suggest the existence of internal compartmentalization, which segments the aquifer into two sectors relatively isolated from each other, as well as the existence of a dual conduit drainage system with differentiated hydraulic properties.

Extrapolation of this information to the possible evolution of contaminants, chemical or bacteriological, from aquifers recharge areas underlines the relevance of knowing these dynamic relationships, especially when the protection and attenuation of risks related to water supplies are concerned.

In any case, the accurate definition of the recharge areas and the characterization of other elements concerning the hydrogeological functioning of the aquifer must be accompanied by more extensive hydrochemical, isotopic, and microbiological studies, in order to achieve a well-planned, for-the-safe, and sustainable utilization of water resources.

**Acknowledgements** This work is a contribution to the project PID2019-111759RB-I00, PCI 2019-103675 (KARMA) and to the RNM-308 Research Group by Andalusia Government (Spain).

## References

Bakalowicz M (2005) Karst groundwater: a challenge for new resources. Hydrol J 13(1):148–160. https://doi.org/10.1007/s10040-004-0402-9

Barberá JA, Mudarra M, Andreo B, De la Torre B (2018) Regional-scale analysis of karst underground flow deduced from tracing experiments: examples from carbonate aquifers in Malaga province, Southern Spain. Hydrol J 26:23–40

Birk S, Geyer T, Liedl R, Sauter M (2006) Process-based interpretation of tracer tests in carbonate aquifers. Groundwater 43(3):381–388

De la Torre B, Mudarra M, Andreo B (2020) Investigating karst aquifers in tectonically complex alpine areas coupling geological and hydrogeological methods. J Hydrol 10 100047. https://doi.org/10.1016/j.hydroa.2019.100047

Doerfliger N, Zwahlen F (1998) Practical guide, groundwater vulnerability mapping in karstic regions (EPIK). Swiss Agency Environ Forests Landscape. Bern, Switzerland, p 56

Field M, Leij F (2012) Solute transport in solution conduits exhibiting multi-peaked breakthrough curves. J Hydrol 440–441:26–35

Fiorillo F, Doglioni A (2010) The relation between karst spring discharge and rainfall by cross-correlation analysis (Campania, southern Italy). Hydrogeol J 18(8):1881–1895

Ford DC, Williams PW (2007) Karst hydrogeology and geomorphology. Wiley, Chichester, United Kingdom, p 562

Hartmann A, Jasechko S, Gleeson T, Wada Y, Andreo B, Barberá JA, Brielmann H, Bouchaou L, Charlier JB, Darling WG, Filippini M, Garvelmann J, Goldscheider N, Kralik M, Kunstmann H, Ladouche B, Lange J, Lucianetti G, Martín JF, Mudarra M, Sánchez D, Stumpp C, Zagana E, Wagener T (2021) Risk of groundwater contamination widely underestimated because of fast flow into aquifers. Proc Natl Acad Sci (PNAS) 118(20): e2024492118. https://doi.org/10.1073/pnas.2024492118

Käss W (1998) Tracing technique in geohydrology. Balkema, Brookfield, MA, p 581

Koit O, Mayaud C, Kogovšek B, Vainu M, Terasmaa J, Marandi A (2022) Surface water and groundwater hydraulics of lowland karst aquifers of Estonia. J Hydrol

Kovačič G (2010) Hydrogeological study of the Malenščica karst spring (SW Slovenia) by means of a time series analysis. Acta Carsologica 39(2):201–215

Lorette G, Lastennet R, Peyraube N, Denis A (2018) Groundwater-flow characterization in a multilayered karst aquifer on the edge of a sedimentary basin in western france. J Hydrol 566:137–149. https://doi.org/10.1016/j.jhydrol.2018.09.017

Mangin A (1984) Pour une meille u reconnaissance des systems hydrologiques a partir des analyses correlatoire et spectrale. J Hydrol 67:25–43

Marín AI, Martín-Rodríguez JF, Barberá JA, Fernández-Ortega J, Mudarra M, Sánchez D, Andreo B (2021) Groundwater vulnerability to pollution in karst aquifers, considering key challenges and considerations: application to the Ubrique springs in Southern Spain. Hydrogeol J 29:379–396. https://doi.org/10.1007/s10040-020-02279-8

Martín-Algarra A (1987) Evolución geológica alpina del contacto entre las Zonas Internas y las Externas de la Cordillera Bética. PhD Thesis. Univ of Granada (Spain), p 1171

Morales T, Uriarte J, Olazar M, Antigüedad I, Angulo B (2010) Solute transport modelling in karst conduits with slow zones during different hydrological conditions. J Hydrol 390:182–189

Mudarra M, Andreo B, Barberá JA, Mudry J (2014) Hydrochemical dynamics of TOC and $NO_3$—contents as natural tracers of infiltration in karst aquifers. Environ Earth Sci 71(2):507–523. https://doi.org/10.1007/s12665-013-2593-7

Padilla A, Pulido-Bosch A (1995) Study of hydrographs of karstic aquifers by means of correlation and cross-spectral analysis. J Hydrol 168:73–89

Sánchez D, Martín-Rodríguez JF, Mudarra M, Andreo B, López-Rodríguez M, Navas MR (2017) Time lag analysis of natural responses during unitary recharge events to assess the functioning of carbonate aquifers in Sierra de Grazalema Natural Park (Southern Spain). In: Renard P, Bertrand C (eds) EuroKarst 2016. Advances in karst science. Springer, Berlin, 157–167

# Quantitative and Geochemical Characterization of the Mokra Karst Aquifer (SE Serbia) by Time Series Analysis and Stochastic Modelling

B. Petrović and V. Marinović

**Abstract**

This paper deals with karst aquifer characterization by using time series and stochastic analysis and modelling. This methodological approach was applied on the example of Mokra karst spring for the period of 2015–2017, which included monitoring and analysis of Mokra karst spring quantity and quality parameters. Quantitative characterization of the Mokra karst spring has shown high karstification degree of the karst aquifer and its very well retention capacity. Also, simulation model has proved that precipitation is the main driving force of the hydraulic behaviour of the Mokra karst groundwater circulation and discharge. Qualitative characterization has shown very small fluctuation of chemical parameters in a hydrologic year and expected karst fingerprinting, while isotopic analysis shown fast water replacement. This case study proved the necessity for karst groundwater characterization in order to manage this natural resource in a sustainable way.

**Keywords**

Karst aquifer characterization • Time series analysis • Stochastic modelling • SE Serbia

## 1 Introduction

Karst groundwater is a vital resource worldwide, primarily as a resource of drinking water for the population. Estimations said that almost 2000 million people in the world use groundwater as a drinking resource, out of which nearly 700 million (9.16% of total world population) use karst groundwater (Stevanović 2019). About 21.6% of the European land surface is covered by the carbonate rocks, including 15.1% of "continuous" and 6.5% of "discontinuous carbonate rocks", with the total area of actual carbonate rock outcrops of 13.8% of total earth surface (Chen et al. 2017). Karst groundwater is significant natural resource in Serbia as well (Fig. 1), considering karst groundwater consumption in Serbia is more than 4 $m^3/s$ or 20% of total groundwater supply of Serbian population (Dimkić et al. 2011; Stevanović and Dokmanović 2015).

Even though karst aquifers are one of the most important resources of drinking water worldwide, the most common problem with them is the seasonal fluctuation of quantity and quality over a hydrological year as well as inherited heterogeneity. For instance, seasonal oscillations of karst groundwater levels and reserves and oscillations in karst groundwater quality (especially during heavy rains) can endanger the water supply of the population. Therefore, characterization and prediction of the karst groundwater regime would be useful in many ways—prediction of extreme values of karst spring discharge (particularly important in the recession period) as well as foreknowledge of deterioration of qualitative parameters (turbidity, bacteria, etc.) couple of days ahead can save large number of people from water scarcity (Stevanović 2015).

Characterization of karst groundwater is based on the definition of geological and hydrogeological characteristics and parameters, preceded by aquifer geometrization, in terms of determining the catchment area and depth of carbonate rocks (Stevanović 2015). After the spatial definition of karst aquifer, the characterization of the quantitative component includes statistical and mathematical methods of evaluating karst spring discharge regimes in terms of analysing recession curve, baseflow and stochastic functions of karst groundwater discharge time series (Kresic 2013). Qualitative characterization of karst groundwater involves the analysis of groundwater quality parameters obtained by chemical analysis, isotopic analysis, etc. (Goldscheider and Drew 2007; Stevanović 2015; Madonia et al. 2020).

B. Petrović (✉) · V. Marinović
Department of Hydrogeology, Faculty of Mining and Geology, The Centre for Karst Hydrogeology, Belgrade University, Belgrade, Serbia
e-mail: branislav.petrovic@rgf.bg.ac.rs


B. Andreo et al. (eds.), *EuroKarst 2022, Málaga*, Advances in Karst Science,
https://doi.org/10.1007/978-3-031-16879-6_8

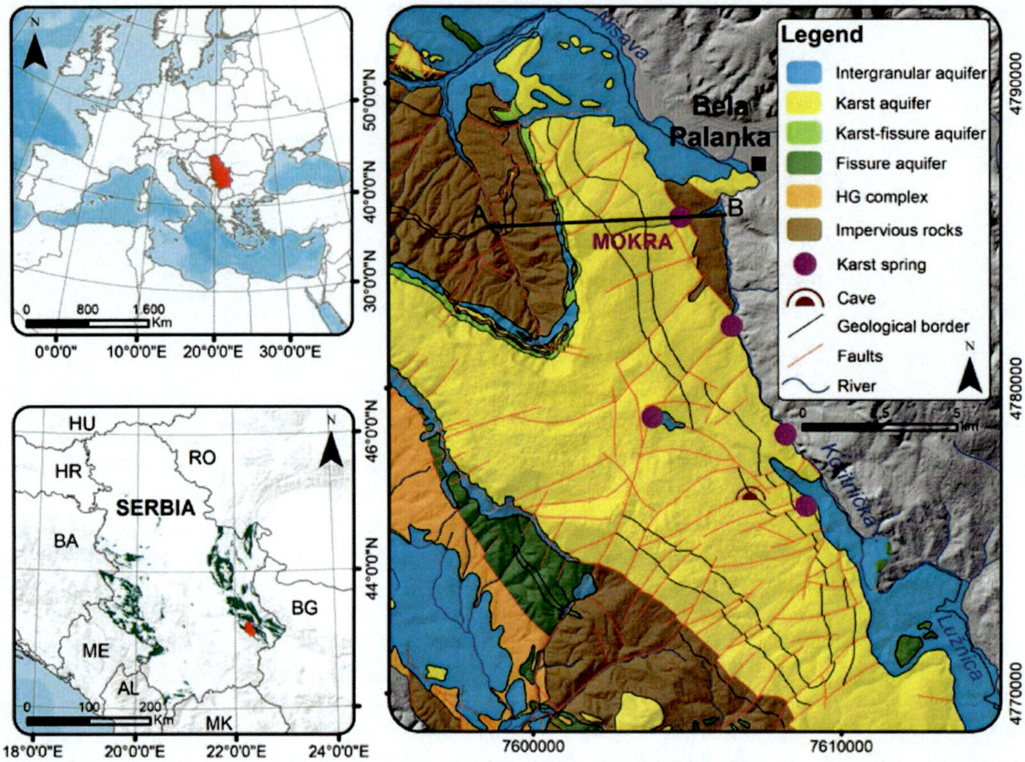


**Fig. 1** **a** Position of Serbia in Europe; **b** location of Mokra karst spring in SE Serbia—red area, while karst aquifers of Serbia are in green; **c** hydrogeology map of the wider area of Mokra karst spring. A–B line corresponds to the geological section of Fig. 2

## 2 Study Area General Characteristics

Mokra karst aquifer is formed on the north-eastern slopes of Suva Planina Mt. which is the part of the Carpatho-Balkanides mountain chain. Suva Planina Mt. is located approximately 230 km SE from Belgrade (Fig. 1a, b). One of the first reports on Mokra karst spring is dated back to the end of XIX century, when the Mokra spring was compared with Mlava spring, another significant karst spring in Eastern Serbia (Cvijić 1896).

Mesozoic karst outcrops of Eastern Serbia cover an area of approximately 3350 km$^2$ (Stevanović 1994), while Suva Planina Mt. catchment area covers about 252 km$^2$ of that surface (Stevanović 1995). Suva Planina Mt. is asymmetrical, normal anticline and its geological setting is very complex due to multiple thrusting and faulting that occurred during the Caledonian Orogeny, the Hercynian orogeny and finally the Alpine orogeny (Dimitrijević et al. 1980). Tectonic movements lead to the uplift of the NW part and the creation of a plunging anticline. The process of erosion in the ascended part caused destruction of limbs of anticline that are created of Cretaceous and Jurassic limestone and dolomite (Fig. 1c), to the core of anticline which is mainly made up of Paleozoic rocks. Local faults are transversal to the NW–SE regional dislocations supposedly causing the enhancement of the karstification process in the limestone and dolomite (Petrović 2020a, b).

Karst aquifers have been created in cracked, fissured and karstified carbonate rocks of the Jurassic and Cretaceous age. The recharge of aquifers is precipitation based with effective infiltration estimated as 55% of total 630 mm precipitation (Stevanović 1991, 1994). Faulting and plunging caused the creation of groundwater flows within the karst aquifer which follow a radial direction from the Suva Planina Mt. anticline axis towards its limbs. However, local faults provoked the enhancement of the karstification process in the limestone and dolomite, and the groundwater flow has been changed to a direction, that is transversal to the original NW–SE direction. Additional karstification provided further and finer division of the groundwater flow and resulted in quite a few orifices of the aquifers in the foothill of Suva Planina Mt. (Stevanović 1991; Petrović and Marinović 2021).

Karst aquifers are mostly discharged via springs as well as through underground pathways into adjacent aquifers.

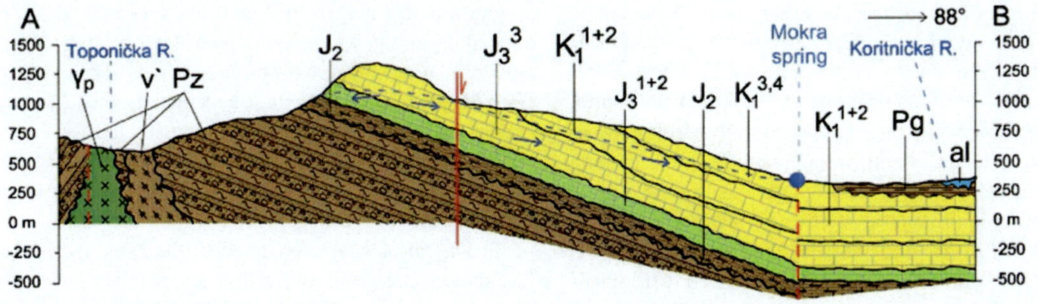


**Fig. 2** Simplified hydrogeological cross section of Suva Planina Mt. (adapted from Petrović 2020a, b). Same vertical and horizontal scales. For cross section location and lithological legend see Fig. 1

Two large karst springs at the NE foothill of the mountain: Mokra (Fig. 2) and Divljana, are a result of a developed process of karstification and a contact of karstified limestone (with dolomite) with less permeable deposits of Neogene age (Petrović 2020a, b).

## 3 Methods

Human curiosity regarding karstic regions has led to the development of specializations of scientific study such as speleology, karst hydrology/hydrogeology, karstology, speleobiology and cave diving (speleodiving) (Petrović 2020a, b). Also, human interest in karst groundwater has been impacted by the usually huge amount of high-quality water that karst aquifer can accumulate and release. Those facts cause the necessity for getting familiar with karst aquifer characteristics in order to manage it in a sustainable way.

Time series analysis of karst groundwater regime stands as an adequate methodological approach to the initial characterization of a karst system (Bonacci 1987; Kresic 2013; Mangin 1994; Jemcov 2008). Time series analysis and stochastic simulation techniques give a general insight into the hydraulic functioning of the entire karst hydrogeological system in different hydrologic periods, although, e.g. precipitation and groundwater discharge, as input and output signals, respectively, are observed at one point. Also, this analysis and techniques must be implemented complementary with the field research which stands as the first step in the characterization of the Mokra karst aquifer in order to define its boundaries, the recharge and discharge zones, as well as to inspect the results of previously conducted (hydro) geological research. Also, permanent monitoring of precipitation, karst groundwater discharge and chemical components are must-have for successfully characterization of a karst system. In this way, time series of precipitation data from Niš station and Mokra spring discharge and quality data for the period 2015–2017 were collected and evaluated.

Quantitative characterization of the Mokra karst system includes several techniques. Firstly, analysis of a recession curve whose shape can provide insight into the behaviour of the karst system in different periods. In other words, extreme maximum values or very steep recession limb may indicate high development of karst conduits, i.e. high karstification degree (Bonacci 1993; Fiorillo 2014). Even though there are several methods for recession curve analysis (Bonacci 1993; Fiorillo 2011, 2014), Maillet's equation was applied for the recession curve analysis of the Mokra karst spring.

Time series analysis were applied to characterize quantitative regime of the Mokra karst spring by autocorrelation and cross-correlation functions. The autocorrelation function implies univariate analysis describing successive members of the same time series and determining its dependence and periodicity (Kresic 1997; Larocque et al. 1998; Jemcov 2008). Slope of the autocorrelogram can point out to the karstification degree of the system. Cross-correlation is a function that encompasses the time domain of time series analysis, but in a bivariate environment. This involves analyzing the correlation of precipitation ($P$) (input signal to the system) and groundwater discharge ($Q$) (output signal from the system) or finding how many days karst system requires to start reacting on a heavy rainy episode (Padilla and Pulido-Bosch 1995; Jemcov 2008).

Stochastic simulation and prediction model of Mokra karst spring discharge was developed based on precipitation and discharge data for the period 2015–2017 (Petrović and Marinović 2021). Stochastic model included a combined autoregressive-cross-regressive moving average model (ARCR-MA), which belong to the group of multivariate time series models implying multiple linear regression. This model included linearized input signal (precipitation) by moving average filter, because the initial model is not able to absorb the intensity of individual rain episodes, which in the simulation model is shown as the white noise (Kresic 1997).

Geochemical characterization of Mokra karst spring groundwater was based on analyses of daily/weekly values of turbidity, specifical electrical conductivity, pH, temperature, and numerous water quality analysis, isotopic ($^{18}O$, $^{2}H$) analyses of rainfall and groundwater (Goldscheider and Drew 2007; Milanović and Vasić 2015; Vasić 2017) as well as simulation of turbidity values for the period 2015–2017.

Turbidity of groundwater was measured daily in situ by Hach Aqua Trend SOM Turbid, while physicochemical parameters, such as temperature (± 0.1 °C), pH (± 0.1 U) and electrical conductivity ($\leq$ 0.5%), were measured weekly, directly in the field, using WTW portable instruments. The sampling campaigns of groundwater from Mokra karst spring were conducted weekly as well. The concentrations of the cations ($K^+$, $Na^+$, $Ca^{2+}$, $Mg^{2+}$) and the anions ($HCO_3^-$, $Cl^-$, $NO_3^-$, $SO_4^{2-}$) were analysed in the laboratory as well nitrites and ammonia, and the total dissolved solids (TDS). A common bench-scale method to estimate natural oxidant demand involves the addition of a permanganate ($KMnO_4$), solution to a sample and monitoring the consumption of permanganate (Xu and Thomson 2008). All chemical analyses were carried out based on the national standards of Serbia (Off. Gazette of FRY, No. 42/98 and 44/99; Off. Gazette of RS, No. 28/2019). Determination of $^{18}O$ and $^{2}H$ contents in water samples was performed in "Institute for Nuclear Research, Hungarian Academy of Sciences—MTA ATOMKI" (Debrecen, Hungary). Analysing of groundwater and rain (snow) samples to determine contents of stable isotope were performed on a Thermo Finnigan Delta plus XP type isotope ratio mass spectrometer. The results were reported in δ‰ versus the V-SMOW standard, with a precision better than ± 1‰ for δD and ± 0.1‰ for $\delta^{18}O$ (Clark and Fritz 1997; Mook 2006; Vasić 2017, Vasić et al. 2019).

## 4 Results and Discussion

Quantitative analysis has shown that discharge varies 0.1–4 $m^3/s$ ($Q_{av}$ = 0.6 $m^3/s$) (Fig. 3a). Recession analysis of Mokra karst spring discharge showed the existence of 3 micro-discharge regimes in 2015, which lasted 98 days, 2 micro-discharge regimes in 2016 (lasted 88 days) and 4 micro-discharge regimes in 2017, which lasted 137 days. The amount of groundwater discharged during recessions was $1.93 \times 10^6$ $m^3$ in 2015, $1.56 \times 10^6$ $m^3$ in 2016 and $4.28 \times 10^6$ $m^3$ in 2017. Recession coefficients varies from $\alpha = 0.124829$ (in 2017) to $\alpha = 0.002292$ (in 2016). Time series analysis of discharge and rainfall data has shown that memory of the Mokra karst system is 22–48 days while system's reaction to heavy rain has a 1–3-days lag period depending on the wetness of a hydrologic year, according to autocorrelation and cross-correlation functions respectively (Fig. 3b, c). Stochastic ARCR MA model of the Mokra karst spring discharge time series has been created for the period 2015–2017 and calibrated with 2018 data (Fig. 3d), which shown accurate fit of simulated with measured values ($R = 0.946$).

Geochemical characterization of the Mokra karst aquifer has shown that groundwater has an "original karst groundwater" fingerprint (Figs. 4 and 5). Mokra spring groundwater temperature varies from 8.7 to 21.1 °C, while the average temperature is 14.2 °C, specifical electrical conductivity (Ec) is in ~100 μS/cm range, from 446 to 542 μS/cm, while the average value is 472 μS/cm (Fig. 4). Groundwater turbidity is usually under 1 NTU, but every year there are periods, at least 3 times per year, when turbidity is higher (Fig. 4), with average value of 0.3 NTU. Karst groundwater is neutral with pH ranging from 7.1 to 7.5, and mean value is 7.4.

Consumption of permanganate is ranging from 1.0 to 3.6 mg/l. Total dissolved solids are in all samples below 348.0 mg/l, while average value is 292.2 mg/l.

The $Ca^{2+}$ and $Mg^{2+}$ are the dominant cations in groundwater (Fig. 5). The concentrations of $Ca^{2+}$ and $Mg^{2+}$ range from 91.0 mg/l to 94.0 mg/l and from 5.9 mg/l to 8.7 mg/l, respectively, with mean concentration of 92.3 mg/l and 7.8 mg/l. All groundwater samples have rather low concentration of $Na^+$ and $K^+$, below 1.7 mg/l and 0.6 mg/l, respectively.

The concentrations of $HCO_3^-$ range from 395.6 to 437.8 mg/l with mean concentration of 413.2 mg/l in this aquifer. Groundwater samples have rather low concentration of $SO_4^{2-}$, $Cl^-$ and $NO_3^-$ below 9.1 mg/l, 5.0 mg/l and 5.4 mg/l, respectively. The Piper diagram and pie chart only confirmed that groundwater belongs to the category of calcium bicarbonate ($Ca\text{-}HCO_3$) groundwater (Fig. 5).

The isotopic composition of stable isotopes ($^{18}O$ and $^{2}H$) of groundwater compared to the local meteoric water line (LMWL) showed that the isotopes are mostly located above the LMWL. Analysis of results of stable isotopes ($\delta^{18}O$ and δD) in groundwater, shows that there is a direct relationship with precipitation. Based on the arrangement of points in the immediate vicinity of LMWL, according to this we may assume that groundwater was accumulated in an open aquifer with rapid water replacement (Petrović 2020a, b), such as those formed in karstified rocks of Suva Planina Mt.

Time series analysis (auto- and cross-correlation) was also performed with turbidity dataset which has shown its short memory (3 days), while response of turbidity to impulse shock caused by heavy rain is 5 days. Also, ARCR MA stochastic model was created for the turbidity time series which shown very well fit of measured and

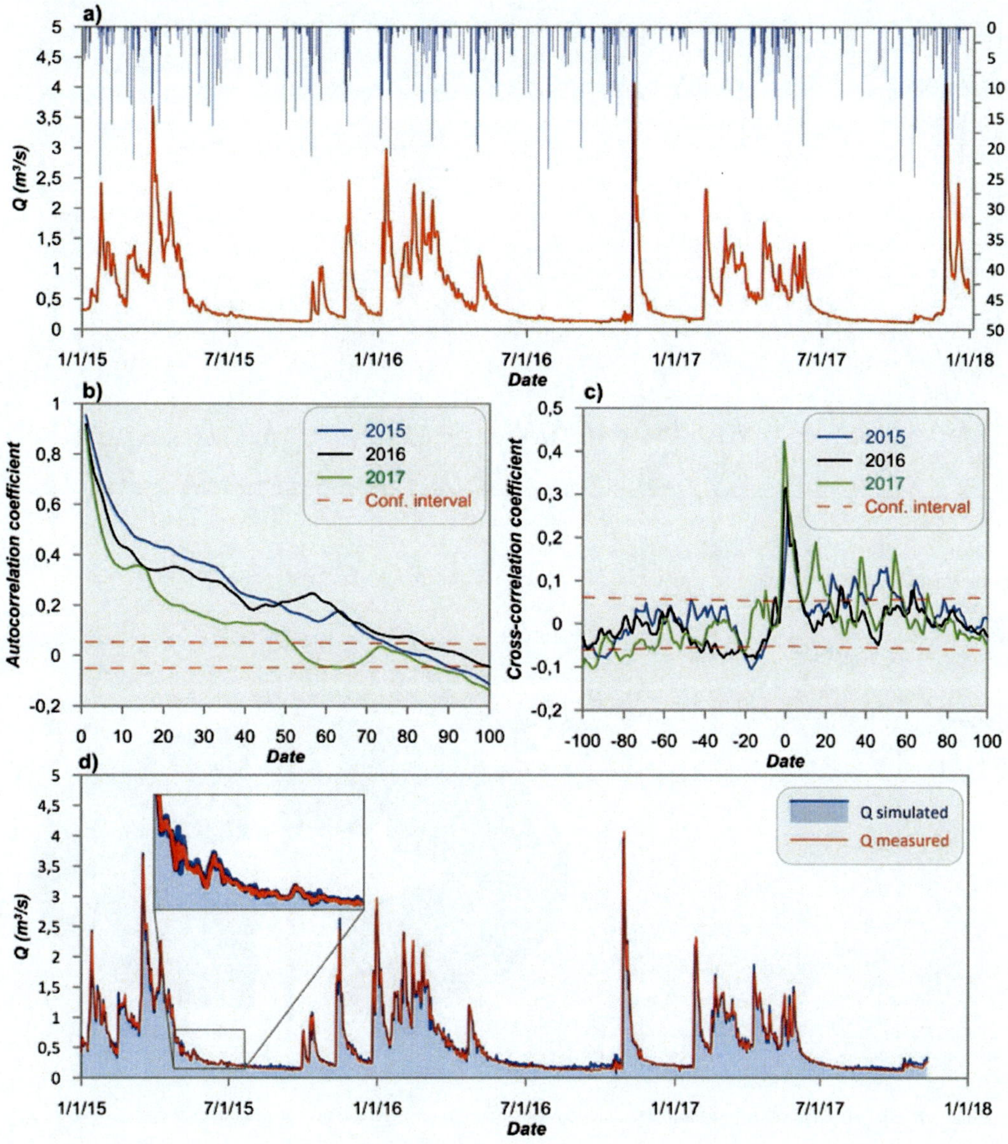


**Fig. 3** **a** Mokra karst spring hydrogram ($Q$) and precipitation ($R$) from Niš station; **b** autocorrelation of Mokra karst spring discharge; **c** cross-correlation of Mokra karst spring discharge and precipitation from Niš station; **d** ARCR MA stochastic simulation model for Mokra karst spring, in the rectangle: groundwater simulation in recession period

simulated values ($R = 0.956$) (Marinović and Petrović 2021) (Fig. 6).

## 5 Conclusions

Quantitative characterization of the Mokra karst aquifer helped us to confirm that after diffuse infiltration of rainfall, groundwater is promptly transferred to the Mokra karst spring via dominant karst conduits. Recession analysis of Mokra karst spring discharge proved very well retention capacity of the aquifer. Stochastic simulation model has confirmed that recharge of this karst aquifer is dominantly by precipitation.

The observed narrow Ec and pH range of groundwater suggest a relatively stable existence of groundwater in Mokra karst aquifer, which is potentially beneficial for long-term water resources exploitation, especially if one take in consideration that spring has already been tapped for water supply of Niš. Consumption of permanganate as an

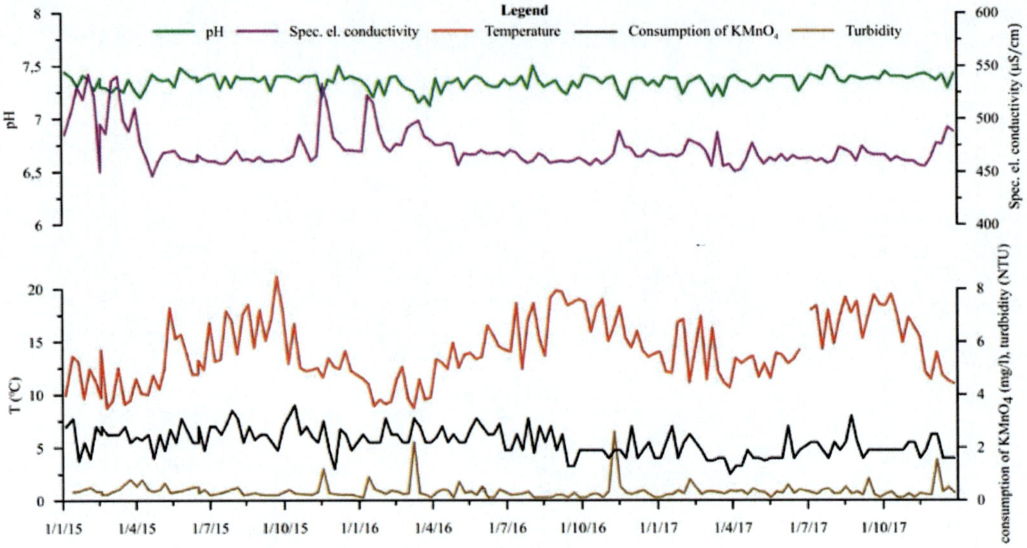


**Fig. 4** Summary diagram of field and laboratory quality parameters of groundwater from Mokra karst spring, 2015–2017

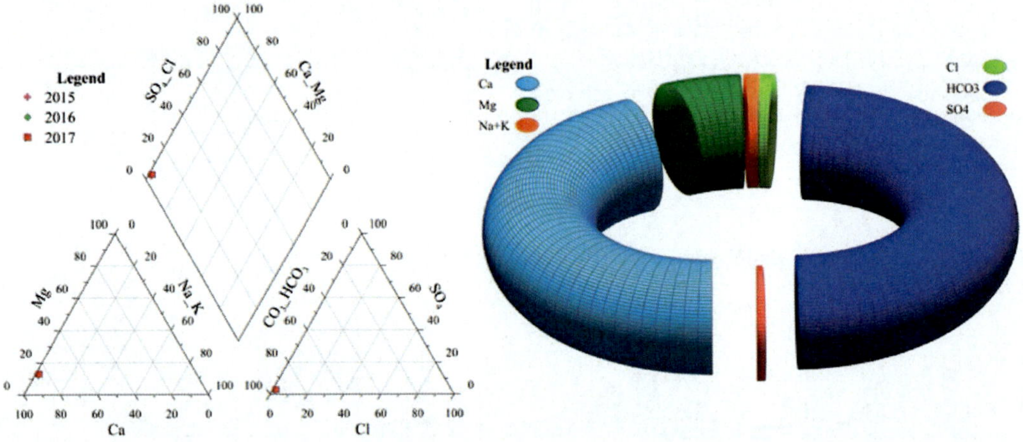


**Fig. 5** Piper diagram (left) and pie chart (right) of basic chemical composition (anions and cations) of Mokra karst spring, 2015–2017

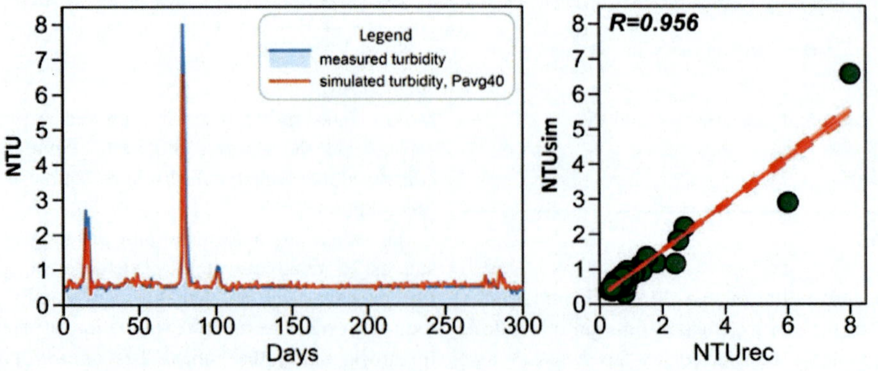


**Fig. 6** Comparative diagram (left) of measured and simulated turbidity values of Mokra karst spring by applying ARCR MA model 2015–2017; correlation (right) of recorded (NTUrec) and simulated (NTUsim) turbidity values

indicator of possible organic pollution showed very low values indicating that quality of groundwater is still pristine. Groundwater is formed in open aquifer with fast water replacement ($^{18}O$, $^{2}H$).

It can be concluded that complex hydrodynamic behaviour related to the system hierarchy is thought to be the origin of turbidity. The fast infiltration through the unsaturated zone causes a flushing effect and a turbulent quick flow in the highly transmissive conduits with very high velocities. In this way, karst aquifer characterization and simulation models within groundwater management may have a function in the early warning system.

## References


Bonacci O (1987) Karst hydrology with special reference to the Dinaric karst. Springer-Verlag, Berlin, p 184

Bonacci O (1993) Karst springs hydrographs as indicators of karst aquifers. J Hydrol 38:51–62

Chen Z, Auler AS, Bakalowicz M, Drew D, Griger F, Hartmann J, Jiang G, Moosdorf N, Richts A, Stevanovic Z, Veni G, Goldscheider N (2017) The world karst aquifer mapping project: concept, mapping procedure and map of Europe. Hydrogeol J 25:771–785

Clark I, Fritz P (1997) Environmental isotopes in hydrogeology. Taylor & Francis Group, Boca Raton, 342 pp

Cvijić J (1896) Izvori, tresave i vodopadi u Istočnoj Srbiji (Sources, peats and waterfalls in East Serbia) (in Serbian). The Voice of Serbian Royal Academy of Sciences, Belgrade

Dimitrijević M, Dragić D, Karamata S, Sikošek B, Petrović B, Veselinović D (1980) Map and explanatory text of the BGM of SFRY, sheet Bela Palanka, F. G. S. of SFRY, 69 pp

Dimkić M, Stevanović Z, Đurić D (2011) Progress and improvement of the status of groundwater in Serbia. In: Proceedings of IWA conference, Belgrade, pp 81–101

Fiorillo F (2011) Tank-reservoir emptying as a simulation of recession limb of karst spring hydrographs. Hydrogeol J 19:1009–1019

Fiorillo F (2014) The recession of spring hydrographs, focused on karst aquifers. Water Resour Manage 28:1781–1805

Goldscheider N, Drew D (eds) (2007) Methods in karst hydrogeology. Taylor & Francis/Balkema, London, 264 pp

Jemcov I (2008) Karst groundwater budget and optimization of its tapping—Serbian examples (in Serbian). PhD thesis, University of Belgrade, Faculty of Mining and Geology, Serbia, 377 pp

Kresic N (1997) Quantitative solutions in hydrogeology and groundwater modeling. CRC Lewis, Boca Raton, 461 pp

Kresic N (2013) Water in karst: management, vulnerability, and restoration. McGraw-Hill, New York, 736 pp

Larocque M, Mangin A, Razack M, Banton O (1998) Contribution of correlation and spectral analyses to the regional study of a large karst aquifer (Charente, France). J Hydrol 205:217–231

Madonia P, Cangemi M, Oliveri Y, Germani C (2020) Hydrogeochemical characters of karst aquifers in Central Italy and relationship with neotectonics. Water 12(7):1926

Mangin A (1994) Karst hydrogeology. In: Gilbert J, Danielopol DL, Stanford JA (eds) Groundwater ecology. Academic Press, San Diego, CA, pp 43–67

Marinović V, Petrović B (2021) Stochastic simulation and prediction of turbidity dynamics in karst systems. Case study: Mokra karst spring (SE Serbia). Rev Bulg Geol Soc 82:222–224

Milanović S, Vasić L (2015) Monitoring of karst groundwater. In: Stevanović Z (ed) Karst aquifers—characterization and engineering. Professional practice in earth sciences. Springer, pp 335–359

Mook WG (2006) Introduction to isotope hydrology-stable and radioactive isotopes of hydrogen, oxygen and carbon. Taylor & Francis, London, 226 pp

Padilla A, Pulido-Bosch A (1995) Study hydrographs of karstic aquifers by means of correlation and cross-spectral analysis. J Hydrol 168:73–89

Petrović B (2020a) Intrinsic groundwater vulnerability assessment by multiparameter methods, a case study of Suva Planina Mountain (SE Serbia). Environ Earth Sci 79:85

Petrović B (2020b) The functioning and impact of epikarst oh the regime, balance and groundwater quality of the eastern part of the Suva Planina Mountain karst system (in Serbian). PhD thesis, University of Belgrade, Faculty of Mining and Geology, Serbia, 305 pp

Petrović B, Marinović V (2021) Application of the discrete autoregressive—cross-regressive moving average model for predicting the daily discharge values of Mokra and Divljana springs. In: Reports of the Serbian geological society for the year 2020, pp 1–15

Rulebook of hygienic safety of drinking water. Official Gazette of FRY, No. 42/98 and 44/99 and Official Gazette of RS, No. 28/2019

Stevanović Z (1991) Hydrogeology of the Carpatho-Balkan karst area of Eastern Serbia and possibilities of water supply (in Serbian, abstract in English). University of Belgrade, Faculty of Mining and Geology, Serbia, 245 pp

Stevanović Z (1994) Karst groundwaters of Carpatho-Balkanides in Eastern Serbia. In: Stevanović Z, Filipović B (eds) Groundwaters in carbonate rocks of the Carpatho-Balkan Mountain range. CBGA, Belgrade, Serbia, pp 203–237

Stevanović Z (1995) Karst groundwater of Serbia—use and potentiality for regional water supply (in Serbian). In: Stevanović Z (ed) Lithospheric water mineral resources of Serbia. University of Belgrade, Faculty of Mining and Geology, Belgrade, Serbia, pp 77–116

Stevanović Z (ed) (2015) Karst aquifers—characterization and engineering. Professional practice in earth sciences. Springer, pp 335–359

Stevanović Z (2019) Karst waters in potable water supply: a global scale overview. Environ Earth Sci 78:662

Stevanović Z, Dokmanović P (2015) Project of groundwater monitoring network expansion in Serbia (in Serbian). University of Belgrade, Faculty of Mining and Geology, Serbia

Vasić L (2017) Genesis and circulation of groundwater of complex karst systems of Kučaj-Beljanica massif (in Serbian). PhD thesis, University of Belgrade, Faculty of Mining and Geology, Serbia, p 393

Vasić L, Palcsu L, Fen H (2019) Groundwater gravitational circulation of karst Veliko Vrelo and Malo Vrelo springs by isotope and the noble gas method: case study of the Beljanica Massif. Environ Earth Sci 78:307

Xu X, Thomson NR (2008) Estimation of the maximum consumption of permanganate by aquifer solids using a modified chemical oxygen demand test. J Environ Eng 134:5 (353)

# Characterization of the Isotopic Signature of Effective Rainfall ($\delta^{18}O$, $\delta^{2}H$) to Constrain the Groundwater Recharge Zones in a Mediterranean Karst Aquifer

T. Garin, B. Ladouche, B. Arfib, B. Dewandel, and J. Gonçalves

**Abstract**

Carbonate aquifers are known as a major source for drinking water in the Mediterranean regions and in other part of the world. Therefore, qualitative and quantitative estimation of the groundwater resource are crucial, especially in area with densely populated areas. Stable water isotopes of surface and ground waters can be used to study water mixing, define recharge area, or identify fast infiltration in karst areas. It relies on (1) the variability of the input signal over space and time at a catchment scale, related to the rainfall isotopic signature, and (2) the flow, storage and mixing within the karst system, related to the soil–epikarst–unsaturated zone–saturated zone hydrodynamic behaviour. Mean groundwater isotopic signature can then be different from the mean total rainfall signature, depending on the effective rainfall amount. The isotopic signature of the recharge was calculated using a water balance model for two sampling sites in south-eastern France. Soil and epikarst are modelled as a first compartment where evapotranspiration occurs, defined by its water capacity. An isotopic mixing balance model uses the calculated monthly effective rainfall and the analysed monthly mean isotopic water values of collected rainfall. This subsurface water capacity has a strong influence on the isotopic signature of the recharge and contributes to deplete the isotopic value of regional groundwater at the scale of a 2-year period.

**Keywords**

Stable water isotopes • Natural tracers • Effective rainfall • Recharge • Carbonate aquifers • Karst

## 1 Introduction

Stable isotopes of rainwater ($\delta^{18}O$, $\delta^{2}H$), because of their conservative characteristics, are used to identify groundwater recharge zones (Clark and Fritz 1997) and thus potentially better constrain hydrogeological catchments. A global approach, by weighted isotope signatures of precipitation over relatively long periods, is often satisfactory for identifying groundwater recharge areas (Darling and Bath 1988; Prada et al. 2016). However, it is also quite common to observe a difference between rainfall and groundwater isotopic ratios (Goldscheider and Drew 2007), related to the hydrogeological functioning of karst aquifers (karst organization, reserve volume, preferential flows, …) (Williams 2008). Soil and epikarst store a part of the precipitation that will never reach the saturated zone and is available for evaporation or/and evapotranspiration. Effective rainfall is the part of the precipitation that will effectively contribute to surface and groundwater flows. Evapotranspiration has no impact on the water stable isotopes signal but reduces the recharge to groundwater by decreasing the effective rainfall.

The goal of this study is to show how water stable isotopes of recharging groundwater can be deviated from rainfall signal, considering evapotranspiration in the soil water capacity. We used a daily water balance and a monthly isotopes balance model, to assess the impact of the effective rainfall amount on the weighted mean isotopic value of the effective rainfall in a Mediterranean case study. Considering six usual values of soil water capacity (SWC), we computed

T. Garin (✉) · B. Arfib · J. Gonçalves
Aix Marseille University, CNRS, IRD, INRAE, CEREGE, Aix-en-Provence, France
e-mail: garin@cerege.fr

B. Arfib
e-mail: arfib@cerege.fr

T. Garin · B. Ladouche · B. Dewandel
BRGM, D3E/NRE-University of Montpellier, 1039 Rue de Pinville, 34000 Montpellier, France



B. Andreo et al. (eds.), *EuroKarst 2022, Málaga*, Advances in Karst Science,
https://doi.org/10.1007/978-3-031-16879-6_9

the deviation between the average water isotopic value of collected rainfall and computed effective rainfall. A global approach over a two-year period is first proposed and then completed with the computation over five hydrological semesters to address the seasonal variability. The process will help to examine the impact of recharge on groundwater isotopic signature measured at karstic springs and boreholes, and thus to better constrain the hydrogeological catchment areas.

## 2 Case Study and Data

The case study is located in south-eastern France, close to Marseille city, on the Port-Miou spring and Huveaune river catchment areas (Fig. 1). The 500 km$^2$ investigated zone is characterized mainly by Triassic to Cretaceous carbonate rocks that have been eroded and karstified during several stages. This zone is also defined by a large range of elevation, from the sea level to 1148 m asl, giving an expected

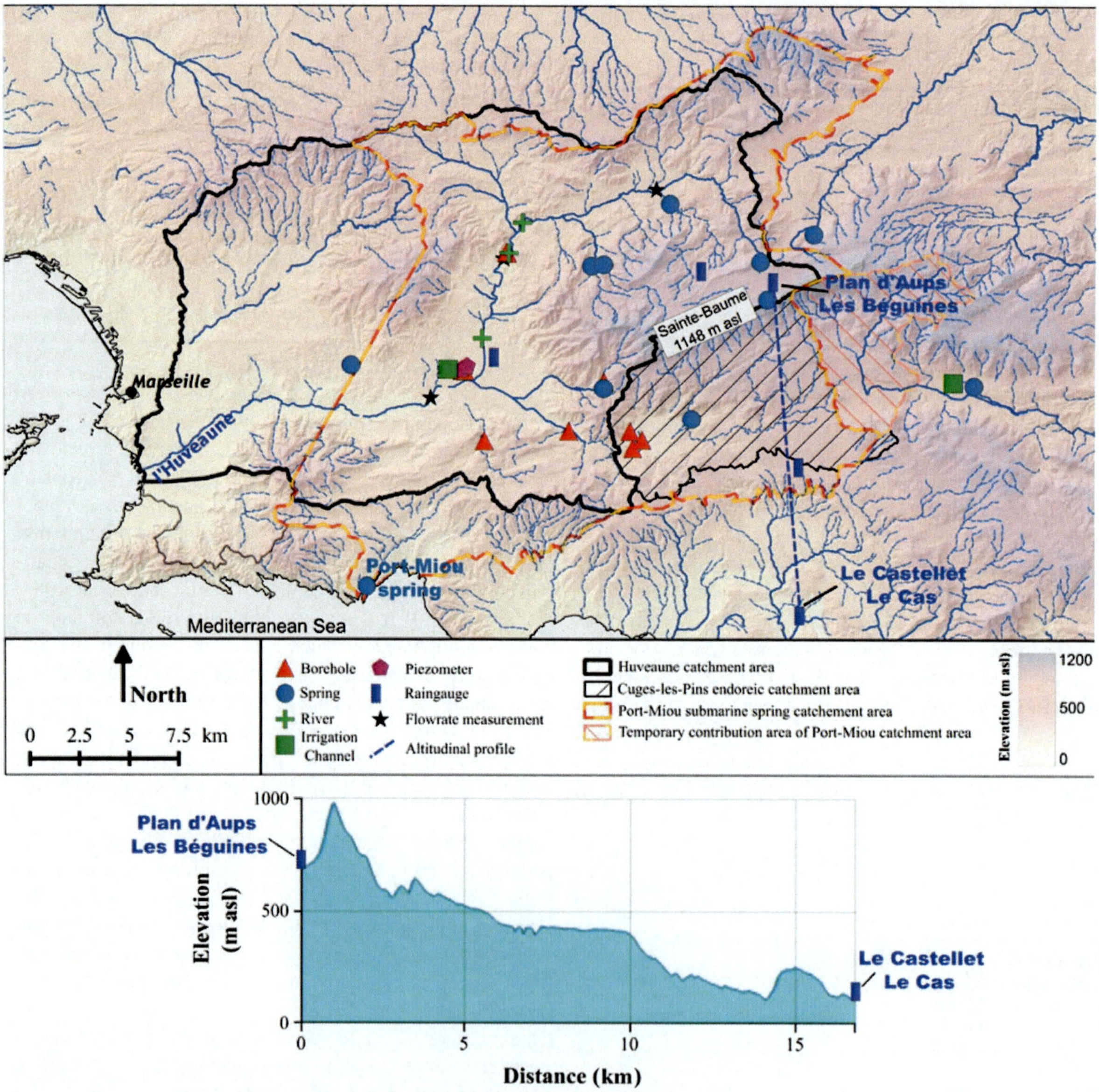


**Fig. 1** Hydrogeological context of the Port-Miou spring and the Huveaune river catchment areas

significant contrast of rainfall isotopic signature. Monthly rainwater samples have been collected during two years at two sampling sites to catch the variability of the stable isotopic signature in the study area: (1) Le Castellet Le Cas station at elevation 103 m, south slope of the Sainte-Baume massif, 3 km from the seashore; (2) Plan d'Aups Les Béguines at elevation 682 m asl, north slope of the Sainte-Baume massif (Fig. 1). These two stations cover the mean elevation of the Port-Miou spring and Huveaune catchment areas (about 360 m asl). Daily precipitation are monitored by automatic rain gauges. Daily potential evapotranspiration is computed by the French National Meteorological Service using Penman–Monteith equation.

## 3 Method

In this study, the time step of precipitation sampling does not allow to work on the scale of one or several daily events. It is a rather global approach that is developed here. An infiltration and isotopic mixing balance model is proposed to calculate the isotopic signature of effective rainfall over time. Figure 2 presents the workflow. The daily effective rainfall, based on daily rainfall, potential evapotranspiration (PE) and six values of SWC, is calculated according to Thornthwaite water balance model (Thornthwaite 1948). The sum of daily effective rainfall led to monthly effective rainfall. Meanwhile, monthly rainfall is collected according to the procedure proposed by Gröning et al. (2012) to avoid evaporation of the rainwater at the two sampling sites. Water stable isotopic composition is measured by laser spectrometry (LAMA Montpellier, France) on the monthly collected rainfall. Then, an isotopic mixing balance model is used to calculate the weighted water stable isotopic composition of effective rainfall over a 2-year period (from April 2019 to March 2021) and over 5 hydrological semesters (starting on September to August). These results can then be compared with the water stable isotopic composition of the cumulated rainfall and the groundwater samples in springs or boreholes. The computation method gives a mean value assuming that the unsaturated and saturated zones are massive waterbodies that buffer the input water geochemical signal. In this paper, we will only compare the results with the isotopic composition of the cumulated rainfall.

## 4 Results

Figure 3a shows the isotopic signal of precipitation and effective rainfall on a 2-year period, according to six values of SWC. The difference between the isotopic signal of rainfall and effective rainfall for a SWC of 100 mm is

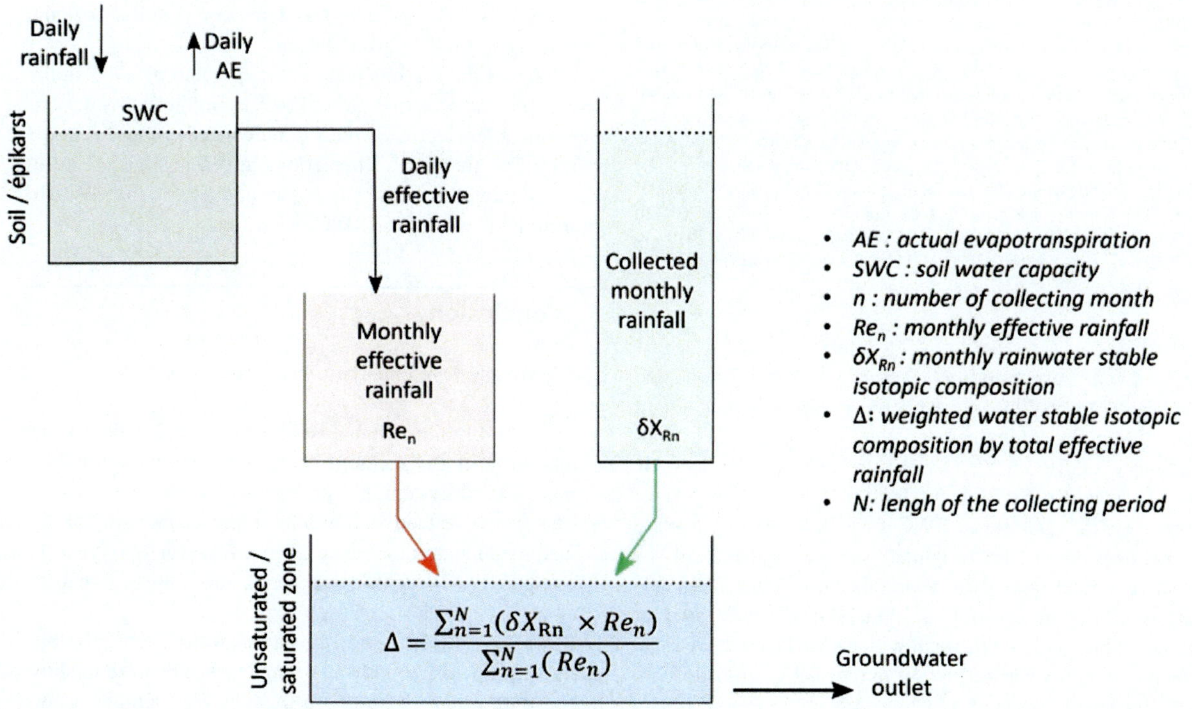


**Fig. 2** Conceptual infiltration model and isotopic mixing balance model

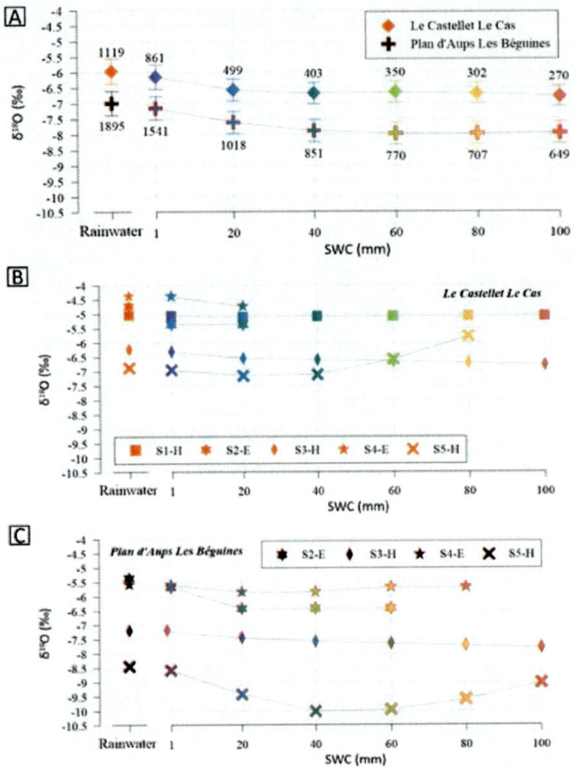


**Fig. 3** **a** Weighted mean of the $\delta^{18}O$ isotopic signature of the rainwater collected at the two collecting stations (Le Castellet Le Cas and Plan d'Aups Les Béguines) for a period of 2 years (April 2019–March 2021), and evolution of the weighted mean isotopic signature of the effective rainfall according to the six values of SWC. Labels give the total rainfall and the total effective rainfall over the studied period. **b**, **c** $\delta^{18}O$ isotopic signature of the rainwater collected at the two collecting stations (respectively, Le Castellet Le Cas and Plan d'Aups Les Béguines) for 5 hydrological semesters and evolution of the isotopic signature of the effective rainfall according to the six values of SWC. The hydrological semesters are defined as: S1—02/11/2018 to 28/02/2019. S2—01/03/2019 to 31/08/2019. S3—01/09/2019 to 28/02/2019. S4—01/03/2020 to 31/08/2020. S5—01/09/2020 to 28/02/2021

obvious with a decrease in $\delta^{18}O$ from −6.99 to −7.96‰ and from −5.97 to −6.79‰, respectively, for the sampling site of Plan d'Aups Les Béguines and Le Castellet Le Cas. Considering a near zero SWC (1 mm) already implies a decrease of 0.16 and 0.19‰ in $\delta^{18}O$ for the two sampling sites. Moreover, the increase in SWC from 1 to 40 mm gives a strong decrease in the weighted isotopic signature of the effective rainfall and then a stabilization between 40 and 100 mm. This observation is linked to the total effective rainfall, which follow this trend with a significant decrease in effective rainfall over the first 40 mm of SWC (from 1119 to 403 mm for Le Castellet Le Cas and from 1895 to 851 mm for Plan d'Aups Les Béguines).

Figure 3b, c show the isotopic signal of precipitation and effective rainfall for 5 hydrological semesters: 3 including autumn and winter (S1-H, S3-H, S5-H) and 2 including spring and summer (S2-E, S4-E). Tables 1 and 2 show the measured rainfall and calculated effective rainfall for the two collecting stations, respectively, and for the different Soil Water Capacity (SWC). The spring–summer semesters have a less total cumulated rainfall than autumn–winter semesters as expected in Mediterranean climate. Results given in tables below Fig. 3b, c show that effective rainfall during summer (S2 and S4) is null starting from a SWC 40 mm at the Castellet station (low elevation, 103 m asl), and very low at the Plan d'Aups station (high elevation, 682 m asl). Major effective rainfall occurs during the autumn and winter seasons. Water stable isotopic composition of rainfall are more depleted during cold autumn–winter semesters than spring–summer semesters for both stations, e.g. at Plan d'Aups station $\delta^{18}O$ of rainfall equal −8.5‰ during S3 and −5.5‰ during S4. Results of effective rainfall water stable isotopic composition calculation for increasing SWC show that the mean value is depleted compared to the total rainfall (expect for the winter S5 with a decreasing slope and then an increasing slope due to one event not detailed in this study). Effective rainfall amount decreases with increasing SWC, that means that only the highest rainfall events are contributing to the effective rainfall (minor events are balance by actual evapotranspiration). Highest events, known as Mediterranean events (usually higher than 50 mm/day) have a strong impact on the isotopic composition of the recharging rainfall to groundwater.

Whatever the seasons and the actual value of SWC in the case study, the effective rainfall recharging groundwater has a water stable isotopic composition depleted compared to the rainfall. The mean $\delta^{18}O$ depletions, given in Fig. 3a, are in the same range for both stations: around −0.5‰ for SWC of 20 mm, and −0.8‰ for SWC of 40 mm.

## 5 Conclusion

The approach developed here highlights:

- The monthly mean stable isotope precipitation data allow the study of the recharge water signature in a Mediterranean karst system at a pluri-annual scale.
- The use of effective rainfall for the calculation of the weighted mean signature leads to a depletion of oxygen 18 compared to the rainfall weighted mean isotopic signature.

Moreover, further comparison of water stable isotope composition with groundwater samples will be used to show that groundwater recharge from effective rainfall occurs primarily during major rainfall events, as expected by the

**Table 1** Measured rainfall and calculated effective rainfall for Le Castellet Le Cas collecting station for the 6 values of soil water capacity (SWC)

| Plan d'Aups Les Béguines | Rainwater (mm) | Effective rainfall depending on SWC (mm) | | | | | |
|---|---|---|---|---|---|---|---|
| Hydrological semesters | | SWC = 1 mm | SWC = 20 mm | SWC = 40 mm | SWC = 60 mm | SWC = 80 mm | SWC = 100 mm |
| S1-H (02/11/2018–28/02/2019) | 193 | 147 | 83 | 81 | 81 | 81 | 81 |
| S2-E (01/03/2019–31/08/2019) | 76 | 41 | 9 | 0 | 0 | 0 | 0 |
| S3-H (01/09/2019–28/02/2020) | 597 | 508 | 370 | 341 | 318 | 294 | 270 |
| S4-E (01/03/2020–31/08/2020) | 189 | 117 | 21 | 0 | 0 | 0 | 0 |
| S5-H (01/09/2020–28/02/2021) | 256 | 195 | 99 | 62 | 32 | 8 | 0 |

**Table 2** Measured rainfall and calculated effective rainfall for Plan d'Aups Les Béguines collecting station for the 6 values of soil water capacity (SWC)

| Plan d'Aups Les Béguines | Rainwater (mm) | Effective rainfall depending on SWC (mm) | | | | | |
|---|---|---|---|---|---|---|---|
| Hydrological semesters | | SWC = 1 mm | SWC = 20 mm | SWC = 40 mm | SWC = 60 mm | SWC = 80 mm | SWC = 100 mm |
| S1-H (02/11/2018–28/02/2019) | nc | nc | nc | nc | nc | nc | nc |
| S2-E (01/03/2019–31/08/2019) | 248 | 154 | 51 | 24 | 4 | 0 | 0 |
| S3-H (01/09/2019–28/02/2020) | 976 | 885 | 707 | 641 | 616 | 592 | 568 |
| S4-E (01/03/2020–31/08/2020) | 268 | 194 | 84 | 40 | 23 | 11 | 0 |
| S5-H (01/09/2020–28/02/2021) | 403 | 307 | 176 | 146 | 128 | 105 | 82 |

large discharge variations observed in-situ. Depletion of the isotopic composition calculated by the balance model will be search in the groundwater isotopic composition. Small rainfall events of a few mm to ten mm may be useful for critical zone ecosystems (soil–plant–atmosphere) but are suspected to be not effective in recharging the regional karst aquifers of the Huveaune–Port-Miou watersheds.

**Acknowledgements** This communication is part of the Karst-Huveaune project funded by Agence de l'Eau Rhône Méditerranée Corse, Région Sud-PACA, Conseil Départemental des Bouches-du-Rhône, Aix-Marseille Provence Métropole, BRGM (French Geological Survey), and Aix-Marseille University. We also thank T. Lamarque (asso. Spélé-H2O) for field assistance, Météo--France for meteorological data, and Syndicat Mixte du Bassin Versant de l'Huveaune (SMBVH). This work was performed within the framework of the Port-Miou observation site, part of the KARST national observatory network (www.sokarst.org) and within the framework of the Castellet observation site, part of the RENOIR observatory network.

## References

Clark I, Fritz P (1997) Environmental isotopes in hydrogeology

Darling WG, Bath AH (1988) A stable isotope study of recharge processes in the english chalk. J Hydrol 101:31–46

Goldscheider N, Drew D (2007) Methods in karst hydrogeology

Gröning M, Lutz OH, Roller-Lutz Z, Kralik M, Gourcy L, Pöltenstein L (2012) A simple rain collector preventing water re-evaporation dedicated for $\delta^{18}O$ and $\delta^{2}H$ analysis of cumulative precipitation samples. J Hydrol 448–449:195–200. https://doi.org/10.1016/j.jhydrol.2012.04.04

Prada S, Cruz JV, Figueira C (2016) Using stable isotopes to characterize groundwater recharge sources in the volcanic island of Madeira Portugal. J Hydrol 536:409–425. https://doi.org/10.1016/j.jhydrol.2016.03.009

Thornthwaite CW (1948) An approach toward a rational classification of climate. Geogr Rev 38(1):55–94. https://doi.org/10.2307/210739

Williams P (2008) The role of the epikarst in karst and cave hydrogeology: a review. Int J Speleol 37(1):1–10. https://doi.org/10.5038/1827-806X.37.1.1

# Application of Statistical Approaches to Piezometry to Improve the Understanding of the Karst Aquifer Hydrodynamic Behaviours at the Cadarache CEA Centre (France)

Manon Erguy, Sébastien Morilhat, Guillaume Artigue, Julien Trincal, Anne Johannet, and Severin Pistre

**Abstract**

Located in the French Mediterranean region, the French Alternative Energies and Atomic Energy Commission (CEA) *Cadarache* Centre is subject to significant rainfall events which can lead to groundwater floods with rapid kinetics and high amplitude. This study aims to improve the understanding of the aquifer's heterogeneity and piezometric responses at the scale of the *Cadarache* centre (900 ha) in order to better apprehend the groundwater floods. The deployed method consists in: (i) taking an inventory of reliable data; (ii) conducing a preliminary study of the spatial variability of rainfall at the centre scale with the method of double-mass curves and the comparison of the rainfall amounts of the different stations; and (iii) conducting a comprehensive statistical study of piezometric data (graphs of sorted piezometric data for three different flood events, definition of an original clustering method using an automatic feature detection). The results are of two kinds: (i) several clusters were identified; three appeared to be representative of specific areas; (ii) for each specific area, piezometers were linked with different plurikilometric NW–SE and NE–SW lineaments identified by previous bibliographic studies. This extensive study on nearly 80 piezometers has thus allowed to define a method of automatic feature extraction of the sorted groundwater levels, and to establish a clustering of piezometers according to their statistical behaviour. Even if the statistical "black box" method was applied without any prerequisite on the location of the boreholes, it allowed establishing a very relevant relation between sorted water height and geological structure constraining the flows.

M. Erguy · G. Artigue · A. Johannet (✉)
HydroSciences Montpellier (Univ. Montpellier, IMT Mines Alès, IRD, CNRS), 6 av. de Clavières, 30100 Alès, France
e-mail: anne.johannet@mines-ales.fr

M. Erguy
e-mail: manon.erguy@cea.fr

G. Artigue
e-mail: guillaume.artigue@mines-ales.fr

M. Erguy · S. Morilhat · J. Trincal
CEA, Centre de Cadarache, 13108 Saint Paul Lez Durance Cedex, France
e-mail: sebastien.morilhat@cea.fr

J. Trincal
e-mail: julien.trincal@cea.fr

S. Pistre
HydroSciences Montpellier (Univ. Montpellier, IMT Mines Alès, IRD, CNRS), Place Eugène Bataillon, 34095 Montpellier, France
e-mail: severin.pistre@umontpellier.fr

**Keywords**

Karst • Groundwater floods • Statistical analyses • Sorted piezometric data

## 1 Introduction

The flood hazard linked to surface water has been greatly studied, given the major economic, social and environmental issues, associated with these disasters (Price et al. 2000; Gaume et al. 2004; Špitalar et al. 2014; Esposito 2015). These studies have been facilitated by the obvious causality between river flooding and associated damage. In contrast, groundwater-related floods are more difficult to observe and their role in surface flooding and the related mechanisms are more difficult to demonstrate. However, groundwater flooding can also cause very significant damage (Kreibich and Thieken 2008; Ascott et al. 2017). The total, or partial, contribution of groundwater to flooding episodes has been demonstrated for many cases, particularly in karst regions due to their rapid groundwater transfers (López-Chicano et al. 2002; Pinault et al. 2005; Jourde et al. 2007; Bailly-Comte et al. 2008). Indeed, karst aquifers are complex hydrosystems particularly sensitive to water table rising (Bonacci 1995; Charlier et al. 2015).


B. Andreo et al. (eds.), *EuroKarst 2022, Málaga*, Advances in Karst Science,
https://doi.org/10.1007/978-3-031-16879-6_10

This study focuses on the Cretaceous karst aquifer of the *Cadarache* centre of the French Alternative Energies and Atomic Energy (CEA) located in the southeast of France. The site is regularly subject to this type of phenomenon. For example, following the major rainy events of November 2011 (cumulative rainfall of 225 mm in 72 h), and November 2019 (cumulative rainfall of 101 mm in 4 days following an episode of 185 mm in 4 days in October), groundwater level rises locally of nearly 40 m. Such rising water tables can submerge sensitive equipment and therefore constitute an important risk to consider and anticipate for the *Cadarache* centre and for many industrial sites or municipalities in the Mediterranean region (Weng and Dörfliger 2002).

In order to forecast groundwater floods, the first stage consists in improving the understanding of the karst aquifer behaviour. The objective of this study is thus to improve the comprehension of the aquifer's piezometric responses, its heterogeneity, and to identify zones having a particular hydrodynamic behaviour.

For this purpose, this paper proposes, in the next section, a presentation of the *Cadarache* site. Then, the third section will present the methods used for the data inventory, their evaluation and their statistical analyses using sorted water level. Afterward, the fourth section presents the results and links them with the underlying geological structures. The conclusion synthetize results and considers future work.

## 2 Presentation of the Study Site: The *Cadarache* CEA Centre

### 2.1 Geological and Hydrogeological Context

The *Cadarache* centre of the French Alternative Energies and Atomic Energy (CEA) is located at *Saint Paul Lez Durance* in the south of France (Fig. 1). The centre takes place on outcropping Cretaceous limestone or covered by Miocene sediments and/or Pliocene and Quaternary alluvium (Guerin 2001). Each formation constitutes an aquifer having a specific behaviour. The Miocene and Quaternary aquifers are not very reactive and their level variations are of low amplitudes (water level rising of few meters in several weeks, or months) (Fig. 1). Consequently, they are not subject to flash groundwater floods; they will not be studied in this paper. The Cretaceous karst aquifer is characterized by an important reactivity, a hydrodynamic spatial heterogeneity and by the absence of highly developed karst conduits (Guerin 2001), thus limiting its capacity for rapid transfer to the outlet. This aquifer exhibits strong fracturing, and a double-diffusivity type behaviour at the local scale (Najib 2007). It is then possible to distinguish: (i) a fracture/conduit continuum, potentially very transmissive, with high amplitude piezometric variations, and significant responses to precipitation, and (ii) a matrix continuum, less reactive. Studies are currently underway on the use of natural tracers, but results are unclear because their variations are not significant enough.

### 2.2 Monitoring Network

The *Cadarache* centre extends over 1600 ha, and has an outstanding hydrogeological monitoring network. The site has eight rain gauges spread over five zones: Z1–Z5, (Fig. 1). In addition, it includes more than 400 piezometers, of which around 230 are equipped with pressure sensors (with 115 piezometers measuring the Cretaceous aquifer) (Fig. 1). This dense monitoring network has enabled the acquisition of a large database of rainfall and groundwater levels over more than 15 years.

## 3 Material and Methods

### 3.1 Data Inventory and Evaluation

The first stage of this study is to make an inventory of the piezometric and pluviometric available data, and then to study their reliability. The first sensors of piezometric data were installed in 2001, but the beginning of measurements is very variable. They do not require a reliability study because data are readjusted by manual measurements.

About rainfall data, daily manual rainfall measurements are available since 1960. Data from automatic rain gauges at 10 or 30 min time-steps are available from 2006. The reliability analysis of the rainfall data was carried out using the following protocol. The periods when the data acquisition rate was less than 100% were identified, then double-mass curves analyses were carried out. This method consists of comparing rainfall accumulations between two rain gauge stations. The aim is to identify potential malfunctions of the rain gauges, or trends in the underestimation or overestimation of rainfall. All the different stations combinations was compared. This allowed to identify the available and reliable data used for the rest of the study.

### 3.2 Analysis of the Spatial Variability of Rainfall at the Centre Scale

The rainfall spatial variability was done using the double-mass curves, and the comparison of cumulative rainfall. The rainfall amounts over different periods were compared to identify possible differences in rainfall at the centre scale. The cumulative rainfalls of the various

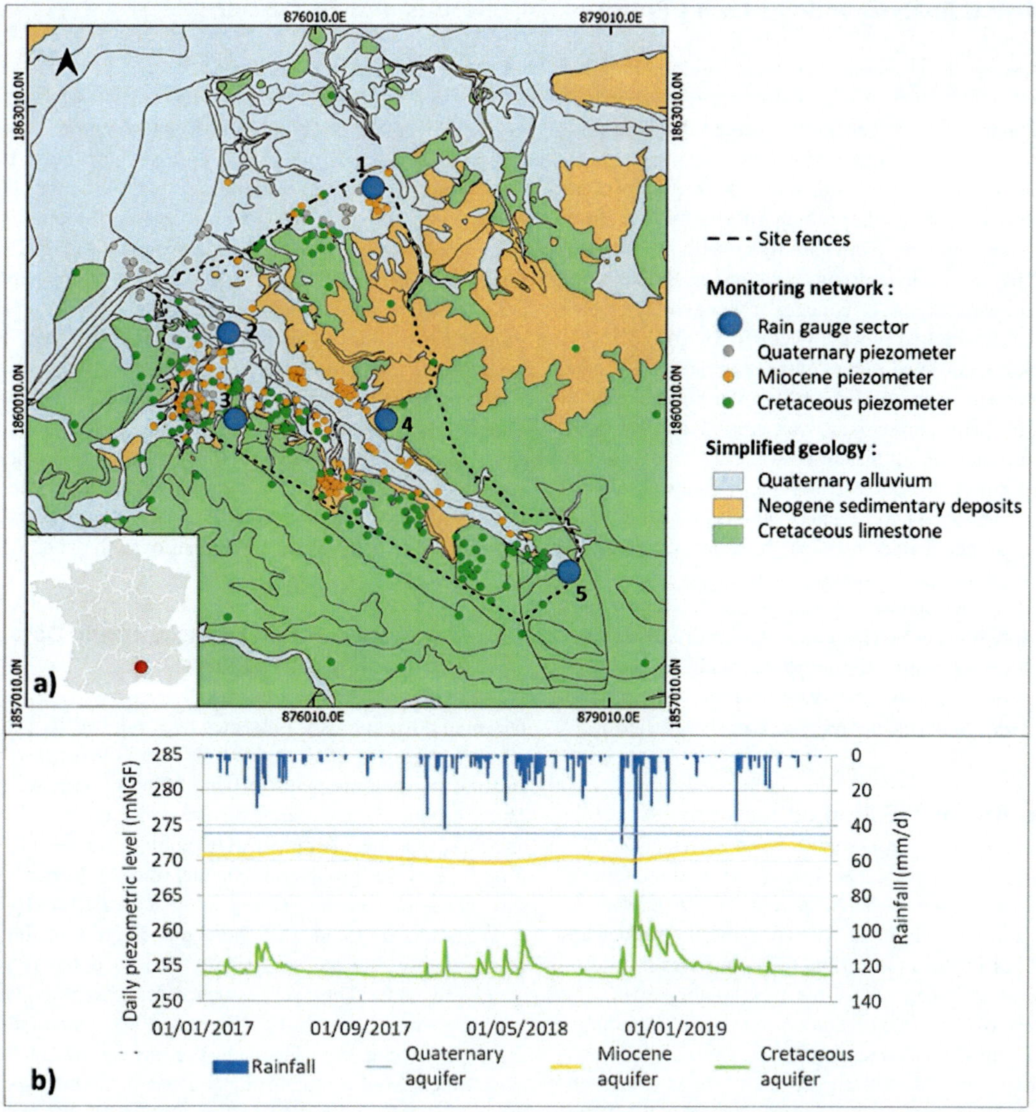


**Fig. 1** Simplified geological map of the *Cadarache* centre with hydrological monitoring network locations (**a**) and piezometric variations of the different aquifers (**b**)

stations were calculated: (i) for the 30 major rainy events, (ii) on a monthly scale since 2017, (iii) annually since 2006. Almost all of the differences greater than 10% can be explained by a temporary malfunction of a rain gauge, identifiable by comparison with the measurement rate for the period. The study of rainfall at the different stations was also done, on a finer time-scale and over more events, in order to identify a possible trend. As the water table is an integrated data of rainfall at the scale of the hydrogeological basin, the limits of which extend beyond the perimeter of the site, it was considered that these small rainfall differences would not have significant impact on the hydrodynamic behaviour of the water table. Based on this hypothesis, it was decided to construct a "reference rainfall chronicle" by averaging data from the two stations, respectively located to the north and to the south of the site (zones Z1 and Z5 in Fig. 1). The advantage is to have a reliable and continuous rainfall chronicle since 2006. This chronicle was used to select the rainfall events studied in the rest of this paper.

### 3.3 Statistical Analysis and Clustering Protocol

The sorted water level curves represents the cumulative empirical probability (in %) versus the piezometric values ranked by classes. As it is generally accepted that rain on a small time-scale behaves like white noise (Mangin 1975), ruptures observed on sorted water levels are only representative of the aquifer behaviour (compartmentalisation, fracturing, connectivity or permeability), without rainfall influence. This allows identifying variations in the slope of the curve as characteristic of changes in behaviour, or particularities of the studied process (Mangin 1975).

The sorted water level curves were compared with each other, for the same time-period and for the boreholes with available data. The comparison was carried out using an automated protocol for all the sorted water level curves. The protocol first detects the curves variations, and the associated piezometric level, using the Ramer–Douglas–Peucker algorithm (Douglas and Peucker 1973). In a second step, the clustering by number of slope variations is made. This makes it possible to identify the main groups having roughly equivalent sorted water levels curves. After this clustering, a visual inter-comparison of the graphs is made by the operator, within each group, and also between the different groups, in order to check the relevance of the clustering.

### 3.4 Selection of the Studied Rain Episodes

The aim is to make statistical analyses on several rain events, in order to characterize the behaviour of the Cretaceous aquifer, according to different types of rainfall episodes and different states of the aquifer (low/high water level).

Three episodes were chosen: (i) November 2011: this event is the most important recorded in terms of cumulative rainfall (212 mm in 72 h) since 1960. It caused the highest rise of the water table (nearly 40 m in some areas) and occurred after a period of pronounced drought; (ii) March 2017: this moderate episode (53 mm in 72 h) generated a response of the aquifer characteristic of medium/low initial water level; (iii) November 2019: in terms of rising water table, this event is comparable to the one observed in November 2011, it is the second most important of the centre. However, it differs by the timing of the rainfall, and the initial state of the aquifer. In 2019, a first rain episode occurred in October (185 mm in 4 days), providing a pre-saturation of the aquifer. The episode in November (101 mm in 4 days) thus occurred when the network of conductive fractures was already largely saturated.

## 4 Results and Discussions

### 4.1 Piezometers Clustering

For the November 2011 episode, 57 piezometric chronicles were considered and the sorted groundwater levels curves were grouped into three main clusters: Group 1 with 5 piezometers, Group 2 with 18 piezometers, Group 3 with 7 piezometers; other data were not groupable (Fig. 2). The groups G1, G2, G3 could be associated with three particular areas. G1 to the northern part of the site: the northwest of the *Mourre Frais* valley. This paleovalley has a Miocene filling and acts as a hydraulic barrier to the flows of the Cretaceous aquifer especially during high groundwater level periods (Guerin 2001). G2 to the central part that is the south of the *Vallée des Piles*. G3 to the southern part: a zone where the water table has a specific behaviour highlighting the compartmentalisation of the aquifer. The piezometers with non-groupable behaviours are spread over the site.

### 4.2 Link Between Sorted Piezometric Data and Geological Features

The method previously presented was applied to the other selected events of 2017 and 2019. These events have more piezometric measurements (around 80 chronicles available). Three groups corresponding to the same three main areas identified for the 2011 episode could also be identified (Fig. 3). Another group could be identified at the north of the site, whose piezometers have been installed after 2011.

We seek: (i) to identify how piezometers in the same group were connected, and (ii) to identify potential preferential flow paths. The Cretaceous karst aquifer does not have very developed conduits (Guerin 2001), and different assumptions can then be made concerning the main flow axes which would explain that distant piezometers present the same sorted water level curves. One possibility is that the fractures (mostly vertical) constitute a preferential axis of circulation.

Based on the following assumptions: (i) lineaments correspond to transmissive subvertical faults (but in reality, some may have an impermeable clay filling), (ii) similar sorted groundwater levels curves indicate a comparable hydrogeological behaviour; we tried to verify that a link between fracturing and groups of sorted groundwater level curves, could be established. In order to test this hypothesis, we overlaid the map showing the extensions of the three classified piezometry groups, and the map with the main

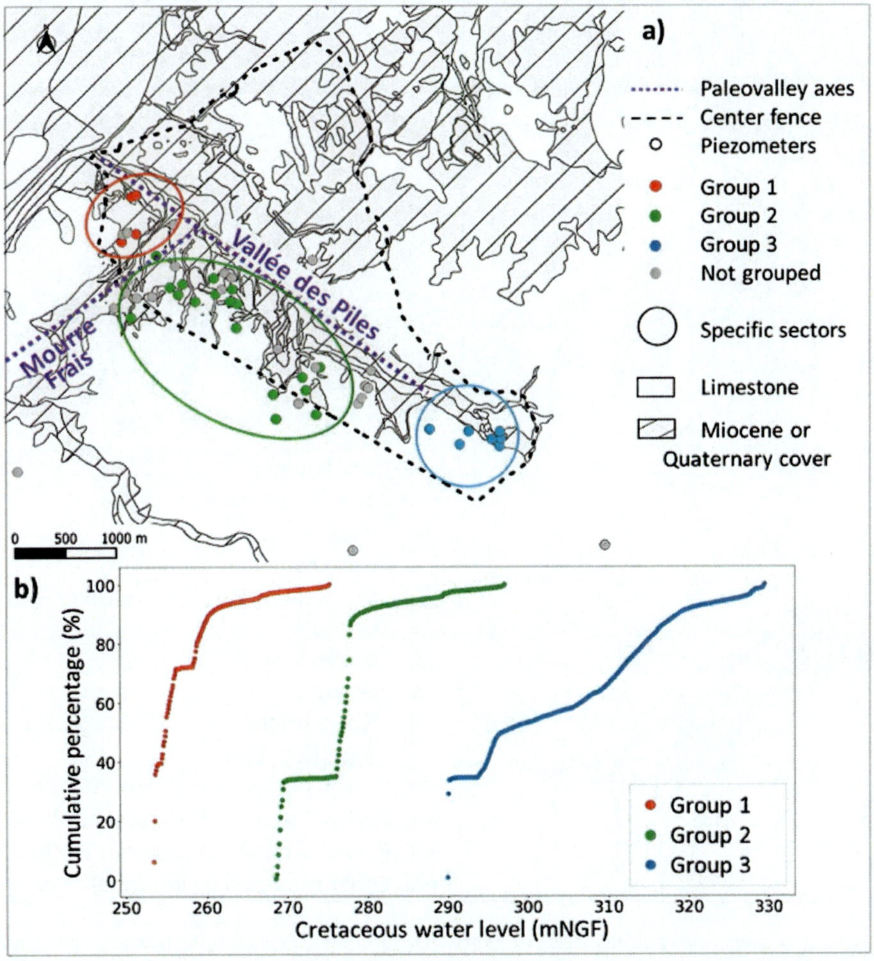


**Fig. 2** Location of the piezometers associated with the different groups of sorted piezometric curves (**a**); typical sorted groundwater levels curves (**b**)

lineaments (identified or supposed faults, visible on the outcrop or hidden from mapping, field studies and geological/hydrogeological models), listed in the bibliography (GEOTER 2017). It appeared that similar lineaments were superimposed on groups of the same category. Within each group, the piezometers can be connected to each other via the nearest lineament. Finally, the lineaments that can potentially play the role of flow axis are identified (Fig. 3).

According to Fig. 3, it is possible to link the majority of the piezometers in each group via the fractures. In the northeast zone, corresponding to the group 1, the N20 subvertical faults seem to play a major role in the groundwater flows. For the central area, associated to the group 2, the N120–N150 and N40–N60 faults, more or less karstified (Guerin 2001), seem to be the preferential flow axes. The flows in the southern sector (group 3) can also be related to the observed lineaments (Fig. 3).

It can be noted, as a first approximation, that the identified preferential flow axes are similar for the three rain episodes (NW–SE and NE–SW faults bordering the paleovalley). However, by comparing the groups formed and the supposed axes, there is a difference in the central part of the site. During the low water period (2017), an additional group, in dark blue, lacking for the high water period (2019) was identified. This difference can be explained by the presence of a connectivity threshold in this area. In low water, the water table is not high enough to reach the altitude where the fractures are connected. This highlights the role of shallow fractures and epikarst on flows in this sector.

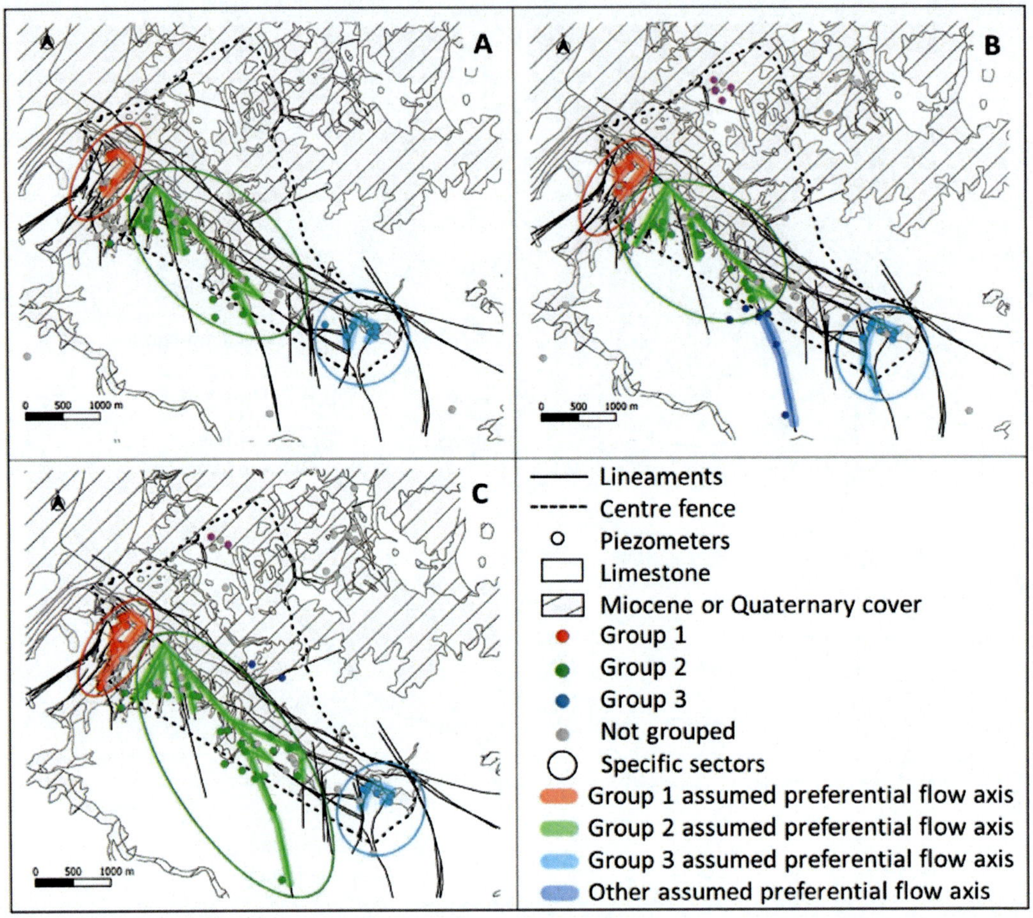


**Fig. 3** Superposition of the piezometers produced from analyses of the sorted piezometric data and the centre fracturing. **a** November 2011; **b** March 2017; **c** November 2019

## 5 Conclusion

The purpose of this work was to enrich the knowledge of the hydrodynamic functioning of the aquifer of the *Cadarache* site, which is sensitive to the rapid rise of water tables. A semi-automated clustering method for sorted groundwater levels was developed and applied to a large amount of data while incorporating visual control of the observer. The following analyse defined three clusters, which could be linked to three distinct zones of hydrogeological behaviour at the site's scale: the north-west, central and southeast areas. It was shown that these areas are morphologically conditioned by structural elements. It is interesting to underline that in the case of this aquifer, which does not have a highly developed karst conduit network, the blind statistical and geological approaches made it possible to identify potential flow paths. The subvertical faults and fractures (NW–SE and NNE–SSW) at the centre scale seems to correspond to the preferential flow axes. Long-term pumping tests coupled with water level measurements on representative piezometers should be carried out in order to confirm these interpretations. Unfortunately, artificial tracers are difficult to implement on the site (aquifer structure, small gradients). In this preliminary study, only the faults were taken in account. It will be complemented by considering other variables (temperature or electrical conductivity) or other parameters such as the nature of the different Cretaceous facies or information on the recharge zones. A study will also be carried out on some boreholes cores to identify fractures and enrich the interpretations presented in this paper. Finally, this study highlights the richness of a multidisciplinary approach.

## References

Ascott MJ, Marchant BP, Macdonald D et al (2017) Improved understanding of spatio-temporal controls on regional scale groundwater flooding using hydrograph analysis and impulse response functions. Hydrol Process 31:4586–4599. https://doi.org/10.1002/hyp.11380

Bailly-Comte V, Jourde H, Roesch A, Pistre S (2008) Mediterranean flash flood transfer through karstic area. Environ Geol 54:605–614. https://doi.org/10.1007/s00254-007-0855-y

Bonacci O (1995) Ground water behaviour in karst: example of the Ombla Spring (Croatia). J Hydrol 165:113–134. https://doi.org/10.1016/0022-1694(95)92769-A

Charlier J-B, Moussa R, Bailly-Comte V et al (2015) How karst areas amplify or attenuate river flood peaks? A response using a diffusive wave model with lateral flows. In: Andreo B, Carrasco F, Durán JJ et al (eds) Hydrogeological and environmental investigations in karst systems. Springer, pp 293–301

Douglas DH, Peucker TK (1973) Algorithms for the reduction of the number of points required to represent a digitized line or its caricature. Cartogr Int J Geogr Inf Geovisual 10:112–122. https://doi.org/10.3138/FM57-6770-U75U-7727

Esposito C (2015) Crue de juin 2010 dans le var: comparaison de la zone inondée et des zones inondables connues. Houille Blanche 34–40. https://doi.org/10.1051/lhb/2015005

Gaume E, Livet M, Desbordes M, Villeneuve J-P (2004) Hydrological analysis of the river Aude, France, flash flood on 12 and 13 November 1999. J Hydrol 286:135–154. https://doi.org/10.1016/j.jhydrol.2003.09.015

GEOTER (2017) Bilan et hiérarchisation des failles cartographiées sur le site de Cadarache. CEA

Guerin R (2001) Synthèse des connaissances géologiques et hydrogéologiques. CEA Cadarache

Jourde H, Roesch A, Guinot V, Bailly-Comte V (2007) Dynamics and contribution of karst groundwater to surface flow during Mediterranean flood. Environ Geol 51:725–730. https://doi.org/10.1007/s00254-006-0386-y

Kreibich H, Thieken AH (2008) Assessment of damage caused by high groundwater inundation. Water Resour Res 44. https://doi.org/10.1029/2007WR006621

López-Chicano M, Calvache ML, Martín-Rosales W, Gisbert J (2002) Conditioning factors in flooding of karstic poljes—the case of the Zafarraya polje (South Spain). CATENA 49:331–352. https://doi.org/10.1016/S0341-8162(02)00053-X

Mangin A (1975) Contribution à l'étude hydrodynamique des aquifères karstiques. PhD thesis, Sciences de la Terre, Université de Dijon, France, 268 pp

Najib K (2007) Phénomènes de remontées de nappe extrêmes en terrain carbonaté fracturé et karstifié: Evaluation de l'aléa et Prévention du risque dans le cadre de la protection de bâti-ments. PhD thesis, Eaux Continentales et Société, Université Montpellier II, France, 376 pp

Pinault, J-L, Amraoui N, Golaz C (2005) Groundwater-induced flooding in macropore-dominated hydrologicalsystem in the context of climate changes. Water Resour Res 41. https://doi.org/10.1029/2004WR003169

Price M, Low RG, McCann C (2000) Mechanisms of water storage and flow in the unsaturated zone of the Chalk aquifer. J Hydrol 233:54–71

Špitalar M, Gourley JJ, Lutoff C et al (2014) Analysis of flash flood parameters and human impacts in the US from 2006 to 2012. J Hydrol 519:863–870. https://doi.org/10.1016/j.jhydrol.2014.07.004

Weng P, Dörfliger N (2002) Projet PACTES. Module: contribution des eaux souterraines aux crues et inondations; site de l'Hérault. BRGM

# Hydrogeological Processes of Karst-Influenced Multi-layered Aquifers of Northern Aquitaine Basin, France

M. Quispe Sihuas, J. B. Charlier, R. Lastennet, O. Cabaret, B. Dewandel, and A. Denis

**Abstract**

This study aims to characterize the regional hydrodynamic and hydrochemical variability of multi-layer carbonate aquifers of the basin edge by considering its karst feature. The study area is located at the northern edge of the Aquitaine Basin, in Southwest France, which is characterized by two main Upper Cretaceous and Jurassic reservoirs. A data set of water levels and physico-chemical analyses in 65 springs/shallow wells and 94 boreholes was collected from French public database. A statistical approach was conducted in order to differentiate aquifer compartments in relation with its hydrogeological context. First results show a contrasted physico-chemical signature between Jurassic and Upper Cretaceous aquifers, which is explained by the lithological properties, the aquifer depth and the residence time. Nevertheless, some locations present temperature and hydrochemical anomalies. Likely hypothesis is the karst influence due to localized recharge from fast river losses, as well as the karstification degree in unconfined and confined compartments.

**Keywords**

Hydrochemistry • Sedimentary basin • Carbonate-rock aquifers

M. Quispe Sihuas (✉) · O. Cabaret
BRGM, Parc Technologique Europarc, 24 Avenue Léonard de Vinci, 33600 Pessac, France
e-mail: m.quispe@brgm.fr

M. Quispe Sihuas · R. Lastennet · A. Denis
UMR 5295 I2M–GCE, Université de Bordeaux, Bordeaux, France

J. B. Charlier · B. Dewandel
BRGM, Université de Montpellier, Montpellier, France

G-eau, INRAE, CIRAD, IRD, AgroParisTech, Institut Agro, BRGM, Montpellier, France

## 1 Introduction

The edge zone of sedimentary basins is the site of interactions involving the recharge of multi-layer aquifers, affecting the quality of the water resource. At the regional scale, the karst feature of the reservoirs is rarely taken into account in hydrogeological processes. However, karst aquifers are characterized by complex recharge, exchanges and mixing between reservoirs, and interaction with surface water. Previous studies on northern Aquitaine Basin (AB) were conducted to understand the geological and hydrogeological functioning of different aquifers (Cabaret et al. 2017; Husson et al. 2016; Lorette 2019; Platel et al. 2010; Von Stempel 1972). Nevertheless, a major question remains about the karstic feature of the carbonate aquifers. This study aims to characterize the regional hydrodynamic and hydrochemical variability of multi-layer carbonate aquifers located in the basin edge by considering their karst features. The study site is the northern AB in Southwest France, characterized by two main Cretaceous and Jurassic reservoirs. First, we present a spatial analysis of the hydrodynamic variability of the two main reservoirs. Then, we investigate the geothermal gradient over the study zone. Finally, we explore the spatial hydrochemical variability of the compartments in link with geological and hydrogeological characteristics. This study focuses on the relationships between the secondary aquifers, Upper Cretaceous and Jurassic, and the role of the edge zone in the water quality. Thus, our objective with this method is to identify representative hydrogeological parameters of the main aquifers, and to identify the processes that best explain their variability.

## 2 Study Area

The Aquitaine Basin (AB) is a regional multi-layer aquifer system located in Southwest France (Fig. 1). In the northern AB, sedimentary terrains from Triassic to Upper Cretaceous


B. Andreo et al. (eds.), *EuroKarst 2022, Málaga*, Advances in Karst Science,
https://doi.org/10.1007/978-3-031-16879-6_11

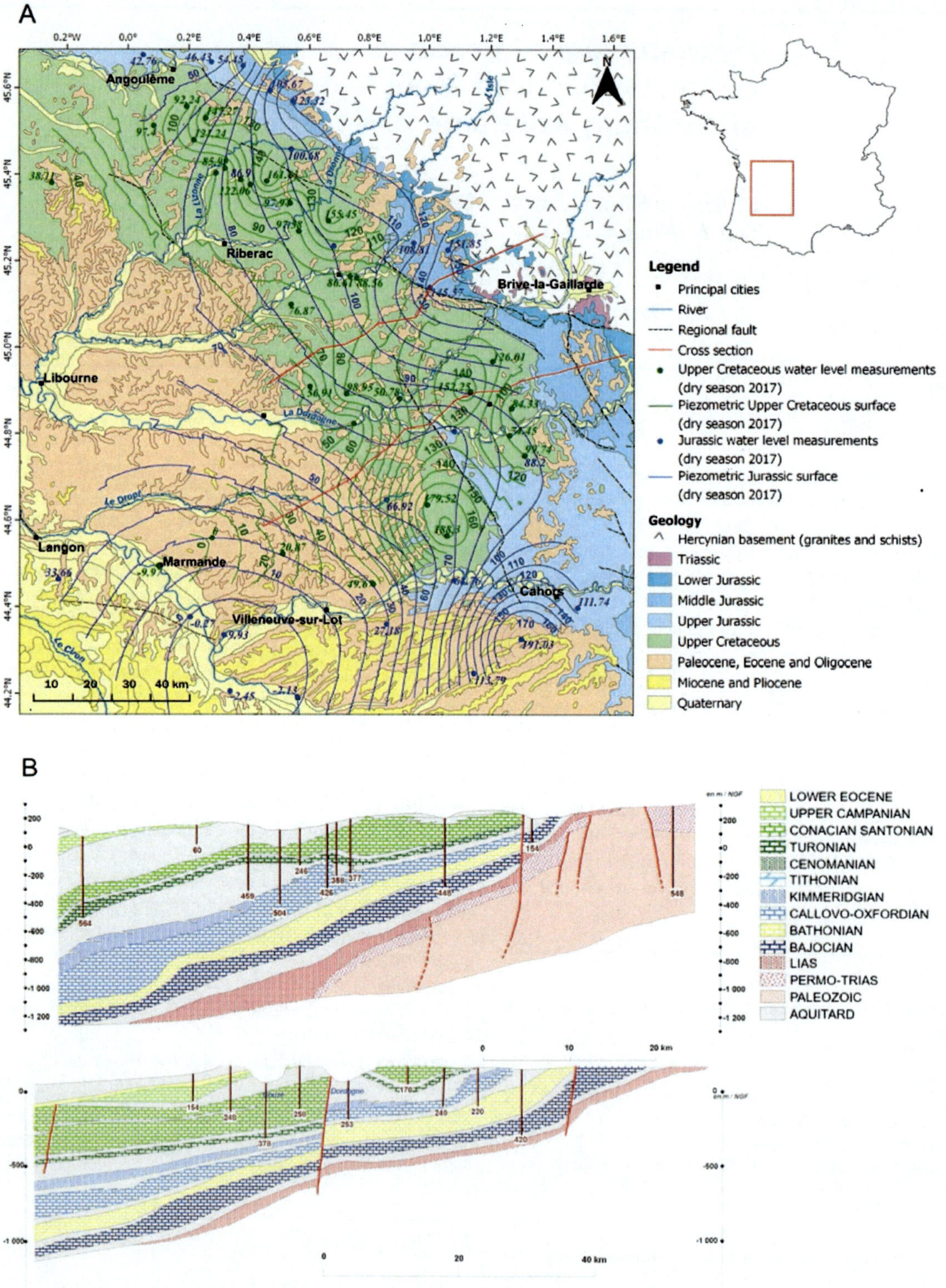


**Fig. 1** **a** Study area in the edge North Aquitaine Basin (southwest France) showing the piezometric map for both Upper Cretaceous and Jurassic aquifers (dry season 2017) and **b** hydrogeological cross sections—red lines in Fig. 1A—(adapted from Platel et al. 2010)

overly a Hercynian granite and schist basement. Because of the variety of sedimentary deposits, the alternation of aquifers and aquitards creates a complex multi-layer aquifer system composed of six majors reservoirs: Jurassic, Upper Cretaceous, Eocene, Oligocene, Miocene, and Plio-Quaternary, which can themselves be subdivided (Platel et al. 2010).

The Jurassic and Upper Cretaceous (at the east of the northern AB) outcrops show evidence of two vast marine transgressions during Mesozoic. These predominantly carbonate formations are karstified, both outcrop and undercover (Platel 1996). The Upper Cretaceous multi-layer system is mainly composed of fractured limestone and contains four aquifers (from the deeper to the shallower): Cenomanian, Turonian, Coniacian-Santonian, and Campanian. The Cenomanian aquifer (50 m thick) is located only in the north of the northern AB. The Turonian (30–60 m) is a fractured and karstified aquifer, which contains predominantly limestone. The Coniacian-Santonian aquifer geometry follows the one of Turonian and is the thickest cretaceous aquifer (30–275 m). It results a complex system with lateral and vertical variations of facies, which develops a secondary porosity in granular limestone (fissures and karstic conduits) and in sandstone layers with interstitial porosity, separated by chalky and marls formations. In the Jurassic unit, it is possible to identify five reservoirs, in some places separated by marly or marl-limestone layers. The Lias aquifer (35–100 m thick) made up of sandstone and dolomitic limestone. The limestone and dolomite of the Bajocian aquifer (30–70 m thick). The limestone of the Bathonian Callovian Oxfordian aquifer (100–650 m thick), which represent the principal Jurassic aquifer in the region. The limestone of the Kimmeridgian aquifer (50–210 m thick). These three aquifers can be considered as a monotonous and homogeneous series formed of limestone with partial secondary dolomitization, highly karstified. However, some marly levels might be present and can delimit local aquitards. Finally, the dolomitic limestone of the Tithonian aquifer, which is located above the marly levels, representing a local aquifer.

## 3 Materials and Methods

Groundwater and hydrological data set were compiled from several public databases. It comprises (1) hydrochemical data (major elements) and water levels from "ADES" (n.d.) and (2) geological and technical information from springs and wells (*La Banque Du Sous-Sol (BSS) | InfoTerre*, n.d).

Regarding water levels, 54 boreholes were selected: (i) 29 in Upper Cretaceous aquifers (Turonian and/or Coniacian-Santonian and/or Cenomanian), (ii) and 25 in Jurassic aquifers (Bajocian and/or Bathonian Callovian Oxfordian and/or Kimmeridgian). Therefore, piezometric maps were established using standard kriging interpolation method for both aquifers during the dry and wet seasons of 2017.

Regarding hydrochemical data, 73 springs and shallow wells (maximum 7 m depth), and 86 boreholes (up to 1.2 km in depth, 330 m in average) were selected. To ensure that each borehole captures only one reservoir, and the boreholes screened both in the Jurassic and Upper Cretaceous aquifers were not selected. Additionally, a minimum of 5 samples per site over the 2000–2019 period was required to integrate the analysis. An ionic balance was carried out to validate the samples, of which a value greater than 5% was not accepted. Physico-chemical parameters (temperature, pH, and electrical conductivity at 25 °C), chemical analyses of major elements, and nitrogen parameters were selected. In this study, the median value of the hydrochemical data set is considered more representative than the mean, since it does not take into account the extreme or aberrant data that can bias the analysis.

To identify patterns in our large data set, multivariate analyses were conducted using a principal component analysis (PCA) on primary variable descriptors of all hydrochemical and physico-chemical data. Linear regression was also used to assess the correlation between variables. The geothermal gradient was calculated from 81 sites (boreholes) with a screen at least at 25 m deep and with flow logs and/or water strikes information. The screened interval must be less than 100 m. For the determination of the gradient, a simple linear regression was used. Additionally, PCA method was used to identify the main hydrochemical components.

## 4 Results and Discussion

### 4.1 Hydrogeological Characteristics

The general pattern of the piezometric map (Fig. 1) in Upper Cretaceous and Jurassic is similar in both the dry and wet seasons. In the Upper Cretaceous aquifers, the seasonal variations are up to 12 m in the unconfined area (outcrops) to a few centimetres in the confined area. The groundwater flow direction is globally from northeast to southwest. These maps highlight recharge zones on the Turonian and Coniacian-Santonian outcrops but do not show evidence of groundwater recharge by river losses. Nevertheless, rivers such as Dordogne, Lizonne, Dronne, and Isle appear to drain the cretaceous aquifer.

In the Jurassic aquifers, the average seasonal water level fluctuation is about 3.1 m and can locally reaches 15 m. Two main flow directions are observed, the first is east–west in the area of Périgueux, and the second is NE–SW between Sarlat-la-Canéda and Cahors in the Causses du Quercy area.

This dividing line seems accentuated by the piezometric depression in the southeast of Sarlat-la-Canéda, which is probably due to the large amount of water abstracted by pumping in this area. Previous Jurassic piezometric surface (Bichot et al. 1997) from 95 to 96 years shows the same piezometric depression but 10 m less deeper and with a main east–west flow direction, which shows the vulnerability of the Jurassic confined aquifers in Villeneuve-sur-Lot.

From the piezometric point of view, it has been shown that stable levels and cyclic seasonal variations, mainly due to aquifer recharge (Platel et al. 2010), characterize the Upper Cretaceous and Jurassic aquifers. Nevertheless, a diminution of the Jurassic groundwater table from years 95 to 96 has been shown, especially at the confined areas.

## 4.2 Geothermal Gradient

Groundwater temperatures from the 81 selected boreholes vary from 13.5 to 40 °C, of which a global regional geothermal gradient of 2.060 °C/100 m ($r^2$ = 0.59) was computed (Fig. 2). This model significantly differs from the theoretical geothermal gradient (3 °C/100 m). This difference is due to cold-water anomalies at important depths, which can be explained by mixing processes with colder superficial waters transported by karstic features. The calculation of the geothermal gradient accounting for the two main Upper Cretaceous and Jurassic reservoirs does not improve so much the model in both cases ($r^2$ of 0.65 and 0.61, respectively). A slightly more efficient model is observed for the Upper Cretaceous aquifers. This could be link to a weaker degree of karstification and an important geological compartmentation of these aquifers, contrarily to the Jurassic aquifers that are strongly karstified.

## 4.3 Hydrochemical Variability

The results of PCA on hydrochemical data are shown in Fig. 3. The two first axes explain 76% of the total variance, where two major poles are differentiated.

Regarding variables, $Mg^{2+}$ is well correlated with water temperature (Tw) and is negatively correlated with $Ca^{2+}$, $HCO_3^-$, and $NO_3^-$. These two groups of variables outline the first axis on Factor 1 of water transit time, increasing from $NO_3^-$ (surface pollution) towards $Mg^{2+}$, which are indicators of fast infiltration and high mineralization (and of dolomite dissolution), respectively. A third group shows that $Na^+$, $Cl^-$, $K^+$, and $SO_4^{2-}$ are well correlated and have a nearly independent pattern compared to tracers determining Factor 1. $Na^+$ and $SO_4^{2-}$ enrichment in Jurassic waters can be explained by vertical leakage from the Lias aquifer because of its evaporitic component (presence of anhydrite). However, an evaporitic and dolomitic signature can be also seen on the confined Upper Cretaceous waters, even if dolomite and anhydrite are not present in its lithology. Following the natural course of $Na^+$ and $K^+$ is difficult because of possible addition from anthropogenic activities.

Regarding observations, a first group is formed mainly by boreholes deeper than 200 m with a few exceptions. This group represents boreholes with elevated values of temperature, and ions such as $Mg^{2+}$ (7–26 mg/L), $Na^+$ (5–72 mg/L), $SO_4^{2-}$ (5–64 mg/L), and $Cl^-$ (9–24 mg/L), that might indicate a higher mineralization (LeGal LaSalle et al. 1996). The mineralization is observed in the deep confined area of Upper Cretaceous and Jurassic aquifers, towards Villeneuve-sur-Lot, which may suggest an enrichment because of the residence time and a vertical drainage up to the upper Cretaceous layers. On the other hand, a second group is formed by springs, shallow wells and boreholes of

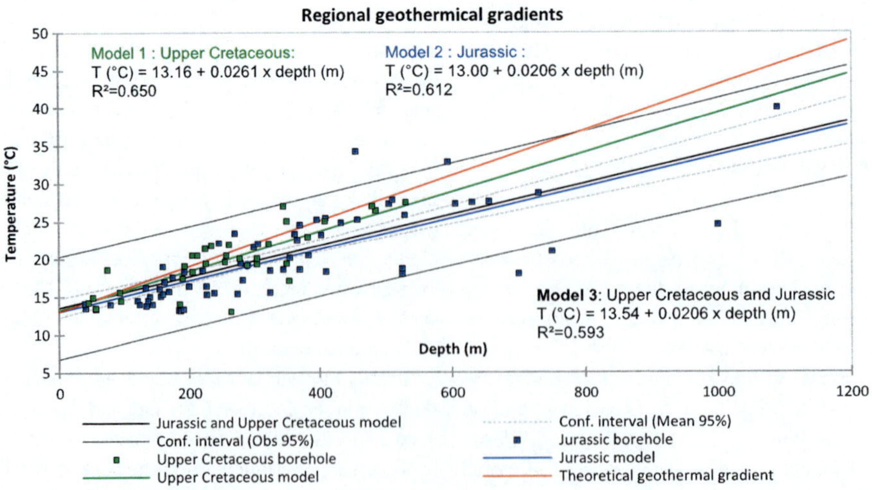


**Fig. 2** Regional geothermal gradient models. Only the confidence intervals for the model 3 are presented

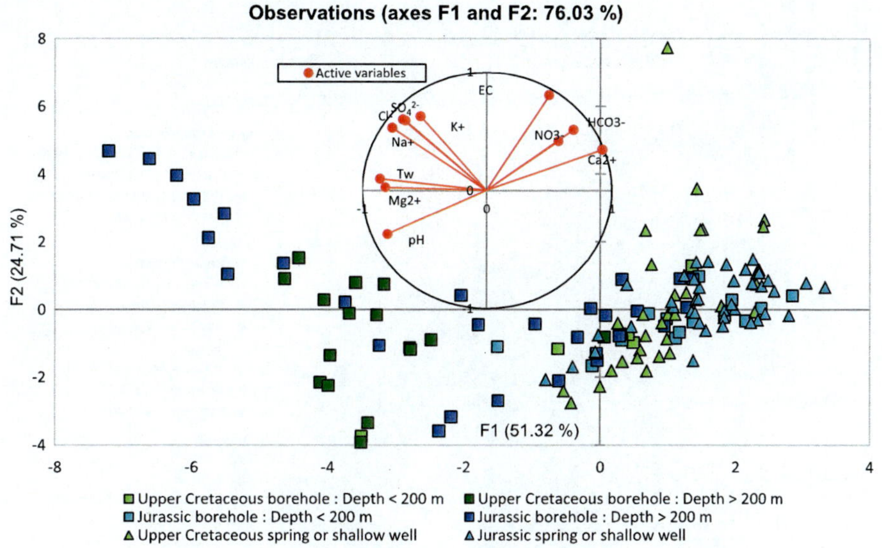


**Fig. 3** Plot based on principal components analysis of hydrochemical sampling from springs and boreholes

shallow depth (<200 m). This group is dominated by high values of calcium (60–140 mg/L) and bicarbonates (220–420 mg/L), which characterize the water–rock interaction in an open carbonate system (Edmunds et al. 1987; Hanshaw and Back 1979). Moreover, higher concentration of nitrates is found in this group (mean value of 12 mg/L and a maximum of 44 mg/L), due to its sensitivity to agricultural practices. However, some deep boreholes are within this group indicating a mixing zone, in which the karstic features may play a major role.

The spatial variability of the main major ions ($Ca^{2+}$ and $Mg^{2+}$) is presented in Fig. 4. High concentrations of $Ca^{2+}$ (around 120 mg/L) are seen in both Upper Cretaceous and Jurassic outcrops, explained by the immediate dissolution of the calcite in an open carbonate system. Rainwater infiltrating the soil overlying the carbonate outcrops quickly becomes saturated with respect to calcite. However, there is a $Ca^{2+}$ decrease from east to west and southwest, which is consistent with the progressive confinement of Upper Cretaceous and Jurassic formations. A similar phenomenon is shown at the chalk aquifer (Edmunds et al. 1987), where it is likely the Mg and Ca exchange takes place during progressive recrystallization of microcrystalline calcite or by incongruent dissolution of the carbonate.

On the contrary, there is an $Mg^{2+}$ increase from east to west and southwest, in both Upper cretaceous and Jurassic aquifers. This pattern is likely explained by the dolomite facies that are only present in Jurassic layers. Additionally, little impurities of magnesium and high residence time in the Cretaceous reservoirs can be the source of the increased concentrations.

## 5 Conclusions and Perspectives

The aim of this study was to characterize the hydrogeological processes of a karst-influenced multi-layered aquifers of basin edge. From a regional analysis in the Aquitaine Basin (AB) of hydrogeological and hydrochemical data, we showed the role of the edge zone into the recharge of multi-layered aquifers. The Upper Cretaceous and Jurassic aquifers are widely heterogeneous. The hydrogeochemistry varies according the lithology facies, the residence time, the depth, and very likely because of the karstic feature. Our results show evidences of mixing zones between the two main Cretaceous and Jurassic reservoirs. This is probably due to the karstic and fractured natures of the carbonate formations that allow a fast infiltration through vertical drainage of water from the surface towards the deep parts of the aquifers. Thus, the groundwater that constitutes the global resource of the AB show a certain vulnerability due to the karstification of the main geological formations. Physical and chemical characteristics of the groundwater data show important variations according to increasing depth, to the progressive confinement from NE–SW and to the geological variations of facies.

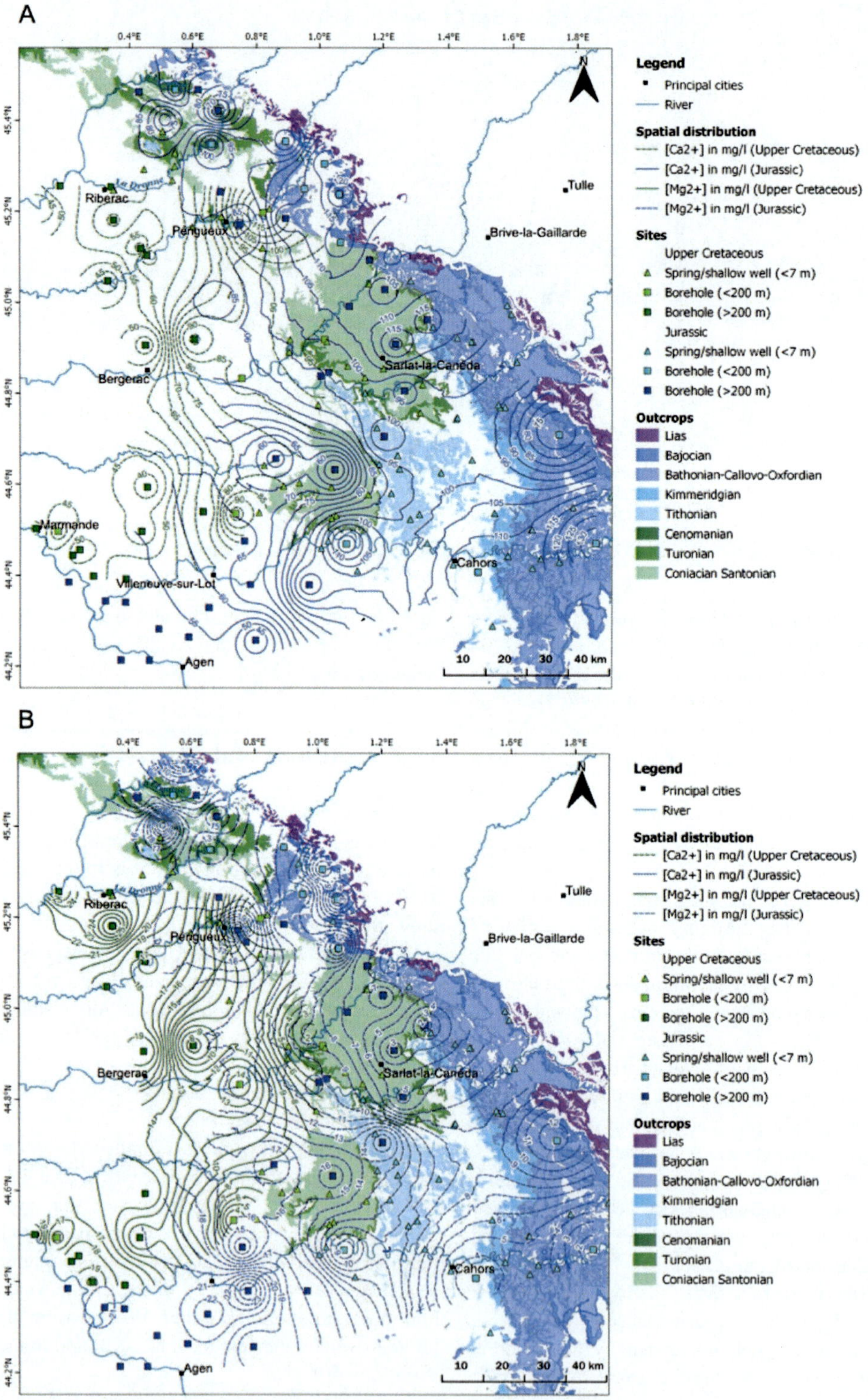


**Fig. 4** Spatial variability of Ca (**a**) and Mg (**b**) concentrations in both Upper Cretaceous and Jurassic aquifers

## References

ADES (n.d.) https://ades.eaufrance.fr/. Accessed 20 Jan 2022

Bichot F, Platel J-P, Benhammouda S, Cournil T, Dufour P (1997) Gestion des eaux souterraines en Aquitaine, année 1. Evaluation des ressources. Extension du modèle Nord-Aquitain: Synthèse des données et concepts (BRGM/RR-39331-FR)

Cabaret O, Perrin J, Wuilleumier A, Nougaret G (2017) Gestion des eaux souterraines en Région Aquitaine—Connaissances des karsts aquitains—Étude des karsts libres et sous couverture du département de la Dordogne—Convention Régionale Eaux Souterraines 2015–2020—Module 3.1—Année 1—Rapport final (BRGM/RP-66488-FR)

Edmunds WM, Cook JM, Darling WG, Kinniburgh DG, Miles DL, Bath AH, Morgan-Jones M, Andrews JN (1987) Baseline geochemical conditions in the chalk aquifer, Berkshire, U.K.: a basis for groundwater quality management. Appl Geochem 2(3):251–274. https://doi.org/10.1016/0883-2927(87)90042-4

Hanshaw BB, Back W (1979) Major geochemical processes in the evolution of carbonate—aquifer systems. J Hydrol 43(1):287–312. https://doi.org/10.1016/0022-1694(79)90177-X

Husson E, Camus H, Lerouge C, Lasseur E, Cabaret O, Saltel M, Pédron N, Coueffé R (2016) Origine, caractérisation et distribution prédictive du Karst sur la bordure Nord-Aquitaine—OKaNA. Rapport final (BRGM/RP-66812-FR)

La Banque du sous-sol (BSS) | InfoTerre (n.d.) https://infoterre.brgm.fr/page/banque-sol-bss. Accessed 20 Jan 2022

LeGal LaSalle C, Marlin C, Savoye S, Fontes JC (1996) Geochemistry and $^{14}C$ dating of groundwaters from Jurassic aquifers of North Aquitaine Basin (France). Appl Geochem 11(3):433–445

Lorette G (2019) Fonctionnement et vulnérabilité d'un système karstique multicouche à partir d'une approche multi-traceurs et d'un suivi haute-résolution: application aux Sources du Toulon à Périgueux (Dordogne, France). These de doctorat, Bordeaux. http://www.theses.fr/2019BORD0116

Platel JP (1996) Stratigraphie, sédimentologie et évolution géodynamique de la plate-forme carbonatée du Crétacé supérieur du nord du bassin d'Aquitaine. Géol Fr 4:33–58

Platel JP, Pédron N, Gomez E, Saltel M (2010) Perspectives de gestion des nappes du Secondaire en Agenais-Périgord. Synthèse géologique et hydrogéologique, modélisation hydrodynamique. BRGM/RP-59330-FR, 217 p., 101 fig., 12 tab., 19 ann. BGRM

Von Stempel C (1972) Etude des ressources en eau de la région de Périgueux (Dordogne). Université de Bordeaux 1, 245 pp

# Implications of Tryptophan-Like-Fluorescence Long-Term Monitoring for Bacterial Detection in a Mountainous Karst Aquifer

J. Fernández-Ortega, J. A. Barberá, and B. Andreo

**Abstract**

Fluorescence spectroscopy approaches like tryptophan-like-fluorescence (TLF) have been presented as a powerful tool to easily detect bacteriological contamination in groundwater used for drinking water supply in rural areas, since bacteria are able to synthetize *L*-Tryptophan. As commonly happens in karst springs, during flooding conditions high turbidity levels (and associated bacteria transport) are observed and thus, the exploitation of groundwater for human consumption is hindered. Hence, TLF has been considered as a potential early warning parameter to prevent polluted groundwater to reach drinking water capture points at Sierra de Ubrique carbonate karst aquifer (S Spain). The obtained results in the studied spring showed low correlation with faecal indicators ($p$ = 0.26–0.75) and suggests the complexity in obtaining reliable continuous measurements mainly due to the multiple potential origins of tryptophan (cheese whey, bacterial activity in soil or karst conduits...) and the biochemical reactions in the soil, epikarst and karst conduits. The suitability test realized in Ubrique test site relegates TLF to a function as a biological risk indicator rather than an early warning parameter.

**Keywords**

Tryptophan • Fluorescence • Karst • Southern Spain

J. Fernández-Ortega (✉) · J. A. Barberá · B. Andreo
Department of Geology and Centre of Hydrogeology, University of Malaga (CEHIUMA), E-29071 Malaga, Spain
e-mail: jaimeortega@uma.es

## 1 Introduction

A wide variety of pathogen microorganisms are transmitted through surface and groundwater systems that supply drinking water to certain populations. The most of them are predominantly of faecal origin, which pose a threat to human health specially in developing countries where water treatment is insufficient (Ashbolt, 2004). Around 9.2% of worldwide population consume karst water (Stevanović 2019) and, in Europe, up to 25.3% of the population lives on karst areas (Goldscheider 2020), which present specific recharge processes that make them highly vulnerable to contamination because low attenuation of contaminants occurs between the recharge area and the springs (Goldscheider and Drew, 2007). During a precipitation event, strong and quick discharge and water quality variations can be observed at the discharge points (Pronk et al. 2007), and thus, polluted water may reach drinking water distribution system or hinder the total use of groundwater resources.

Hence, drinking water quality should be frequently controlled to assure and protect the health of the final users. However, the detection of enteric pathogens in groundwater is—usually—indirectly made through the analysis of indicator organisms such as thermotolerant (faecal) coliforms (TTCs) or E. coli. Moreover, the analysis of these indicators requires 24 h culture in laboratories and specific reagents (Mikell et al. 1996), which may delay the detection of a potential contamination episode. Besides these methods, fluorescence spectroscopy approaches like tryptophan-like-fluorescence (TLF) (Fig. 1) have been presented as a powerful tool to easily detect bacteriological contamination in groundwater used for drinking water supply in rural areas, given that different bacteria families (such as *Escherichia Coli*) are able to synthetize L-Tryptophan, which is an essential amino acid for biological processes.

Thus, given that the accuracy and reliability of this method has been little tested in karst aquifers, this work aims to test the suitability of TLF for contamination detection



B. Andreo et al. (eds.), *EuroKarst 2022, Málaga*, Advances in Karst Science,
https://doi.org/10.1007/978-3-031-16879-6_12

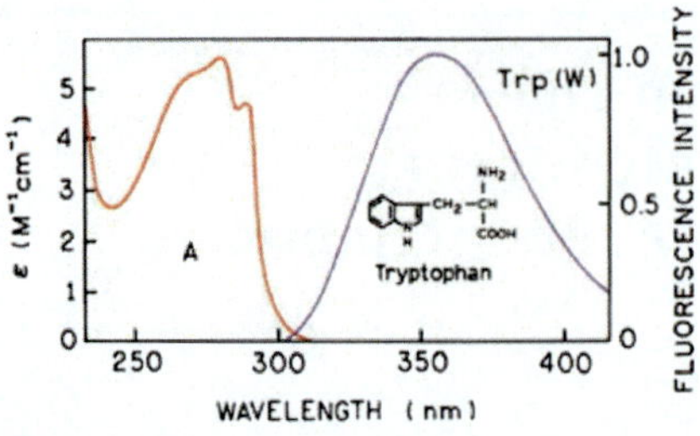


**Fig. 1** Absorption (*A*)—emission spectrum of *L*-Tryptophan (modified from Lakowicz, 2006)

after 7 months (October to May) of continuous record of as a potential early warning parameter at Sierra de Ubrique system (S Spain), where episodic high turbidity levels (and associated bacteria detection) partly hinder the exploitation of groundwater during flooding conditions in this mountainous karst aquifer that supplies drinking water to 17.000 inhabitants of Ubrique town.

## 2 Site Description

The study area composes a 26 km$^2$ mountainous karst aquifer within the boundaries of Sierra de Grazalema Natural Park (Cádiz province, S Spain), approximately 80 km north-east from Cádiz city. This zone is characterized by a humid Mediterranean climate and rainy period mainly occurs in autumn and winter (mean annual rainfall is 1350 mm) (Andreo et al. 2014). Abrupt elevations (400–1500 m a.s.l.) with steep slopes and NE-SW direction define the orography of Sierra de Ubrique (Marín et al. 2021).

The study area is—geologically—situated within the Betic Cordillera and characterized by Triassic evaporites with clays, Jurassic dolostones (lower) and limestones (upper) -500 m thick-, and Cretaceous-Paleogene marly-limestones (Martín-Algarra 1987) (Fig. 2). The geological structure is determined by anticline folds, in which core Jurassic formations can be found, and synclines matching with depressions constituted by more recent Cretaceous

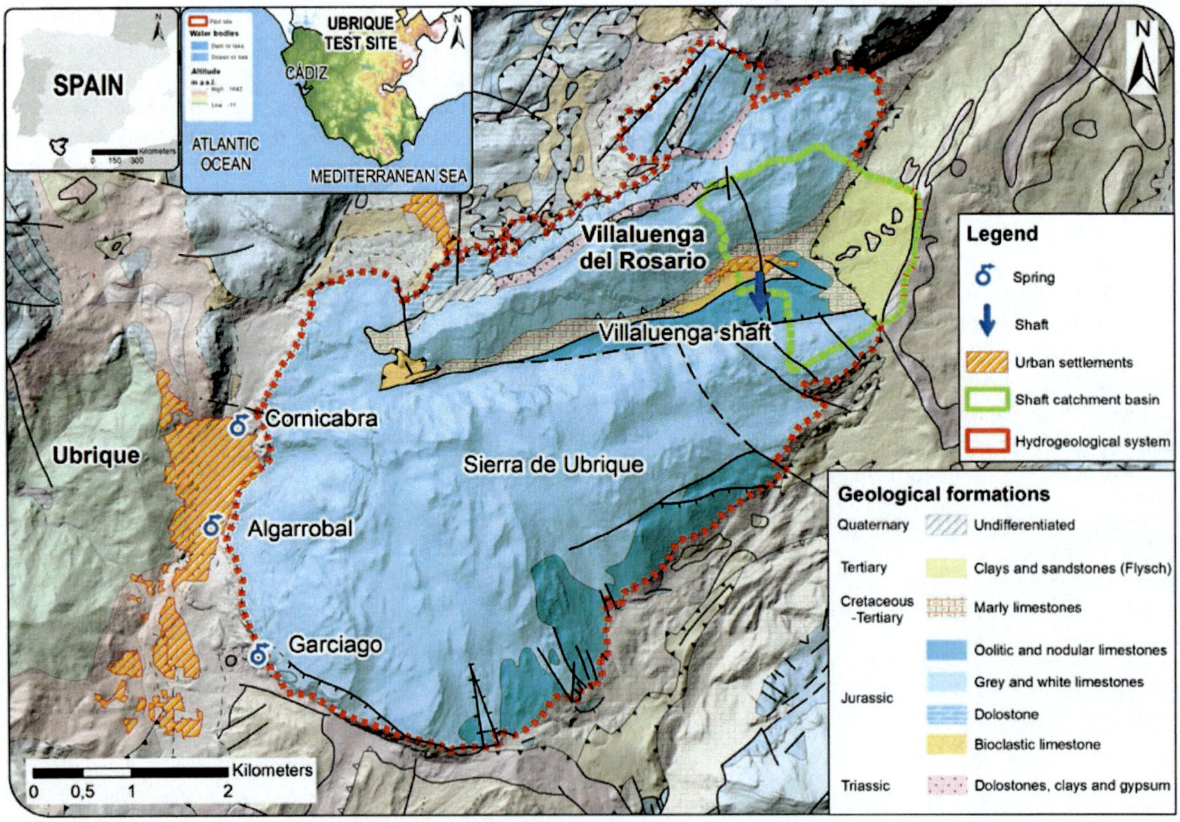


**Fig. 2** Hydrogeological setting of the study area

materials. Furthermore, some tertiary flysch clays and sandstones formations overthrust over the previous rocks.

Sierra de Ubrique test site constitutes a fractured and karstified aquifer which geometry is mainly determined by the anticline folds that define, respectively, the Sierra de Ubrique and the Sierra del Caillo (Martín-Algarra, 1987). Recharge in this binary karst system is produced by diffuse infiltration from rainfall through carbonate outcrops (autogenic) and concentrate flow from a neighbour catchment which enters the system through the Villaluenga swallow hole (allogenic). This 3 km$^2$ catchment developed in Cretaceous-Tertiary materials feeds the Albarrán intermittent stream, which can rapidly variate from 0 to 6.000 l/s in a few minutes during a storm event. Natural discharge occurs through the SW border of the aquifer, the most significant springs are the Cornicabra (349 m a.s.l.) and Algarrobal (317 m a.s.l.) perennial outlets, which present discharge rates of 10–2460 l/s (average 406 l/s) and 10 to 2,625 L/s (average 157 l/s), respectively. Additionally, Garciago overflow spring (422 m a.s.l.) whose discharge rate is, on average, 311 l/s but ranges from 0 to 6059 l/s, appears only during large flooding events (Martín-Rodriguez et al. 2016; Sánchez et al. 2016).

Moreover, different karst groundwater pressures related to rural activities developed in small populations were identified in the recharge area. In this study case, those are goat and sheep livestock, small cheese factories and the presence of a wastewater treatment plant (WWTP) which pours the—untreated—sewage into the Albarrán stream. Hence, recharge features in this system together with the well-developed karst network and human activities certainly increase groundwater vulnerability and risk to contamination.

## 3 Methods

### 3.1 Water Sampling and Monitoring Network

Different physical–chemical parameters of groundwater samples were measured: water temperature and electrical conductivity (EC) with WTW™ Cond 3110 (± 0.1 °C and ± 1 µS/cm resolution, respectively), (Xylem, USA) and turbidity with HACH turbidity pocket metre 2100Qis (±0.1 NTU) (HACH, USA). Furthermore, samples were taken to the laboratory for Total Organic Carbon (Shimadzu mod. TOC-VCSN) and semi-quantitative analysis of total coliforms and E. coli (IDEXX ColiSure, USA). Continuous measurements of several parameters (spring discharge, temperature, EC, turbidity and TLF) were realized at the two permanent springs (used for drinking water supply). Water level was recorded through the use of Odyssey® data loggers (Dataflow Systems LTD, New Zealand) (±0.8 mm resolution). Water temperature (± 0.1 °C), electrical conductivity (± 0.1 µS/cm), turbidity (± 0.01 NTU) and TLF (± 0.01 ppb) were measured with the thermometer, conductivimeter and fluorometer integrated within the GGUN-FL30 probe (Albilia Co, Switzerland). The device which holds specific optic channels for $\lambda_{EX}/\lambda_{EM}$ around 600/550 nm for turbidity and 280/350 nm for TLF with a sampling interval of 15 min.

## 4 Results

### 4.1 Time Series Analysis

Spring response to four effective recharge events has been registered at the two main discharge points. Maximum discharge values between 1200 and 1800 l/s were observed in Cornicabra spring (Fig. 3A) and turbidity up to 63.7 NTU during the first flooding event and not higher than 15 NTU at the following ones. Regarding TLF, this spring shows a baseline concentration between 3 and 5 equivalent ppb and a maximum value of 36.5 eq. ppb which was as well registered along the first event. The successive events display smoother peaks with maximum concentrations close to 20 eq. ppb which also coincide with greatest turbidity intensity lapses. On the other hand, Algarrobal spring showed a maximum discharge around 800 l/s and maximum turbidity values up to 334.2 NTU in the first event and 200 NTU at the following ones (Fig. 3B). TLF baseline was also found between 3 and 5 eq. ppb and maximum record of TLF of 122.1 eq. ppb during the third event. However, common maximum values of TLF records are displayed around 80 eq. ppb and tend to reproduce the shape of turbidity record.

Some minor TLF peaks (red triangles, Fig. 3) are observed in both springs coinciding with little precipitation events where no turbidity or spring discharge increase is observed. Furthermore, fast increases of TLF record in Algarrobal spring are observed mostly coinciding with the first moments of a recharge event.

### 4.2 Correlation with Contamination Indicators

Flooding events provoke an increase of spring discharge and turbidity, and thus, TLF should also show any correlation to these parameters. In Algarrobal spring, it is

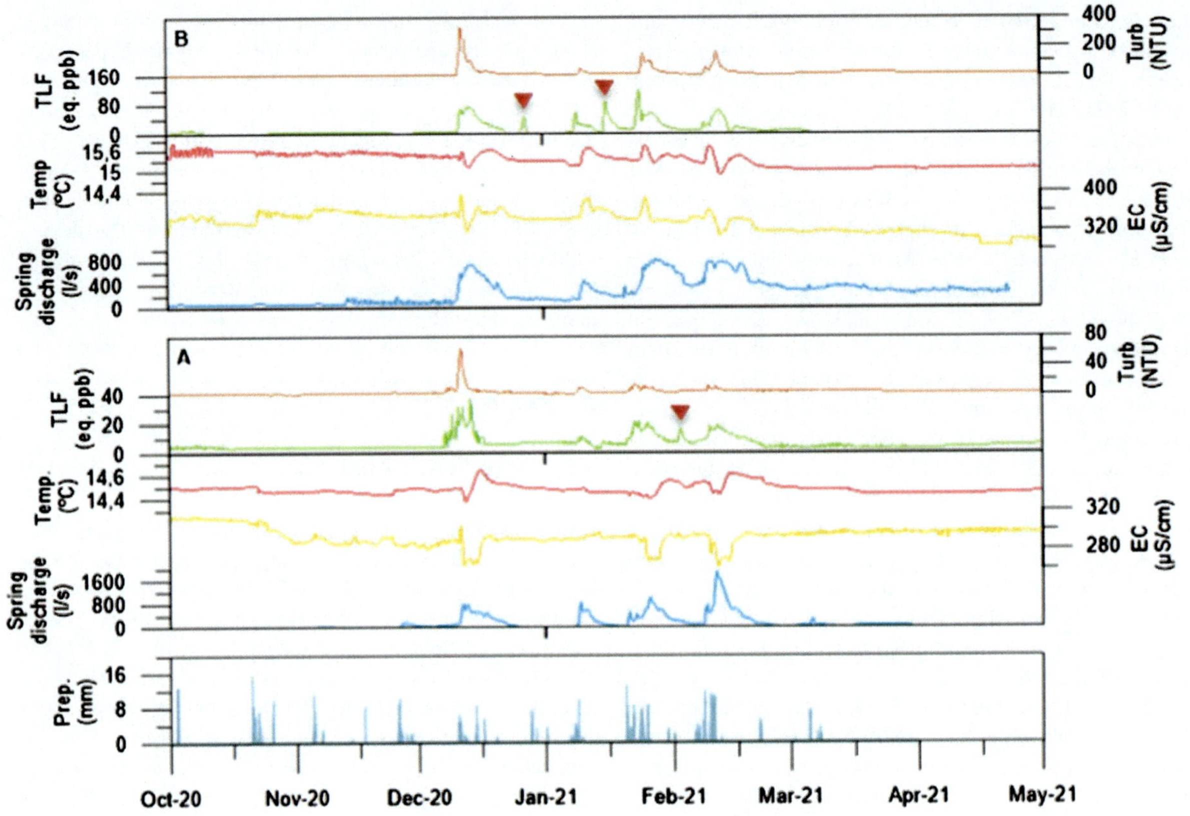


**Fig. 3** Precipitation and continuous raw data time series of groundwater parameters recorded at Cornicabra (*A*) and Algarrobal (*B*) springs from October 2020 to May 2021

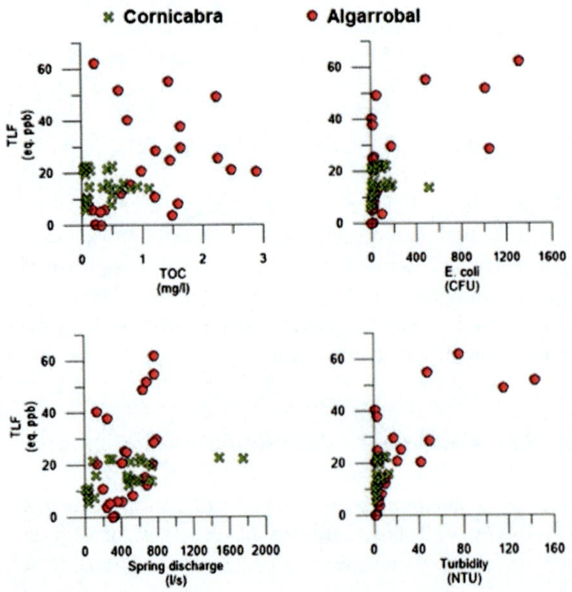


**Fig. 4** Statistic correlations between TLF and other water parameters (TOC, E. coli and Turbidity)

observed that maximum TLF values are related to maximum discharge values ($p = 0.47$), however, in Cornicabra spring TLF is reached during diverse hydrodynamic conditions (Fig. 4). This might be explained because of intense TLF response during low water conditions, nevertheless, Cornicabra spring shows a higher Pearson correlation coefficient ($p = 0.59$).

Turbidity and TLF show slightly better correlation at both springs ($p = 0.45$ at Cornicabra and $p = 0.71$ at Algarrobal) (Fig. 4), nonetheless, still high TLF are measured when turbidity is below 5 NTU (Spanish national law maximum threshold for drinking water supply). Despite that tryptophan or TLF substances are organic compounds, there is no direct correlation between it and TOC values ($p = -0.11$ for Cornicabra and $p = 0.20$ for Algarrobal) (Fig. 4) and a high dispersion is observed in both springs. On the other hand, an apparent correlation has been found between biological activity indicators (E. Coli) and the equivalent concentration of TLF at Algarrobal spring ($p = 0.75$), however, Cornicabra spring does not show a good correlation between them ($p = 0.26$) (Fig. 4).

# 5 Discussion

## 5.1 Method Technical Limitations

Minor technical issues (such as the decay of light intensity or interferences) arose during the study period. Interferences and quenching processes are well known in these techniques as multiple substances may cause the decrease of TLF intensity (Lakowicz 2006). Among them, substances used as pharmaceutical products or medicines such as acrylamide, halogen anaesthetics, hydrogen peroxide, imidazole, nicotinamide or succinimide are found. Furthermore, recent investigations (Markechova et al. 2013; Bridgeman et al. 2015; Ward et al. 2020) described the hardness to detect low levels of microbial contamination in some cases due to closeness of the TLF peak with the humic-like fluorescence (HLF) peak.

As the two springs studied in this research are now integrated within the village, some quenching substances may appear from (e.g.) losses from sewage systems and nearby roads washed by runoff. This situation may explain some of the TLF peaks with significant concentration ($\approx$ 15 eq. ppb) which do not coincide in time with effective rain episodes. The spring that most evidences these problems is Algarrobal, as extreme turbidity values might suppose a potential TLF interference during specific moments of a flooding event by enhancing or saturating the protein-like optical of the used device. Iron precipitates were also observed over the probe in this spring, what can lead to incorrect fluorescence intensity readings. Nevertheless, the effect of these substances is still poorly understood and difficult to estimate in our test site, and thus increasing the uncertainty of the data.

Previous works (Khamis et al., 2015) described the effect of turbidity and temperature variations in two different specific TLF devices and suggested to implement a correction algorithm in order to obtain reliable continuous measurements. The obtained results together with preceding studies suggest the complexity in obtaining reliable continuous measurements as this technique shows some limitations related to the measurement environment and could limit the use of TLF for reliable contamination.

## 5.2 Tryptophan Origin

Previous researches (Determann et al. 1998; Cumberland et al. 2012; Quiers et al. 2014; Sorensen et al. 2015, 2016; Frank et al. 2018 and Nowicki et al. 2019) showed that protein-like fluorescence is directly related to the activity of bacteria. However, it has also been proved that if tryptophan is available in the environment, E. coli (among other bacteria families) is able to assimilate it in order to synthetize proteins (Yanofsky 2007), this means they stop synthesizing it to consume the tryptophan present in the groundwater. In our test site, the presence of the faecal waters from WWTP and lixiviates from cheese factories provides a continuous source of tryptophan that could feed bacterial colonies during low water conditions.

Thus, despite the non-conservative behaviour of tryptophan, the results of this study suggest that TLF baseline in the springs might be caused by biological processes (Martens and Frankenberger 1993) in the soil and within the system. On the other hand, diluted protein-like substances in concentrated recharge (washing the remains from the WWTP and cattle activities areas) might contribute, to a greater extent, to the increase of TLF in the springs. The increase of TLF signal in the springs could serve as a tracer of human activities, and this is reflected in the higher TLF values in Algarrobal spring, which also has a higher contaminant load.

## 5.3 TLF as Indicator of Water Quality

Sorensen et al. (2015) proposed a precautionary threshold of 0.33 μg/L to infer faecal contamination and 0.43 μg/L on a new research in 2016. However, these thresholds may be adapted to each study site and consider its geological features and contamination sources, as the TLF baseline in our karst study site is approximately 5 μg/L.

The behaviour of TLF record is slightly different at each spring regarding maximum values and correlation with water parameters. The data acquired in Algarrobal spring shows an apparent correlation with faecal contamination indicators such as E. coli; however, it does not always show a good correlation in Cornicabra spring. Similar results were also found by Bridgeman et al. (2015) and Carstea et al. (2020) suggested that further work is required to understand TLF in groundwater and for assessing drinking water quality.

# 6 Conclusions

The first application of TLF long-term monitoring in Ubrique test site resulted successful in Algarrobal and Cornicabra springs and showed proportional values according to contaminant load according to other pollution indicators (turbidity, E. coli). However, TLF approaches showed some limitations related to the measurement method (fluorescence interferences), intrinsic features of the system (turbidity) and the location of springs (urban area). In that way, monitoring tasks focused on detecting the arrival of contaminated groundwater to the capture point should be complemented with other easy-to-measure parameters (such as turbidity or electrical conductivity).

Due to the uncertainty about the weight that the different tryptophan sources produce in the recorded TLF signal and the poor correlation between TLF and faecal indicators, the use of this technique in Ubrique test site springs is more suitable as a biological risk indicator rather than an early warning parameter. Nonetheless, TLF approach has proved to be a promising tool in order to rapidly obtain information about the water quality.

**Acknowledgements** This work is a contribution to the European Project "KARMA", PRIMA call, ANR -18-PRIM-0005- and to PiD2019-111759RB-IOO project done in the frame of the Research Group RNM-308 funded by the Autonomous Government of Andalusia (Spain). The authors thank the local government of Ubrique town.

## References

Andreo B, Sánchez D, Martín-Algarra A (2014) Caracterización hidrogeológica y evaluación de los recursos hídricos de la Sierra de Grazalema (Cádiz) para su potencial implementación como reserva estratégica de agua en la cabecera de la Demarcación Hidrográfica del Guadalete-Barbate. Andalusian Water Agency Technical report, Spain, p 128

Ashbolt NJ (2004) Microbial contamination of drinking water and disease outcomes in developing regions. Toxicology 198(1–3):229–238

Bridgeman J, Baker A, Brown D, Boxall JB (2015) Portable LED fluorescence instrumentation for the rapid assessment of potable water quality. Sci Total Environ 524–525:338–346

Carstea EM, Popa CL, Baker A, Bridgeman J (2020) In situ fluorescence measurements of dissolved organic matter: a review. Sci Total Environ 699:134361

Cumberland S, Bridgeman J, Baker A, Sterling M, Ward D (2012) Fluorescence spectroscopy as a tool for determining microbial quality in potable water applications. Environ Technol 33(6):687–693

Determann S, Lobbes JM, Reuter R, Rullkötter J (1998) Ultraviolet fluorescence excitation and emission spectroscopy of marine algae and bacteria. Mar Chem 62:137–156

Frank S, Goeppert N, Goldscheider N (2018) Fluorescence-based multi-parameter approach to characterize dynamics of organic carbon, faecal bacteria and particles at alpine karst springs. Sci Total Environ 615:1446–1459

Goldscheider N, Drew D (eds) (2007) Methods in karst hydrogeology. International Contribution to Hydrogeology, IAH, vol 26. Taylor and Francis/Balkema, London.

Goldscheider N, Chen Z, Auler AS, Bakalowicz M, Broda S, Drew D, Hartmann J, Jiang G, Moosdorf N, Stevanovic Z, Veni G (2020) Global distribution of carbonate rocks and karst water resources. Hydrogeol J 28:1661–1677. https://doi.org/10.1007/s10040-020-02139-5

Khamis K, Sorensen JPR, Bradley C, Hannah DM, Lapworth DJ, Stevens R (2015) In situ tryptophan-like fluorometers: assessing turbidity and temperature effects for freshwater applications. Environ. Sci Process Impacts 17:740–752

Lakowicz JR (2006) Principles of Fluorescence Spectroscopy, 3rd edn. Springer, New York

Marín AI, Martín Rodríguez JF, Barberá JA, Fernández-Ortega J, Mudarra M, Sánchez D, Andreo B (2021) Groundwater vulnerability to pollution in karst aquifers, considering key challenges and considerations: application to the Ubrique springs in southern Spain. Hydrogeol J 29:379–396

Markechova D, Tomkova M, Sadecka J (2013) Fluorescence excitation-emission matrix spectroscopy and parallel factor analysis in drinking water treatment: a review. Pol J Environ Stud 22:1289–1295

Martens DA, Frankenberger WT (1993) Metabolism of tryptophan in soil. Soil Biol Biochem 25–12:1679–1687

Martín-Algarra M (1987) Evolución geológica alpina del contacto entre las Zonas Internas y Externas de la Cordillera Bética. PhD Thesis, University of Granada, Spain, p 1171

Martín-Rodriguez JF, Sánchez-García D, Mudarra-Martínez M, Andreo-Navarro B, López-Rodríguez M, Navas-Gutiérrez MR (2016) Evaluación de recursos hídricos y balance hidrogeológico en acuíferos kársticos de montaña.Caso de la Sierra de Grazalema (Cádiz, España). In: Las aguas subterráneas y la planificación hidrológica. Spanish-Portuguese Congress. IAH Spanish Chapter. Madrid (Spain), Nov 2016. pp 163–170

Mikell AT, Smith CL, Richardson JC (1996) Evaluation of media and techniques to enumerate heterotrophic microbes from karst and sand aquifer springs. Microb Ecol 31:115–124

Nowicki S, Lapworth DJ, Ward JST, Thomson P, Charles K (2019) Tryptophan-like fluorescence as a measure of microbial contamination risk in groundwater. Sci Total Environ 646:782–791

Pronk M, Goldscheider N, Zopfi J (2007) Particle-size distribution as indicator for faecal bacteria contamination of drinking water from karst springs. Environ Sci Technol 41(24):8400–8405

Quiers M, Batiot-Guilhe C, Bicalho CC, Perrette Y, Seidel JL, Van Exter S (2014) Characterisation of rapid infiltration flows and vulnerability in a karst aquifer using a decomposed fluorescence signal of dissolved organic matter. Environ Earth Sci 71:553–561

Sánchez D, Martín-Rodríguez JF, Mudarra M, Andreo B, López M. Navas MR (2016) Time-lag analysis of natural responses during unitary recharge events to assess the functioning of carbonate aquifers in Sierra de Grazalema Natural Park (Southern Spain). In: 2016 Eurokarst Neuchâtel, pp 157–167

Sorensen JPR, Lapworth DJ, Marchant BP, Nkhuwa DCW, Pedley S, Stuart ME, Bell RA, Chirwa M, Kabika J, Liemisa M, Chibesa M (2015) In-situ tryptophanlike fluorescence: a real-time indicator of faecal contamination in drinking water supplies. Water Res 81: 38–46

Sorensen JPR, Sadhu A, Sampath G, Sugden S, Dutta Gupta S, Lapworth DJ, Marchant BP, Pedley S (2016) Are sanitation interventions a threat to drinking water supplies in rural India? An application of tryptophan-like fluorescence. Water Res 88:923–932

Stevanović Z (2019) Karst waters in potable water supply: a global scale overview. Environ Earth Sci 78:662. https://doi.org/10.1007/s12665-019-8670-9

Ward JST, Lapworth DJ, Read DS, Pedley S, Banda ST, Monjerezi M, Gwengweya G, MacDonald AM (2020) Large-scale survey of seasonal drinking water quality in Malawi using in-situ tryptophan-like fluorescence and conventional water quality indicators. Sci Total Environ 140674

Yanofsky C (2007) RNA-based regulation of genes of tryptophan synthesis and degradation, in bacteria. RNA 13(8):1141–1154

# What Microbial Signature Means in Terms of Groundwater Dynamics, Vulnerability and Residence Time—Comparison of Shallow and Deep Karst Resources

M. Sinreich and A. Pochon


## Abstract

The microbial signature is a fundamental characteristic of karst groundwater resources, both with respect to the natural state and potential pollution. Karst springs are regularly affected by water quality derogation in the course of recharge events. Deeper resources with longer groundwater residence times, on the other hand, may be free of fecal pollution and turbidity problems, what puts them into the focus for future exploration strategies. Both aquifer types are however still few understood regarding their microbiological constitution. Novel flow-cytometry data (TCC, Live/Dead cells and LNA/HNA cells) were assessed at shallow and deep resources of a karst system in the Swiss Jura Mountains. While spring results followed hydrological variations, the well data evidenced microbial stability, linked to the increased flow time of several years, and approaching equilibrium with the natural biocenosis of the groundwater ecosystem. Spring and well monitoring considers the end-members of karst groundwater dynamics and vulnerability. This allowed for (i) assessing the hydrogeological meaning of the flow-cytometry parameters, (ii) deducing general conclusions from the signature regarding karst functioning, and (iii) gaining deeper insight into microbial transport and fate processes. The findings provide a conceptual framework for the evolution of diverse microbiological parameters with residence time, contributing to the protection and management of karst resources.



## Keywords

Microbiology • Flow-cytometry • Karst • Vulnerability • Residence time


## 1 Introduction

Although the microbiological constitution represents an important feature for the characterization of karst aquifers, it is still poorly understood. Given fecal pollution being of major concern for the use of karst groundwater, microbiological investigation classically restricts to the analysis of fecal indicator bacteria. More recent approaches allow for complementing such standard analyzes.

Flow-cytometry (FCM) is a novel, easy and insightful tool to characterize microbiological parameters in water. Applying this innovating approach to groundwater started about 10 years ago in the framework of the Swiss National Groundwater Monitoring with an overview of ranges for different aquifer types (Koetzsch and Sinreich 2014). Following work examined that the Total cell count (TCC) is an appropriate criterion for investigating and characterizing karst aquifers with respect to their microbiological state and vulnerability (Sinreich et al. 2014). Additional FCM parameters have the potential for making further progress in this direction by specifying the microbiological status in terms of a cell fingerprint.

The present study compares spring and well data in the same hydrogeological setting of a karst system located in the Swiss Jura Mountains, notably highlighting the importance of deeper karst resources for water supply (Hessenauer et al. 2001; Jeannin et al. 2021). While emerging tools such as FCM have a potential to improve the understanding of karst aquifer hydrogeology and microbiology, a comprehensive examination of the FCM parameters is still missing to date in this context. The present study wants to fill this gap.


M. Sinreich (✉)
Federal Office for the Environment FOEN, Hydrology Division, Bern, Switzerland
e-mail: michael.sinreich@bafu.admin.ch

A. Pochon
Aquagéo Sàrl, Moudon, Switzerland




B. Andreo et al. (eds.), *EuroKarst 2022, Málaga*, Advances in Karst Science,
https://doi.org/10.1007/978-3-031-16879-6_13

## 2 Investigation Design

### 2.1 Study Area

The Dou-Torrent-Raissette karst system, located in the folded Jura Mountains in the Northwestern part of Switzerland, discharges an estimated 50 km$^2$ catchment amounting to 2 m$^3$/s of mean annual flowrate. The main aquifer consists of late Jurassic karstified limestones, which are outcropping on both flanks of the St-Imier valley syncline (Fig. 1).

The Torrent and Raissette springs are important perennial drinking water supplies for the region and exhibit water quality and quantity fluctuations typical for shallow karst groundwater resources, strongly correlated with recharge events. The Dou spring acts as the overflow of the system falling almost totally dry during very low water stages.

Hydrogeological prospection for additional and less vulnerable resources in the Valley of St-Imier recently involved the successful implementation and commissioning of the Sauges well, which taps the same Jurassic aquifer at a depth between 400 and 600 m under a thick cover of low permeable tertiary sediments.

The extracted groundwater is sub-thermal (~18 °C), with very stable physico-chemical characteristics, no trace of bacteriological or chemical pollution and an estimated mean residence exceeding 40 years based on tritium analysis sampled in the first exploratory well in the mid-nineties.

### 2.2 Sampling and Analyses

The Torrent and Raissette springs as well as the Sauges pumping well were monitored over a more than 1 year period from end of 2020 to beginning of 2022. A campaign of 12 samplings targeted on differing hydrological situations in order to specifically cover periods of both low and high water stages, with a focus on the flood and recession period of summer/autumn 2021 (Fig. 2). Data were set in context with continuous discharge and physico-chemical measurements, which is a primary information to be checked with microbiological occurrence and abundance.

Water samples were filled into prepared containers, subsequently kept cooled at about 4 °C during transport and storage, and mostly analyzed within 24 h. Flow-cytometry (FCM) technique was used to obtain cell count parameters, conducted according to the method described by Hammes et al. (2008). Additionally, classical microbiological parameters, i.e. aerobic mesophilic bacteria (AMB) and fecal indicator bacteria (FIB: *Escherichia coli* (*E. coli*) and Enterococci), were analyzed by culture methods according standard protocols.

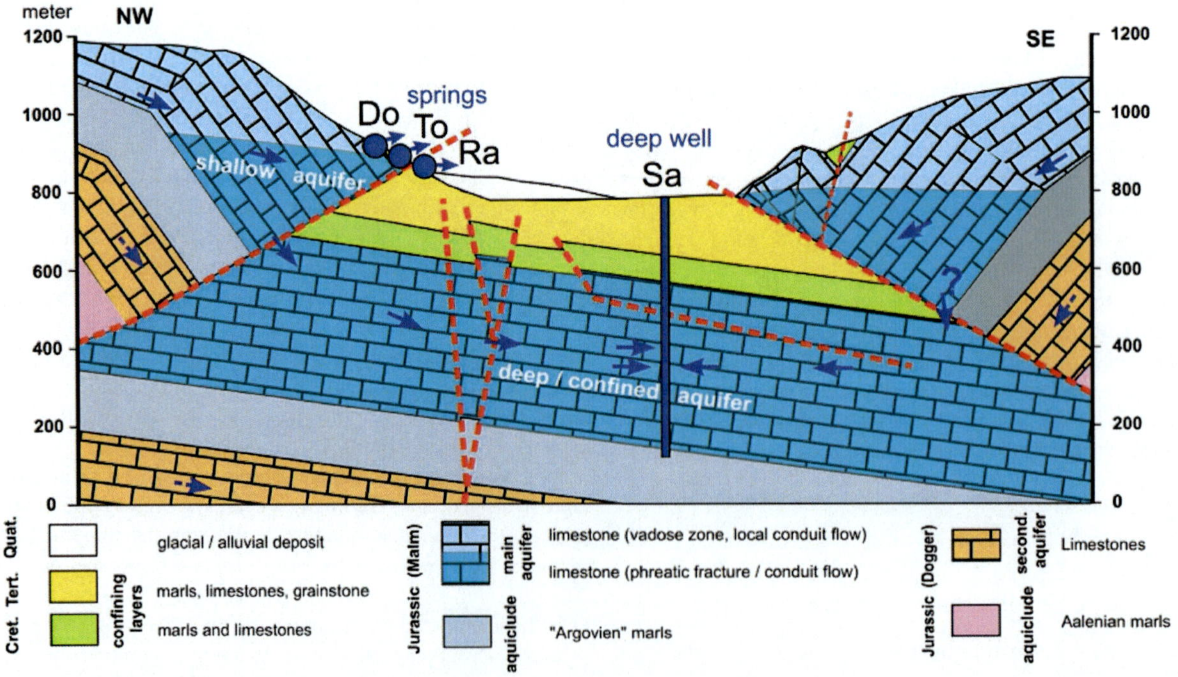


**Fig. 1** Hydrogeological cross section through the St-Imier valley displaying the groundwater conditions at the study area including sampling sites (adapted from Della Valle et al. 1998)

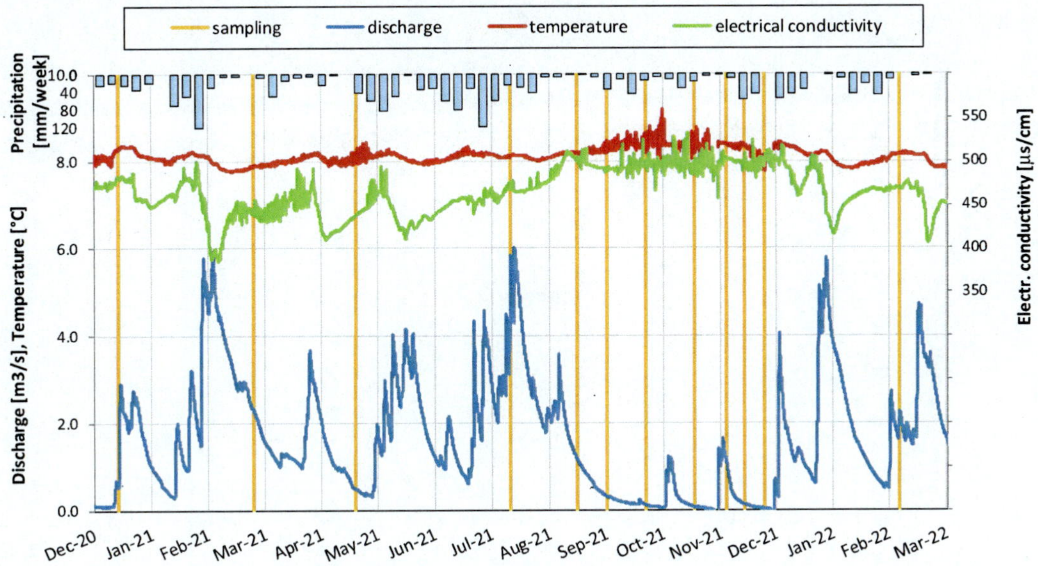


**Fig. 2** Continuous physico-chemical and discharge data for the shallow groundwater resource (Dou overflow spring) over the study period with related sampling dates

Microbial signature using FCM comprises Total cell count (TCC) as well as specific subsets for Live (intact) vs. Dead (permeabilized) cell count and Low nucleic acid (LNA) vs. High nucleic acid (HNA) cell count. Ratios of living cells and LNA cells with respect to TCC are used for illustrating their relative importance.

# 3 Results

## 3.1 Shallow Groundwater

Figure 2 displays the evolution of discharge and electrical conductivity for the Dou overflow spring as representative for the whole shallow groundwater system. Typically for dynamic karst aquifers, sharp peaks linked to rainfall and snowmelt events followed by recession phases dominate the hydrograph. The study period was characterized by an extremely wet summer and a relatively dry autumn.

For the shallow groundwater resource, captured at the Torrent and Raissette springs, microbiological monitoring parameters (Figs. 3 and 4) evidenced similar variability as physico-chemical parameters (i.e. electrical conductivity, temperature, and turbidity), correlated to discharge variations. As typically known from dynamic karst system, FIB levels strongly rose over flood events, temporarily causing water quality derogation. *E. coli* variability ranged over three orders of magnitude (from 1,000 cfu/100 mL to almost absent). Enterococci showed systematically lower values, but with similar pattern as *E. coli* or AMB.

All FCM cell types (TCC, Live, Dead, LNA, HNA) followed an evolution comparable to FIB, though to differing degrees. TCC reached more than 1,000,000 but never declined below 200,000 cells/mL, i.e. within one order of magnitude. Live cell and LNA cell ratios varied between 70 and 90%, and 50 and 80% of TCC, respectively.

## 3.2 Deep Groundwater

The dynamics and microbial signature of the deep karst resource, exploited at the Sauges well, differ significantly from the shallower groundwater as it remained very stable over time, independent of the prevalent hydrological conditions (Figs. 3 and 4). No significant reaction has been observed with respect to surface water conditions, and spring discharge variations, respectively (Table 1).

No fecal indicators were detected in the well water throughout the whole year, confirming the excellent quality of extracted groundwater. TCC remained permanently around 10,000 cells/mL, which is more than one order of magnitude below spring levels during baseflow conditions, without any rainfall-driven or seasonal variation observed. All cell types had lower counts in the deep aquifer than in the shallow aquifer. However, their relative importance toward TCC exceeded spring values for Live cell ratio (> 90%), and LNA cells ratio (75–85%).

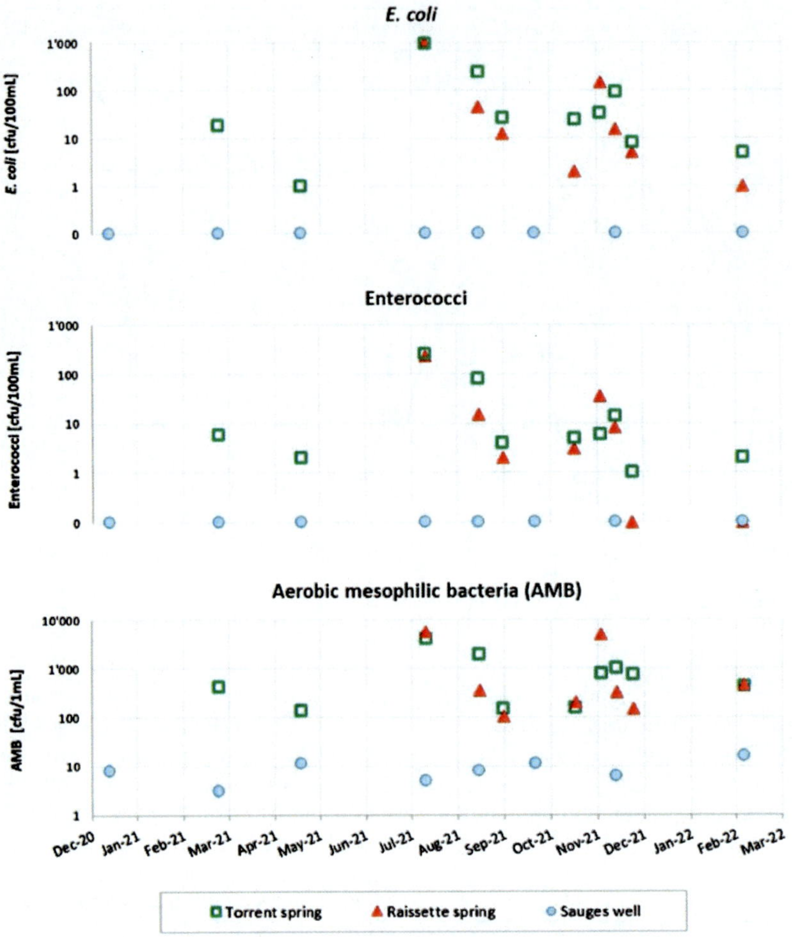


**Fig. 3** Fecal indicator parameters for the three sampling sites

## 4 Discussion

### 4.1 Microbiological Parameters

Table 1 provides an overview of the typical abundances for microbiological parameters analyzed. These are consistent with data from numerous other Swiss karst sites (Sinreich et al. 2014; Pronk et al. 2020) and similar hydrogeological settings (Koetzsch and Sinreich 2014; Schönher et al. 2021). Both FIB and FCM parameters differed significantly between the shallow and deep karst groundwater resources, and are discussed in the following.

#### 4.1.1 Fecal Indicator Bacteria (FIB)

Occurrence of FIB, such as *E. coli* and Enterococci, mainly results from fecal pollution sources (e.g. sewer or livestock), transported via fast preferential flow components in the course of recharge events.

During baseflow conditions, higher proportion of slow flow components are feeding the springs, and groundwater linked to the preceding recharge events exhibits long enough residence times for sufficient FIB die-off. In a similar vein, water extracted from the well, having even much longer residence times, is free of fecal bacteria at any time. This confirms the FIB approach as indicator for mainly fast microbial/fecal input not exceeding several weeks. It goes along with AMB, which are not specifically of fecal origin, but may indicate near surface influence. AMB are nonetheless permanently found also in the deep aquifer, but at very low levels.

#### 4.1.2 Total Cell Count (TCC)

TCC values measured at the springs are in agreement to what had been defined by Sinreich et al. (2014) as characteristic for dynamic and vulnerable karst systems. Variations for this overall parameter were related to the prevalent

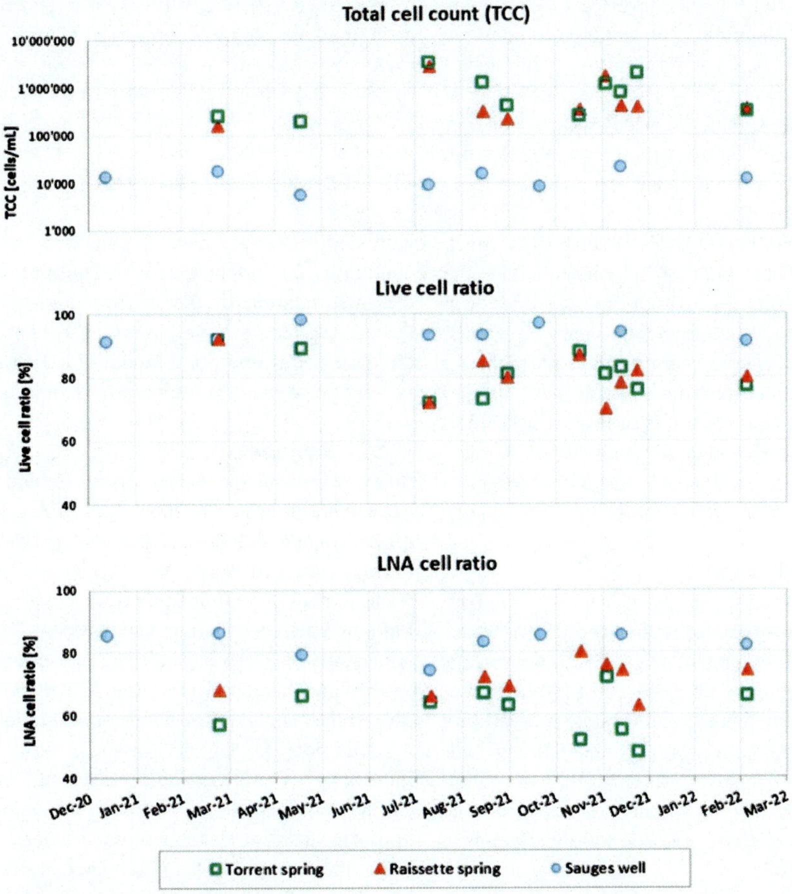


**Fig. 4** Flow-cytometry parameters for the three sampling sites

**Table 1** Aggregated monitoring results (mean values and standard deviation) for eight samplings at each of Torrent spring, Raissette spring, and Sauges well sampling sites (microbial signature in bold)

| | | | Shallow resource | | | | Deep resource | |
|---|---|---|---|---|---|---|---|---|
| | | | Torrent spring | | Raissette spring | | Sauges well | |
| | | | Mean | std-*d* | Mean | std-*d* | Mean | std-*d* |
| | El. conductivity | µS/cm | 499 | 13 | 469 | 11 | 414 | 7 |
| | Temperature | °C | 8.3 | 0.1 | 8.1 | 0.1 | 18.6 | 0.5 |
| | Turbidity | NTU | 3.6 | 5.4 | 11.3 | 29.2 | 0.3 | 0 |
| FIB | *E. coli* | cfu/100 mL | **174** | 327 | **158** | 360 | **0** | 0 |
| | Enterococci | cfu/100 mL | **47** | 90 | **37** | 79 | **0** | 0 |
| | AMB | cfu/mL | **1,161** | 1,316 | **1,534** | 2,365 | **9** | 4 |
| FCM | TCC | cells/mL | **1,165,908** | 1,023,106 | **806,714** | 926,912 | **12,600** | 5,121 |
| | Live cell ratio | % | **79** | 5 | **79** | 6 | **94** | 3 |
| | LNA cell ratio | % | **61** | 8 | **72** | 6 | **82** | 4 |

hydrological conditions, with higher levels at flood events and lowest levels during baseflow. This pattern, comparable to FIB, suggests that TCC peaks at the springs are dominated and triggered by similar mechanisms. High loads of bacterial cells may derive from sinking surface water, soil, or microbial pollution, mainly during recharge events. Bottom values reached during baseflow are considered characteristic for each karst aquifer system.

In contrary, well groundwater is characterized by much lower and stable TCC, one order of magnitude below the lowest values at the springs. For both shallow and deep resources, TCC is a factor 1,000 higher compared to AMB, and thus a much more sensitive parameter. This makes it particularly useful for characterizing systems with limited surface influence, such as karst aquifers of low vulnerability, or porous aquifers (Koetzsch and Sinreich 2014). Consequently, TCC may be a good proxy for microbiological variability when AMB are too few (Van Nevel et al. 2017).

### 4.1.3 Live/Dead Cell Count

Live cells predominated in all groundwater samples (>70% of TCC), with highest ratios related to the deep groundwater. The high amount of both living and dead cells at the springs during floods, in conjunction with a reduced Live cell ratio, suggests that the input signal is the main source of dead (not viable) cells. Also the increasing proportion of living cells with discharge recession suggests that dead cells are rather of allochthonous origin than the result of bacterial fate during transport. In this respect, Live cell ratio has opposite evolution compared to TCC.

During baseflow, living cells are reaching values up to 90% of TCC at the springs. They are thereby approaching levels of the deep resource, while TCC still remains one order of magnitude above. This may represent the natural situation with lowest influence of fresh water in the system, and with living cells dominating. A stable and high Live cell ratio seems to be typical for the deep system, protected from rapid surface influences and changes. Whilst low ratio can be associated with surface influence and vulnerable systems, high ratio may occur for either systems of low vulnerability, or highly vulnerable systems at baseflow.

### 4.1.4 LNA/HNA Cell Count

LNA cells were the dominant fraction in almost all groundwater samples (> ca. 50% of TCC). They indicate the presence of smaller cells, while HNA cells are expected being more abundant at elevated nutrient availability, such as surface water. Although absolute counts of LNA and HNA cells were both fluctuating with spring discharge, there was no evidenced impact of flood events causing increased importance of HNA cells in groundwater, as might be assumed. Overall, there was no clear relationship with time confirmed between LNA cell ratio and discharge or TCC in the course this study.

Nonetheless, LNA cell ratio was generally higher in the deep than the shallow resource. Consequently, microbiological processes may influence the fate of bacteria in terms of LNA and HNA, additionally to transport.

## 4.2 Conceptual Framework

Knowledge about the catchment and the hydrogeological conditions, in conjunction with continuous data, allows for the interpretation of the microbiological results regarding spatial and temporal variations in the karst system. This is the basis for a conceptual model of karst microbial signature in terms of dynamics, vulnerability, and residence time.

### 4.2.1 Dynamics

Sharp fluctuations of spring discharge and physico-chemical parameters indicate the high dynamics of the shallow karst groundwater and the fast arrival of freshly infiltrated rainwater during recharge events. As for microbial pollution, groundwater dynamics were identified for governing the input of all kind of bacteria into the karst system. It rises the overall abundance of bacteria, i.e. TCC, and drastically changes the quality and microbial signature of the groundwater. This is despite the mainly diffuse nature of infiltration in the catchment.

After long-lasting dry periods, slow flow components increasingly dominate and represent the baseflow signature of the system. Such conditions are most appropriate to compare different systems to each other.

At the deeper karst groundwater, no fresh input is to be considered. Such lack of hydrological dynamics goes along with stable microbial signature and low abundances. It is noteworthy that the dynamic shallow groundwater reaches during baseflow similar high Live cell ratios as the stable deep resource, though always maintains much higher TCC.

### 4.2.2 Vulnerability

The overall amount of fast flow components determines the global vulnerability of a groundwater system or spring. Microbial signature can provide information about the vulnerability of a catchment. Sinreich et al. (2014) proposed a microbiological classification for different karst settings and vulnerabilities, postulating higher TCC values for more vulnerable systems, even for baseflow. The present findings strongly confirm this hypothesis, including the respective TCC ranges.

This effect highlights most between the shallow and the deep resources. Those represent end-members in terms of vulnerability, and natural protection, respectively. Global vulnerability of the sampling points is consistent with values from Sinreich et al. (2014), e.g. around 10′000 cells/mL for naturally protected aquifers of low vulnerability. Slight differences in the mean values for TCC, and LNA cell ratio, observed for Torrent and Raisette springs in this study (see Table 1) are also expected to reflect small contrasts in global vulnerabilities of both sub-catchments.

Recharge events load the system in a way that there is a sustainable microbial signature even for baseflow, allowing for global vulnerability evaluation on the basis of one single sample, as mentioned in Sinreich et al. (2014). FIB alone are not able to do so. The absence of FIB and a permanently low TCC, in contrary, are a reliable indication for a resource protected towards non-persistent pollution.

### 4.2.3 Residence Time

Freshly infiltrated meteoric water following recharge events is associated with the arrival of fast and preferential flow at springs within hours to days. In the course of discharge recession, these components are more and more replaced by slower pathway and storage components. These changes are highlighted by electrical conductivity (Fig. 1), with negative peaks indicating marked dilution due to fresh, lowly mineralized water. At these events, important release from the surface causes high fluxes in all kind of bacterial cells creating a strong input signature.

With ongoing recession, mean residence time of groundwater increases. Tracing tests performed in the catchment during low flow conditions showed transit times up to several weeks to reach the springs. Mean residence time in the deeper resource, however, is up to several years given the long transfer in the phreatic zone until reaching the well.

In the present study, mean residence time showed a clear relationship with several microbiological parameters, more dynamic/vulnerable systems having higher baseline abundances and higher variability ranges. Elevated TCC and only moderate Live cell ratio are then due to lower overall residence time compared to less vulnerable, i.e. more naturally protected, aquifers. FIB die-off processes are time-dependent, too. Figure 5 illustrates the conceptual framework for the microbiological signature of karst groundwater.

### 4.2.4 Microbial Biocenosis

Following the above interpretation, shallow resources are governed by the input signature following recharge events whilst deeper resources can be considered being adapted to the intrinsic environmental conditions. This latter case is representing the microbial biocenosis of pristine karst groundwater. The adaption with time occurs when allochthonous bacteria are either eliminated or had become part of the natural system. At these times, fecal bacteria are not surviving any more. The deep system, with low organic nutrients, and elevated LNA ratio, respectively, is thus characterized by both low TCC and a high Live cell ratio. In

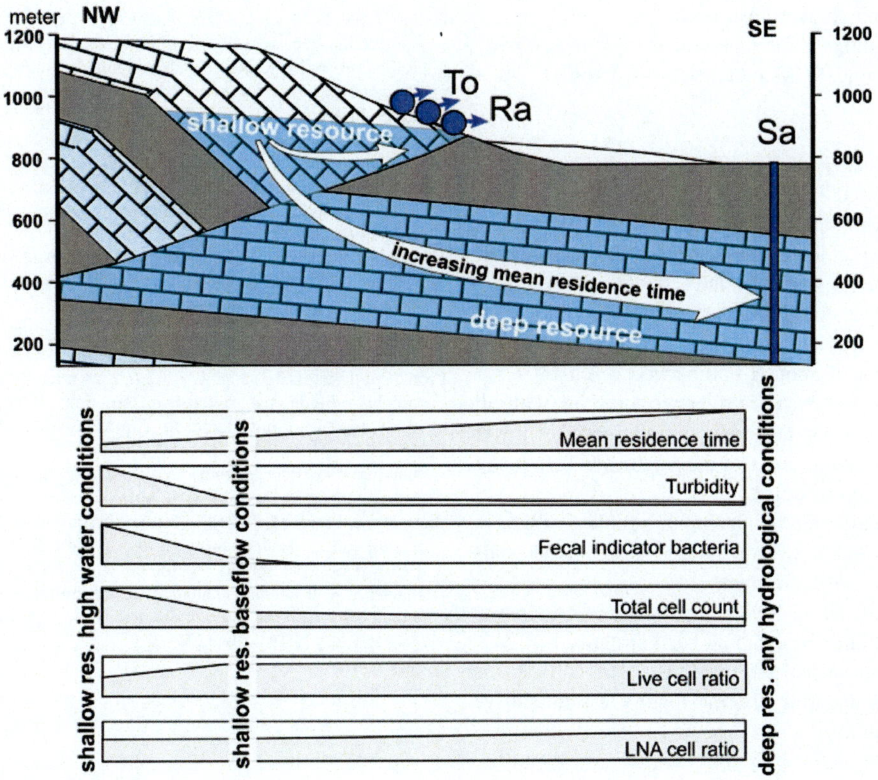


**Fig. 5** Conceptual model for the evolution of microbiological parameters in shallow and deep resources with respect to karst groundwater mean residence time (explanation see text)

the case of the Sauges well, it appears that potentially high-loaded input does not play a role any more. Signature has instead adapted to the natural biocenosis after several years of residence time, in equilibrium to the autochthonous microbial community.

Also shallow systems may reach such pristine state when slow flow components clearly dominate. During low flow conditions, the two springs exhibit high Live cell ratios comparable to the well, suggesting this parameter being an indication for a natural biocenosis. Thereby, TCC always remained at higher level at the springs suggesting specific signature for a nutrient-rich environment, in conjunction with lower LNA cell ratio.

## 5 Conclusions

The present study provides a conclusive interpretation of the meaning of FCM parameters for karst groundwater, clearly evidenced at end-members of shallow (vulnerable) and deep (low vulnerability) groundwater resources and related to spring variability. Classical microbiology targets fast flow components with *E.coli* appearance and dynamics, and being limited to FIB survival times. In deep resources of low vulnerability, and high mean groundwater residence time, respectively FIB and AMB are not useful discriminates any more, whereas FCM parameters are. They clearly characterize both deep and shallow resources.

Some basic findings of the present study may provide principles in applying FCM parameters as a tool in karst research:

- There are typical levels of TCC for shallow (about 100,000 to 1,000,000 cells/mL) and deep karst systems (about 10,000 cells/mL).
- All bacteria occur at lower abundances in the deep aquifer than in the shallow one (absolute counts), though does not account for relative importance (ratios).
- Overall abundances of bacteria are TCC > AMB > *E. coli* with a factor of about 1,000 between each.
- Differing signatures exist for dynamic systems, with all FCM cell types rising at floods and approaching lower levels during baseflow, some of the parameters then being similar to the deep resource.
- Living cells clearly predominate in all samples, with Live cell ratio increasing with residence time; dead cells mainly derive from flood input.
- Rising TCC and falling Live cell ratio are in correlation with FIB occurrence in dynamic karst systems.
- LNA cells form the majority of TCC. LNA cell ratio is elevated for low dynamic systems, though less conclusive regarding variability.
- Pristine karst groundwater consists of a typical microbial biocenosis with differences in shallow and deep resources, and Live cell ratio as an indicator.
- Transport processes may rule microbial signature during high water stages, while microbial fate becomes more important for longer residence times.
- A dynamics–vulnerability–residence time approach, building the conceptual framework of the present findings, may explain the constitution and changes of the microbial signature in karst groundwater.

There are hence typical cell count values for each resource type and in relation with mean residence time, but with temporal changes for shallow/dynamic systems. Therefore, FCM parameters give more insight into the functioning of karst aquifers, going beyond fecal pollution issues and potentially complementing FIB as indicators. Sound understanding of the meaning of the diverse FCM parameters amplifies the actual knowledge on karst groundwater microbiology. It provides an overall picture and a means of characterization, not only for contrasted cases as in the present study. This may represent an important step forward in karst hydrogeology.

## References


Della Valle G, Rieben Ch, Adatte P (1998) Prospection d'eau souterraine par forages profonds dans le Vallon de St-Imier. Rapp. bureau MFR, Office de l'économie hydraulique et énergetique du canton de Berne, 60 pp

Hammes F, Berney M, Wang Y et al (2008) Flow-cytometric total bacterial cell counts as a descriptive microbiological parameter for drinking water treatment processes. Water Res 42:269–277

Hessenauer M, Rieben C, Flury F (2001) Prospection d'eau souterraine par forages profonds à Muriaux (Canton de Jura). Bull Géolog Appl 6(2):147–164

Jeannin PY, Malard A, Sinreich M (2021) Karstwater as drinking water—A way to use karst groundwater for Managed Aquifer Recharge (MAR). 48th IAH congress, Brussels

Koetzsch S, Sinreich M (2014) Zellzahlen zum Grundwasser—Bestimmung mittels Durchflusszytometrie. Aqua Gas 3:14–21

Pronk M, Urfer D, Bonnet C (2020) Etat microbiologique de l'eau potable—Etude par cytométrie en flux alimentant la région d'Yverdon-les-Bains-Grandson. Aqua Gas 6:22–29

Schönher CH, Proksch P, Kerschbaumer D et al. (2021) Every cell counts—Experiences with flow cytometry for Austrian drinking water supply. Österr Wasser u Abfallwirtschaft 73:501–511

Sinreich M, Pronk M, Kozel R (2014) Microbiological monitoring and classification of karst springs. Environ Earth Sci 71(2):563–572

Van Nevel S, Koetzsch S, Proctor CR et al (2017) Flow cytometric bacterial cell counts challenge conventional heterotrophic plate counts for routine microbiological drinking water monitoring. Water Res 113:191–206

# Deriving Major Ion Concentrations at High Resolution from Continuous Electrical Conductivity Measurements in Karst Systems

B. Richieri, D. Bittner, A. Hartmann, P. Benettin, B. M. van Breukelen, D. Labat, and G. Chiogna

**Abstract**

Time series of hydrochemical parameters support the investigation of dominant karst hydrological processes and conceptual model structures. Nevertheless, high costs for sample collection and analyses cause hydrochemical data to be rarely available at a sufficiently high temporal resolution, e.g. hourly. The electrical conductivity (EC), however, can be cheaply and continuously measured by an EC sensor. To overcome this issue, a method is proposed to derive continuous major ion concentrations from continuous EC using both low-frequency (i.e. weekly) and high-frequency (i.e. every 3 h) ionic measurements. Due to the large ion concentrations and complex speciation characterizing karst springs, the concentrations of each element as free ion and as aqueous complexes are computed separately. The first is computed based on their contributions to the total EC, whereas the concentrations as aqueous complexes are obtained from speciation calculations. The method is tested in two karstic watersheds with different types of bedrock and temporal resolution of the available hydrochemical datasets, i.e. the Kerschbaum dolostone system in Austria and the Baget limestone system in France. The results show that complexation needs to be considered for $SO_4^{2-}$, $Ca^{2+}$, $Mg^{2+}$ and $HCO_3^-$, whose neglection would lead to an underestimation of the total concentrations. The proposed methodology allows us to obtain high-resolution major ion concentrations without performing costly laboratory analyses. The investigation of the proper temporal resolution, required to apply this method to different geologies, can support future fieldwork and sampling campaigns.

B. Richieri (✉) · G. Chiogna
Chair of Hydrology and River Basin Management, TUM School of Engineering and Design, Technical University of Munich, Arcisstr. 21, 80333 Munich, Germany
e-mail: beatrice.richieri@gmail.com

G. Chiogna
e-mail: gabriele.chiogna@tum.de

D. Bittner
Erftverband, Department for River Basin Management, Am Erftverband 6, 50126 Bergheim, Germany
e-mail: Daniel.Bittner@erftverband.de

D. Bittner · A. Hartmann
Institute of Groundwater Management (IGW), Technical University of Dresden, Bergstr. 66, 01069 Dresden, Germany
e-mail: andreas.hartmann@tu-dresden.de

P. Benettin
Laboratory of Ecohydrology ENAC, IIE/ECHO, École Polytechinque Fédérale de Lausanne, 1004 Lausanne, Switzerland
e-mail: paolo.benettin@epfl.ch

B. M. van Breukelen
Department of Water Management, Faculty of Civil Engineering and Geosciences, Delft University of Technology, 2628 CN Delft, The Netherlands
e-mail: b.m.vanbreukelen@tudelft.nl

D. Labat
Université Toulouse 3-Géosciences Environnement Toulouse - CNRS-UPS-IRD, 14 Avenue Edouard Belin, 31400 Toulouse, France
e-mail: david.labat@get.omp.eu

G. Chiogna
Institute of Geography, University of Innsbruck, Innrain 52, 6020 Innsbruck, Austria

## 1 Introduction

Hydrochemical data of karst springs can provide significant information about the subsurface structures and relevant karst hydrological processes. Several studies have used hydrochemical data to investigate the subsurface functioning of karst systems, e.g. to characterize different kinds of karst systems or to estimate the ratio between water from the vadose and the phreatic zones under varying hydrometeorological conditions (Aquilina et al. 2006; Dreiss 1989; Liu


B. Andreo et al. (eds.), *EuroKarst 2022, Málaga*, Advances in Karst Science,
https://doi.org/10.1007/978-3-031-16879-6_14

et al. 2004; Mudarra and Andreo 2011; Ravbar et al. 2011; Torre San et al. 2020). When used in a multi-objective model calibration approach, hydrochemical time series can support the identification of realistic model structures and the reduction of the parameter uncertainty (Chang et al. 2021; Charlier et al. 2012; Hartmann et al. 2017; Hartmann et al. 2013a, b; Husic et al. 2019; Perrin et al. 2003; Zhang et al. 2019). However, high cost and time requirements limit the applicability of major ion time series, which are rarely available at a temporal resolution sufficiently high to investigate the transport processes of a system. On the contrary, the electrical conductivity (EC) signal, which represents an integrated signal of major ion concentrations, can be easily and cheaply measured by means of automatic instruments and thus shows a huge potential in model development and evaluation (Chang et al. 2021; Jourde et al. 2018; Meus et al. 2014; Cano-Paoli et al. 2019).

Benettin and van Breukelen (2017) proposed a method to retrieve, from continuous EC signal and discrete measurements of major ions, the individual ion concentrations at the same temporal resolution as the EC signal. The method is based on the decomposition of the EC signal measured at a river gauge into the individual major free ions that conduct current in water. However, due to the large ion content in water from karst systems and more complex water speciation as compared to surface streams, the method of Benettin and van Breukelen (2017) may not be directly applicable to karst systems.

This study presents a modified framework of the methodology from Benettin and van Breukelen (2017). The modified approach is tested in two study areas characterized by different bedrocks and different temporal resolution of available hydrochemical datasets. In particular, we apply the modified EC decomposition approach for the Kerschbaum dolostone karst system in Austria and the Baget limestone karst system in France.

## 2 Materials and Methods

In the following, we present the characteristics of both investigated karst systems, the available datasets as well as the modified methodology of Benettin and van Breukelen (2017).

### 2.1 Study Areas

This work considers two study areas, i.e., the Kerschbaum springshed in Austria and the Baget springshed in France (Fig. 1a). The two karst systems are characterized by a different degree of karstification and discharge variability, which result from the different geology of the karst formations (Fig. 1b, c). Kerschbaum is a pre-alpine, small-scale (2.5 km$^2$) dolostone karst system with a mean discharge of 34 l/s (Bittner et al. 2021). Despite the developed network of fractures and conduit, no sinkholes are present in the catchment and the spring is mainly fed by diffusive infiltration processes (Narany et al. 2019). The response of the spring discharge to precipitation events is characterized by a piston and flushing effect of the dissolved elements stored in the saturated zone. The Baget springshed is located in the Pyrenees and is a small-scale (13 km$^2$) limestone karst system characterized by a mean annual discharge of 477 l/s. Las Hountas is the only perennial spring of the Baget catchment and displays a complex response characterized by different mechanisms, i.e. dilution processes during high discharge periods and a piston and flushing effects with the first heavy rains after long dry periods (Ulloa-Cedamanos et al. 2020).

The water quality data available for the two study areas have different temporal resolutions and durations. The dataset for the Kerschbaum spring is provided by the waterworks of Waidhofen a.d. Ybbs. It covers the year from 10/17/2018 to 10/3/2019 and consists of hourly EC (µS/cm) and weekly concentrations (mg/L) of the major ions $Ca^{2+}$, $Mg^{2+}$, $HCO_3^-$, $SO_4^{2-}$, $NO_3^-$, $Cl^-$, $Na^+$ and $K^+$. Conversely, the dataset for the Las Hountas spring results from an event-based sampling campaign and covers the eleven days from 10/4/2021 to 10/14/2021, with EC (µS/cm) and major ion concentrations (mg/L) measured every fifteen minutes and 4–5 h, respectively.

### 2.2 Electrical Conductivity (EC) Decomposition Method

Due to the large ionic content and complex water speciation characterizing the discharge of karst springs, the occurrence of aqueous complexes is not negligible. Thus, the concentrations of each element occurring as free ions and as aqueous complexes are computed separately. The concentrations as free ions are computed based on the contribution of each element to the total measured EC signal, whereas the concentrations as aqueous complexes are obtained from speciation calculations. All model equations are implemented using Phreeqc (Version 3.6.4).

The total EC (µS/cm) of water is equal to the sum of the contributions of each individual ion i (Eq. (1)), which is calculated from the molar conductivity $\Lambda_m^{\circ}$ (S/m/ (mol/m$^3$)), the molar concentration m (mol/m$^3$) and the electrochemical activity coefficient $\gamma_{EC}$ dimensionless (−) of the individual ion as shown in Eq. (2).

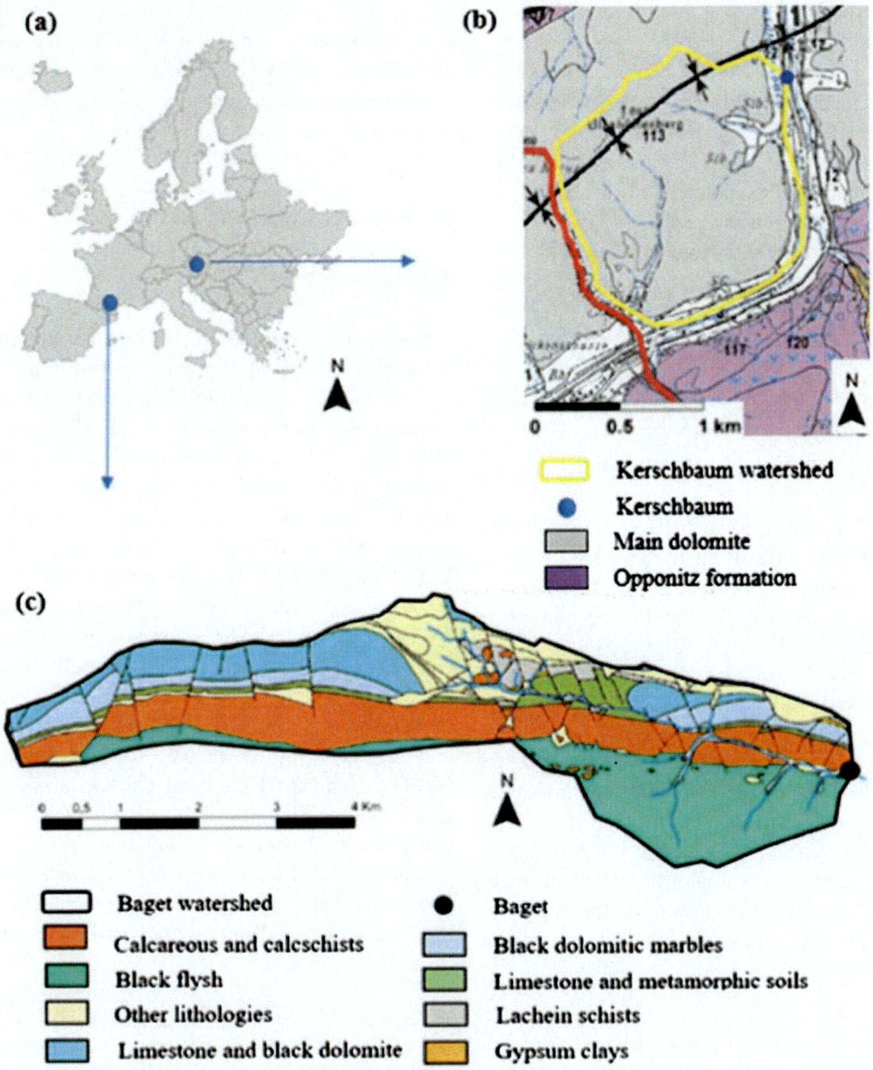


**Fig. 1** Geological maps of the study areas showing the location of the springs and recharge areas. **a** The locations of the study areas in Europe. **b** Kerschbaum dolostone karst system in Austria. **c** Baget limestone karst system in France

$$\mathrm{EC} = \sum_i \mathrm{EC}_i \tag{1}$$

$$\mathrm{EC}_i = \Lambda^\circ_m * m * \gamma_{EC} \tag{2}$$

The molar conductivity and the diffusion coefficient $D_W$ ($m^2/s$) at 25 °C are correlated, as shown in Eq. (3), by a coefficient of proportionality which is a function of the charge number z (−) of the ion, the Faraday´s constant F (96,485.33 C/mol), the gas constant $R$ (8.31 J/°K/mol) and the absolute temperature $T$ (°K). The diffusion coefficient at a specific temperature different from 25 °C needs to be adjusted considering the viscosity of water $\eta$ (Pa s) at that temperature (Eq. (4)).

$$\Lambda^\circ_m = \frac{z^2*F^2}{R*T} D_W \tag{3}$$

$$(D_W)_T = (D_W)_{298} * \frac{T}{298} * \frac{\eta_{298}}{\eta_T} \tag{4}$$

The electrochemical activity coefficient $\gamma_{\mathrm{EC}}$ (−) is computed as shown in Eq. (5) based on the charge number $z$ (−) and the Debye-Hückel activity coefficient $\gamma_{\mathrm{DH}}$ (−) considering an ionic strength $I < 0.36|z|$.

$$\log(\gamma_{\mathrm{EC}}) = \log(\gamma_{\mathrm{DH}}) * \frac{0.6}{|z|^{0.5}} \tag{5}$$

The chemical properties of each ion i can be expressed by a single coefficient $a_i$ ((μS/cm)/(mg/L)) as in Eq. (6), with M the ion molar mass (g/mo(l)).

$$a_i = \frac{\Lambda^{\circ} * \gamma_{EC}}{M} \tag{6}$$

Given as input the low-resolution ion concentrations, Eqs. (1–6) are implemented on Phreeqc to derive the individual electrical conductivity $EC_i$ of the free ions $Ca^{2+}$, $Mg^{2+}$, $HCO_3^-$, $SO_4^{2-}$, $NO_3^-$, $Cl^-$, $Na^+$ and $K^+$ at low resolution. For a more detailed explanation of the procedure, one can refer to the Phreeqc manual (Parkhurst and Appelo 2013).

The total EC can be written as function of the coefficients $a_i$, concentrations $C_i$ (mg/l) and time $t_k$ as in Eq. (7), where the low-resolution contribution of each ion i to the total EC can be quantified with the weight factor $f_i$ (−) (Eq. (8)).

$$\mathrm{EC}(t_k) = \sum_i \mathrm{EC}_i(t_k) = \sum_i a_i(t_k) * C_i(t_k) \tag{7}$$

$$f_i(t_k) = \frac{\mathrm{EC}_i(t_k)}{\mathrm{EC}(t_k)} \tag{8}$$

As proposed by Benettin and van Breukelen (2017), the low-resolution $f_i$ are then linearly interpolated to the same resolution as the measured EC, and the high-resolution concentrations $C_i$ of the individual free ions are computed with Eq. (9).

$$C_i(t_k) = \frac{\mathrm{EC}(t_k) * f_i(t_k)}{a_i(t_k)} \tag{9}$$

In addition to the procedure described in Benettin and van Breukelen (2017), to account for complexation processes, we consider that the total concentration of each element is equal to the sum of its concentration as free ion and as part of aqueous complexes. The concentrations of the elements in aqueous complexes are, first, computed at the temporal resolution of the ionic measurements by speciation calculation on Phreeqc as difference between the total molality of an element (mol/kgw) and the molality of the solution (mol/kgw). Then, they are linearly interpolated to the resolution of the measured EC and summed to the corresponding concentration as free ion.

## 3 Results and Discussion

### 3.1 Speciation Calculation

The speciation calculations performed with Phreeqc allow us to determine the equilibrium composition of each water sample and thus give information about the distribution in concentration of the different chemical species of each element. Table 1 shows the average percentage of each element present as free ion and as ion involved in aqueous complexes. The results of the speciation calculations are consistent between the two study areas. For both the Kerschbaum and Las Hountas springs, $SO_4^{2-}$ has the largest tendency to form aqueous complexes, with 16.3% and 14.2% of its total concentration involved in the formation of $CaHSO_4^+$, $NaSO_4^-$ and $HSO_4^-$. For both the springs, $Ca^{2+}$, $Mg^{2+}$ and $HCO_3^-$ are, in decreasing order, involved in complexes, i.e. $CaOH^+$, $CaHSO_4^+$, $CaHCO_3^+$, $MgHCO_3^+$ and $MgOH^+$. Only about 0.2% of $Na^+$ forms $NaSO_4^-$ and $NaCO_3^-$, whereas $NO_3^-$ and $Cl^-$ are present only as free ions. For Las Hountas, 0.1% of $K^+$ forms $KSO_4^-$, whose concentration is zero for Kerschbaum.

The total EC derives mainly from the presence of free ions in solution, whereas charged complexes have a minor role. As shown in Table 1, the contribution of the elements

**Table 1** Elements occurring as free ions (%) and aqueous complexes (%) expressed, for each element, as mean percentage of the total ion concentration, together with their percentage contribution to the total electrical conductivity EC. The percentages result from speciation calculation on Phreeqc based on the samples collected at the Kerschbaum and Las Hountas springs for the periods 10/2018–10/2019 and 10/4/2021–10/14/2021, respectively

| | Kerschbaum spring 10/2018–10/2019 | | Las Hountas spring 10/4/2021–10/14/2021 | |
|---|---|---|---|---|
| | *Percentage of the total ion concentration* | | | |
| | Elements as free ions (%) | Elements as aqueous complexes (%) | Elements as free ions (%) | Elements as aqueous complexes (%) |
| $Ca^{2+}$ | 95.0 | 5.0 | 93.9 | 6.1 |
| $Mg^{2+}$ | 95.4 | 4.6 | 94.5 | 5.5 |
| $HCO_3^-$ | 96.9 | 3.1 | 94.8 | 5.2 |
| $SO_4^{2-}$ | 83.7 | 16.3 | 85.8 | 14.2 |
| $NO_3^-$ | 100.0 | 0.0 | 100.0 | 0.0 |
| $Cl^-$ | 100.0 | 0.0 | 100.0 | 0.0 |
| $Na^+$ | 99.8 | 0.2 | 99.78 | 0.2 |
| $K^+$ | 100.0 | 0 | 99.9 | 0.1 |
| | *Percentage contribution to the total* EC | | | |
| | Elements as free ions (%) | Elements as aqueous complexes (%) | Elements as free ions (%) | Elements as aqueous complexes (%) |
| EC | 99.4 | 0.6 | 98.8 | 1.2 |

present as free ions is 99.4% and 98.8% of the total EC for the Kerschbaum and Las Hountas springs, respectively.

For both the study areas, the total EC is mainly correlated to $Ca^{2+}$, $Mg^{2+}$ and $HCO_3^-$ and thus to the dissolution of carbonate rocks, with a total percentage contribution of 92.2% and 90.7% for Kerschbaum and Las Hountas, respectively. The different geology of the studied systems is noticeable by looking at the relative contributions of $Ca^{2+}$ and $Mg^{2+}$. Kerschbaum is dominated by dolomitic formations and shows a percentage contribution to the total EC of $Ca^{2+}$ and $Mg^{2+}$ of 30.2% and 20.2%, respectively. Baget, as a limestone system, shows a lower contribution of $Mg^{2+}$, which is equal to 6.2%. At the Kerschbaum and Las Hountas springs, $SO_4^{2-}$ is responsible for 2% and 6.3% of the total EC, respectively. The larger contribution of $SO_4^{2-}$ observed at the Las Hountas spring results from gypsum dissolution and carbonate dissolution by sulfuric acid (Ulloa-Cedamanos et al. 2020). The contribution of $NO_3^-$ is approximately the same for both the springs, i.e. 1.1%. As typical for dolomite and limestone systems, $Na^+$, $Cl^-$ and $K^+$ barely contribute to the total EC. In the case of Kerschbaum, $Na^+$ and $Cl^-$ show an increase in contribution during the winter season, most likely due to the salt flushed from the streets.

## 3.2 Computed Electrical Conductivity (EC)

Figure 2 shows the continuous observed EC (solid line) together with the low-resolution computed EC (dashed line), which is calculated on Phreeqc from the given ionic measurements as the sum of the contribution of each element present as free ion and as aqueous complexes. For both the study areas, Phreeqc slightly underestimates the observed EC, with a mean relative error (−) of 0.02 and 0.035 for Kerschbaum and Las Hountas, respectively. Due to the low temporal resolution of hydrochemical data in Kerschbaum, the computed EC does not capture the dynamic of the system (Fig. 2a). For example, no water samples were collected in Kerschbaum in December 2018 and January 2019 and thus, the computed EC does not represent the observed high peak. Considering the Las Hountas dataset with a higher temporal resolution of hydrochemical measurements, the computed EC time series better matches the dynamics of the system (Fig. 2b). However, changes in EC at a lower temporal resolution than the hydrochemical measurements cannot be matched by linear interpolation, e.g. the drop in EC on October 7th.

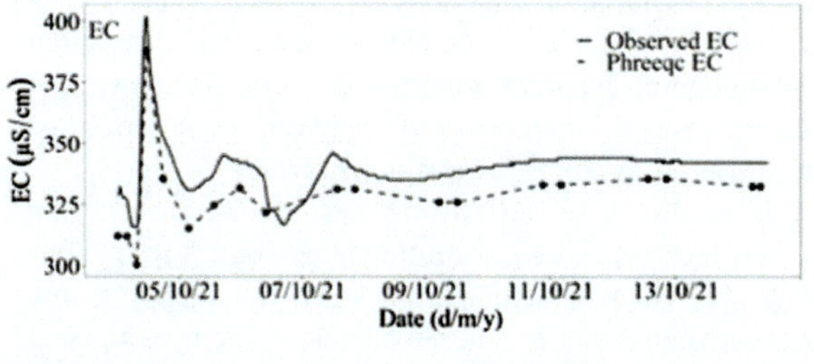


**Fig. 2** Continuous observed EC (μS/cm) (solid line) and the EC computed on Phreeqc (dashed line) at the same temporal resolution as the ionic measurements. **a** Kerschbaum spring for the period 10/2018–10/2019. **b** Las Hountas spring for the period 10/4/2021–10/14/202

## 3.3 Computed Ion Concentrations at High Frequency

The calculation of the high-resolution ion concentration time series is based on the linear interpolation of the weight factors $f_i$ (−) (Eqs. (8, 9)), rather than on the direct interpolation of the low-resolution ionic measurements. Figure 3 shows the resulting concentration time series at high frequency (black line) together with the point measurements (red points) for the Kerschbaum (a) and Las Hountas (b) springs. For Las Hountas, we show the computed high-resolution time series for the event occurred from 10/4/2021 to 10/9/2021, without showing the results related to the baseflow conditions (10/10/2021–10/14/2021), during which no significant changes in ion concentrations are observed. The results show more high-frequency fluctuations than the direct linear interpolation. The high-resolution ion concentrations are obtained by multiplying the linearly interpolated weight factors $f_i$ (−) to the continuously observed total EC. In this way, the computed concentration time series are, at each time step, calculated based on the actual total ion content, rather than being function of only the two closest ionic measurements.

## 4 Conclusions

From continuous time series of EC and point ionic measurements, the proposed methodology makes it possible to obtain major ion concentrations at the same temporal resolution as the EC signal. Due to the large ionic content and complex water speciation of the discharge of karst springs, complexation processes cannot be neglected. Thus, the concentration of each element is calculated as the sum of that

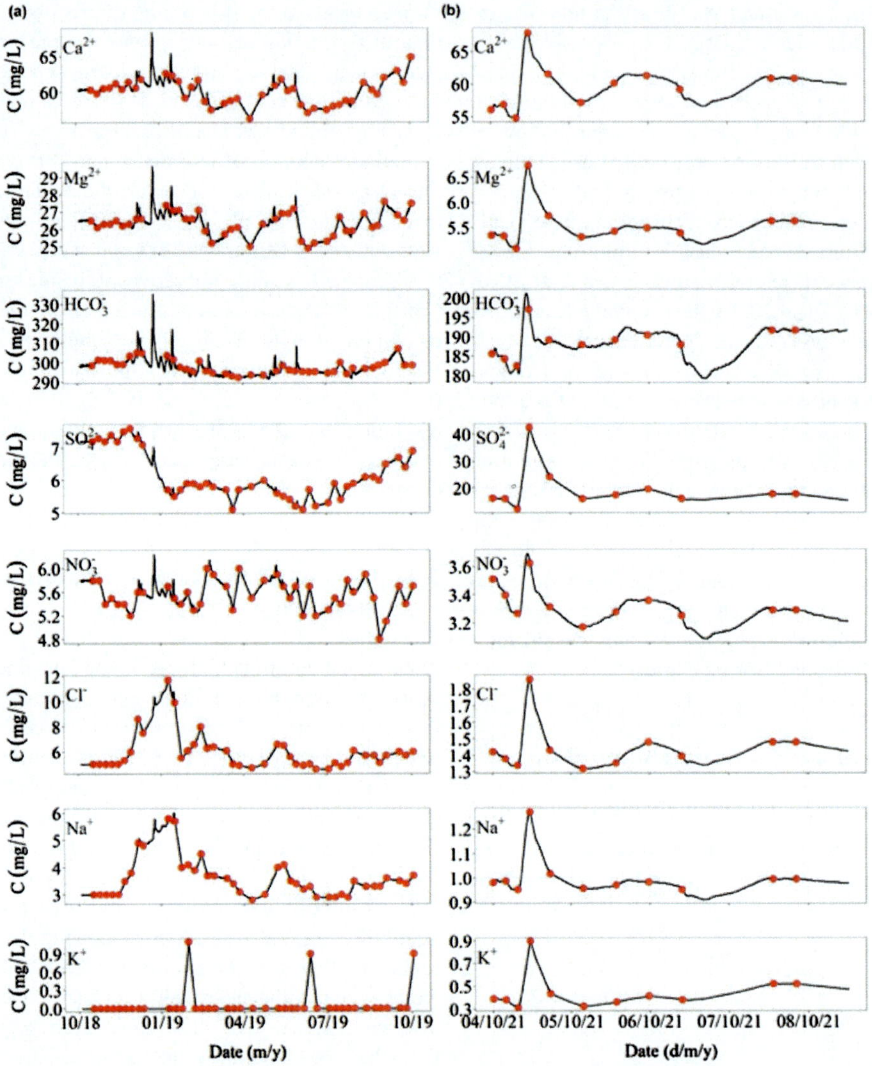


**Fig. 3** High-resolution individual ion concentrations *C* (mg/L) computed with the EC decomposition method (black line) together with the observed ionic measurements (red points) for **a** the Kerschbaum spring and **b** the Las Hountas spring

element present as free ion and as ion involved in aqueous complexes. The concentration of each element as free ion is computed based on its contribution to the total observed EC (Benettin and van Breukelen 2017), whereas its concentration involved in aqueous complexes with speciation calculation as the difference between the total molality of an element and the molality of the solution.

Our results highlight that the EC signal of karst springs can be used to obtain major ion concentrations at high temporal resolutions, when both the elements present as free ions and as ions involved in aqueous complexes are considered. Indeed, the neglection of complexation processes would lead to an underestimation of the total concentrations of $SO_4^{2-}$, $Ca^{2+}$, $Mg^{2+}$ and $HCO_3^-$. The uncertainties related to each computed individual ion need to be estimated to test whether the presented method can be used for the prediction of the temporal dynamics of the major ions. Thus, we will further investigate the correlation between the contribution of each ion to the total EC and its uncertainty.

The possibility to retrieve high-resolution concentration time series of major ions, for different geological formations and resolution of the available datasets, can support future works on karst systems conceptualization. The investigation of the proper temporal resolution, required to apply this method to different karst formations, can additionally guide fieldworks and sampling campaigns. We will further study how the different hydrological behaviors of dolostone and limestone karst systems influence the application of the method and the adequate temporal resolution of the input data.

## References


Aquilina L, Ladouche B, Dörfliger N (2006) Water storage and transfer in the epikarst of karstic systems during high flow periods. J Hydrol 327(3–4):472–485. https://doi.org/10.1016/j.jhydrol.2005.11.054

Benettin P, Van Breukelen BM (2017) Decomposing the bulk electrical conductivity of streamflow to recover individual solute concentrations at high frequency. Enviro Sci Technol 4(12):518–522. https://doi.org/10.1021/acs.estlett.7b00472

Bittner D, Richieri B, Chiogna G (2021) Unraveling the time-dependent relevance of input model uncertainties for a lumped hydrologic model of a pre-alpine karst system. Hydrogeol J 29:2363–2379. https://doi.org/10.1007/s10040-021-02377-1

Cano-Paoli K, Chiogna G, Bellin A (2019) Convenient use of electrical conductivity measurements to investigate hydrological processes in Alpine headwaters Sci. Total Environ 685:37–49. https://doi.org/10.1016/j.scitotenv.2019.05.166

Chang Y, Hartmann A, Liu L, Jiang G, Wu J (2021) Identifying more realistic model structures by electrical conductivity observations of the karst spring. Water Resour Res 57 e2020WR028587. https://doi.org/10.1029/2020WR028587

Charlier JB, Bertrand C, Mudry J (2012) Conceptual hydrogeological model of flow and transport of dissolved organic carbon in a small Jura karst system. J Hydrol 460:52–64. https://doi.org/10.1016/j.jhydrol.2012.06.043

Dreiss SJ (1989) Regional scale transport in a Karst Aquifer: 1. Component separation of spring flow hydrographs. Water Resour Res 25(1): 117–125. https://doi.org/10.1029/wr025i001p00117

Hartmann A, Wagener T, Rimmer A, Lange J, Brielmann H, Weiler M (2013a) Testing the realism of model structures to identify karst system processes using water quality and quantity signatures. Water Resour Res 49:3345–3358. https://doi.org/10.1002/wrcr.20229

Hartmann A, Weiler M, Wagener T, Lange J, Kralik M, Humer F et al (2013b) Process-based karst modeling to relate hydrodynamic and hydrochemical characteristics to system properties. Hydrol Earth Syst Sci 17(8):3305–3321. https://doi.org/10.5194/hess-17-3305-2013

Hartmann A, Barberá JA, Andreo B (2017) On the value of water quality data and informative flow states in karst modelling. Hydrol Earth Syst Sci 21(12):5971–5985. https://doi.org/10.5194/hess-21-5971-2017

Husic A, Fox J, Adams E, Ford W, Agouridis C, Currens J et al (2019) Nitrate pathways, processes, and timing in an agricultural karst system: Development and application of a numerical model. Water Resour Res 55:2079–2103. https://doi.org/10.1029/2018wr023703

Jourde H, Massei N, Mazzilli N, Binet S, Batiot-Guilhe C, Labat D et al (2018) SNO KARST: a French network of observatories for the multidisciplinary study of critical zone processes in karst watersheds and aquifers. Vadose Zone J 17(1):1–18. https://doi.org/10.2136/vzj2018.04.0094

Liu Z, Groves C, Yuan D, Meiman J (2004) South China Karst aquifer storm-scale hydrochemistry. Ground Water 42(4):491–499. https://doi.org/10.1111/j.1745-6584.2004.tb02617.x

Meus P, Moureaux P, Gailliez S, Flament J, Delloye F, Nix P (2014) In situ monitoring of karst springs in Wallonia (southern Belgium). Environ Earth Sci 71(2):533–541. https://doi.org/10.1007/s12665-013-2760-x

Mudarra M, Andreo B (2011) Relative importance of the saturated and the unsaturated zones in the hydrogeological functioning of karst aquifers: the case of Alta Cadena (Southern Spain). J Hydrol 397(3–4):263–280. https://doi.org/10.1016/j.jhydrol.2010.12.005

Narany TS, Bittner D, Disse M et al (2019) Spatial and temporal variability in hydrochemistry of a small-scale dolomite karst environment. Environ Earth Sci 78:273. https://doi.org/10.1007/s12665-019-8276-2

Parkhurst DL, Appelo CAJ (2013) Description of input and examples for PHREEQC version 3-A computer program for speciation, batch-reaction, one-dimensional transport, and inverse geochemical calculations. U.S. Geological Survey Techniques and Methods, book 6, chap. A43, 497 p. http://pubs.usgs.gov/tm/06/a43/

Perrin J, Jeannin P-Y, Zwahlen F (2003) Epikarst storage in a karst aquifer: A conceptual model based on isotopic data, Milandre test site. Switzerland J Hydrol 279(1–4):106–124. https://doi.org/10.1016/s0022-1694(03)00171-9

Ravbar N, Engelhardt I, Goldscheider N (2011) Anomalous behavior of specific electrical conductivity at a karst spring induced by variable catchment boundaries: The case of the Podstenjšek spring. Slovenia Hydrol Process 25(13):2130–2140. https://doi.org/10.1002/hyp.7966

Torre San F, Fabbri P, Piccinini L, Dalla Libera N, Pola M, Zampieri D (2020) Defining the hydrogeological behavior of karst springs through an integrated analysis: a case study in the Berici Mountains area (Vicenza, NE Italy). Hydrogeol J 28:1229–1247. https://doi.org/10.1007/s10040-020-02122-0

Ulloa-Cedamanos F, Probst J-L, Binet S, Camboulive T, Payre-Suc V, Pautot C, Bakalowicz M, Beranger S, Probst A (2020) A forty-year karstic critical zone survey (Baget Catchment, Pyrenees-France): lithologic and hydroclimatic controls on seasonal and inter-annual variations of stream water chemical composition, pCO2, and carbonate equilibrium. Water 12(5):1227. https://doi.org/10.3390/w12051227

Zhang Z, Chen X, Cheng Q, Soulsby C (2019) Storage dynamics, hydrological connectivity and flux ages in a karst catchment: conceptual modeling using stable isotopes. Hydrol Earth Syst Sci 23(1):51–71. https://doi.org/10.5194/hess-23-51-2019

# Hydrogeological Characterization and Modeling at Two Test Sites of the Apulian Karst (Southern Italy)

I. S. Liso, C. Cherubini, and M. Parise

**Abstract**

Differently from the classical hydrogeological methods applied for porous or fractured aquifers, karst groundwater needs specific adaptation in the investigating techniques. Monitoring and surveying actions, performed both at the surface and within the subsoil, have to consider the optimal way in order to properly characterize karst environment. In this sense, the caves offer an unvaluable opportunity for scientists. Taking advantage from the possibility to explore two cave systems of Apulia (Southern Italy) and to reach the water table by means of speleological techniques, geological and hydrogeological surveys have been performed at the surface and within the cave systems. All available data have been used for the first attempt to implement numerical and computational models, to predict groundwater flow and solute transport in the Apulian karst. Numerical modeling is a powerful tool to understand the hydrodynamics of karst systems. The final aim of this research, still ongoing, is to integrate and use surface and subsurface geological data in order to obtain more realistic model outputs.

**Keywords**

Hydrogeology • Karst • Cave • Modeling • Apulia

## 1 Introduction

Karst environments around the globe host precious freshwater resources for humans, as one quarter of the global population is completely or partially relies on drinking water from karst aquifers (Hartmann et al. 2014; Chen et al. 2017a, b; Stevanović 2018; Goldscheider et al. 2020). The geological and hydrogeological characterization certainly represents a necessary starting point to protect and safeguard groundwater, to build conceptual models (Castany 1985; Gunn 2007; Parise et al. 2018), and to implement the numerical and computational ones. Uncertainties in the results of the simulations, compared to the heterogeneity of the natural system, are definitely greater in karst than in porous aquifers (Kovacs and Sauter 2007). The physical heterogeneity and anisotropy of karst groundwater, including its triple porosity (porous matrix, fractures and conduits) requires specific adaptation in the investigating techniques, which need to be different from the classical hydrogeological methods applied for porous or fractured aquifers.

The collection of geological and hydrogeological field data and high temporal resolutions monitoring (ideally, continuous), both at the surface and directly within the subsoil, can be considered the keys to properly characterize a karst environment. In this sense, the presence of caves, and the direct access inside the karst systems, represents an unvaluable opportunity for scientists, especially when the cave allows cavers to arrive at water table. Building a karst conceptual model, using all available data and in line with them, is an essential step in understanding the hydrodynamics and hydrology and represents the basis of numerical model implementation, a powerful tool in predicting the groundwater behavior in karst. Taking advantage from the possibility to explore two cave systems in Apulia (Southern Italy) and to reach the water table by means of speleological techniques, geological and hydrogeological surveys have been performed both at the surface and within these cave systems. The final aim of this research, still ongoing, is to

C. Cherubini
Department of Physics and Earth Sciences, University of Ferrara, Via Saragat 1, 44122 Ferrara, Italy

I. S. Liso (✉) · M. Parise
Department of Earth and Environmental Sciences, University Aldo Moro, Via Edoardo Orabona, 4, 70125 Bari, Italy
e-mail: i.serena.liso@gmail.com


B. Andreo et al. (eds.), *EuroKarst 2022, Málaga*, Advances in Karst Science,
https://doi.org/10.1007/978-3-031-16879-6_15

integrate surface and subsurface data in order to obtain more realistic model outputs.

## 2 Site Description

Apulia (S Italy) is an almost entirely karst region surrounded by the Adriatic and Ionian Seas for most of its extent. The Apulian karst districts are: Gargano (the promontory to the north), the Murgia plateau (center Apulia), and the Salento peninsula to the south (Fig. 1).

The development of karst processes in Gargano and Murge started before than in Salento, where karst phenomena appeared only during Quaternary. The study will focus on Murge and Salento zones, where the two only Apulian caves through which it is possible to directly reach the water table are located.

From a hydrogeological standpoint, in the Murge the recharge generally occurs through a series of endorheic basins and dolines, hosting a large number of swallow holes (Zumpano et al. 2019; Pisano et al. 2020). The hydraulic head reaches the highest values in correspondence of the hydrogeological watershed (Maggiore and Pagliarulo 2004), and rapidly decreases toward the sea. Along the coast, some springs may be encountered, most of them being submarine.

In the Salento peninsula, freshwater floats as a lens above the saltwater, forming a wedge that extends inland from the shoreline (Cotecchia 1955; 2014; Zorzi and Reina 1955; Margiotta and Negri 2004), so that the Adriatic and Ionian seas are hydraulically connected. The hydrogeological watershed drains groundwater toward the Adriatic and the Ionian seas and its position is not fixed, but it moves in function of recharge and abstraction from wells. In Salento, karst springs are mostly located along the Adriatic coast, the only exceptions being the Santa Cesarea Terme springs (D'Angeli et al. 2021) and the Chidro and Boraco on the Ionian side (Liso and Parise 2020).

Murge is subdivided into High (or NW) Murge, the upper portion of the plateau, and Low (or SE) Murge, the portion of the karst plateau closer to the sea. The most important karst landform in Low Murge is definitely the Canale di Pirro polje, a tectono-karst valley with flat bottom, 12 km long (Parise 2006; Pisano et al. 2020). It reaches on its eastern side the Murgia escarpment, which is the main NW–SE tectonic line separating inland Murge from the Adriatic coastline. As typical of poljes in other karst regions of the world (Gams 1978, 2005; Lopez Chicano et al. 2002; Gracia et al. 2003; Nicod 2003; Bonacci 2004; Milanovic 2004), it often presents flooded sectors, becoming a temporary lake after severe storms (Dogan 2003; Gracia et al. 2003), due to the hydraulic functioning of the swallow holes at its bottom. At Canale di Pirro polje, the two main swallow holes are: the Gravaglione and the Inghiottitoio di Masseria Rotolo Cave, the deepest cave in the Region (Rotolo Cave in the following; Parise and Benedetto 2018; Liso et al. 2020; Parise et al. 2020), objects of this study. It opens at 300 m a.s.l., presents large underground environments, reaches groundwater at about 260 m from the surface, and so far, it has been explored down to a depth of −324 (−60 m below the groundwater surface, thanks to diving explorations). Inghiottitoio di Masseria Rotolo cave is totally developed within Cretaceous limestones.

Moving farther southward, the Salento peninsula is separated by the Low (SE) Murge through a tectonic structure located between Brindisi and Taranto, named *Soglia Messapica* (Messapian Threshold). In this area, dolines and swallow holes are definitely the most widespread karst features (Bruno et al. 2008; Del Prete et al. 2010; Festa et al. 2012; Pepe and Parise 2014; Sansò et al. 2015; Margiotta et al. 2016, 2021). In the center of the peninsula, the second cave object of this study (Vora Bosco) opens at 64 m a.s.l. within an E-W doline. Vora Bosco essentially develops within Miocene and Plio-Pleistocene calcarenites, with only the last 10 m in limestone; it reaches the water table about 60 m below the topographic surface.

## 3 Methods

The two studied caves are very different in terms of (1) development, (2) space and geometry of underground voids and (3) geological environments; this led us to choose appropriate investigation methods and techniques, adapting them to the karst setting and, also, to the specific geological, geomorphological and hydrological features of the sites.

Research actions have been carried out at both the surface and within the cave systems, in order to fully characterize the karst environment. At the surface, a karst-geomorphological map was built by means of aerial photograph interpretation using digital stereoscopic view (Pisano et al. 2020). Surveying of the rock mass discontinuities was performed following the classical procedure of structural geology, at the available outcrops located within the polje. Further, the preferential directions of the main karst systems in Low Murge were investigated through analyses of the cave maps. Rainfall time series were collected through the regional rain-gauge network website, managed by the Civil Protection Authority (https://protezionecivile.puglia.it/). In addition, in October 2017, a new meteorological station was installed about 2 km from the Rotolo Cave entrance, and is still working. Isotope, chemical-physical and microbiological parameters of the aquifer have been assessed for samples from the caves, but also from a network of private wells within the study areas. Furthermore, bio-speleological surveys have been carried out to evaluate the ecological status of groundwater and, more in general, of the underground environment. Within the cave

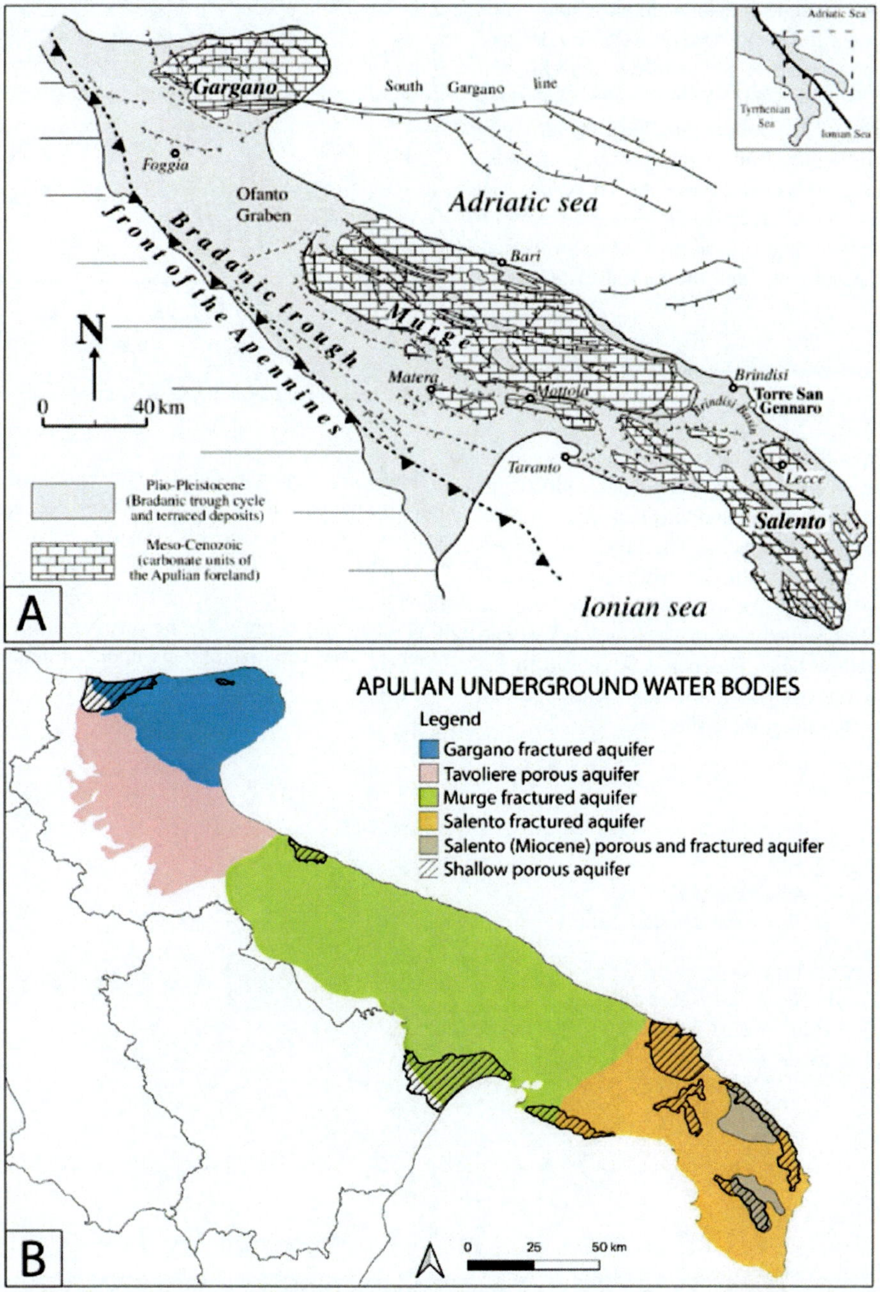


**Fig. 1** A: geological sketch of Apulia (modified after Pieri et al. 1997) showing (1) Bradanic Trough sediments and terraced deposits (Pliocene–Pleistocene); (2) carbonate units of the Apulian Foreland (Mesozoic-Cenozoic). The three main karst areas of the region (Gargano, Murge, and Salento), as described in Parise (2011), are also shown; B: schematic representation of Apulian underground water bodies (original drawing, data from Tiziano Project available at www.sit.puglia.it)

systems, other activities have been performed, including the topographic survey, and a characterization of the stratigraphic succession of the rock mass by means of thin sections analyses. In addition, we collected data series on the cave air relative humidity and temperature, and on groundwater level, temperature and electrical conductivity.

All collected data have been used to implement two different numerical models, in function of the available data and the specific hydrogeological features, at each study site. In the Low Murge area a discrete fracture (DF) model has been chosen to simulate the groundwater flow, deriving the statistics on fracture data from pumping-well tests and geological field surveys. The model simulates the groundwater flow velocity, direction and its preferential pathways. At Vora Bosco, a lumped model has been chosen to simulate functioning of the karst aquifer. In detail, a simple reservoir model has been built with the time series of water level data, derived directly from the cave, and the rainfall data (Leins et al. 2022).

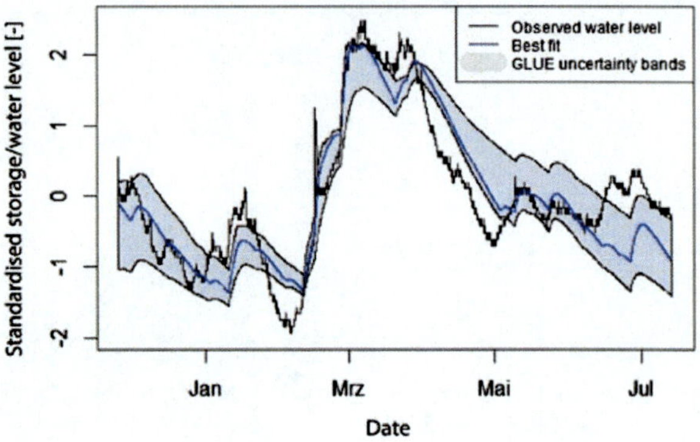


**Fig. 3** Vora Bosco model output (after Leins et al. 2022)

## 4 Results

The final model output at the Canale di Pirro polje is shown in Fig. 2, where the arrows represent the water velocity in terms of magnitude and direction, and the black straight lines are the preferential water flowpaths. The latter are toward NE, from the polje to the Adriatic Sea, with the mean flow velocity of 34.5 m/d. Away from the domain borders, the arrow tips are mainly oriented toward the Rotolo Cave and the Gravaglione swallow hole. This could be related to the large underground voids hosted immediately below the two main absorption points within the polje.

The Vora Bosco model did not reproduce entirely some quick dynamics of the aquifer system, especially its steep recessions (Fig. 3), probably because of the short time series of available data. However, it was able to describe the general trend of the observed water level. The calibrated model parameters proved to be very sensitive, consistent and reliable. The very simplified model structure therefore is well-suited, even for the short time series available at the cave.

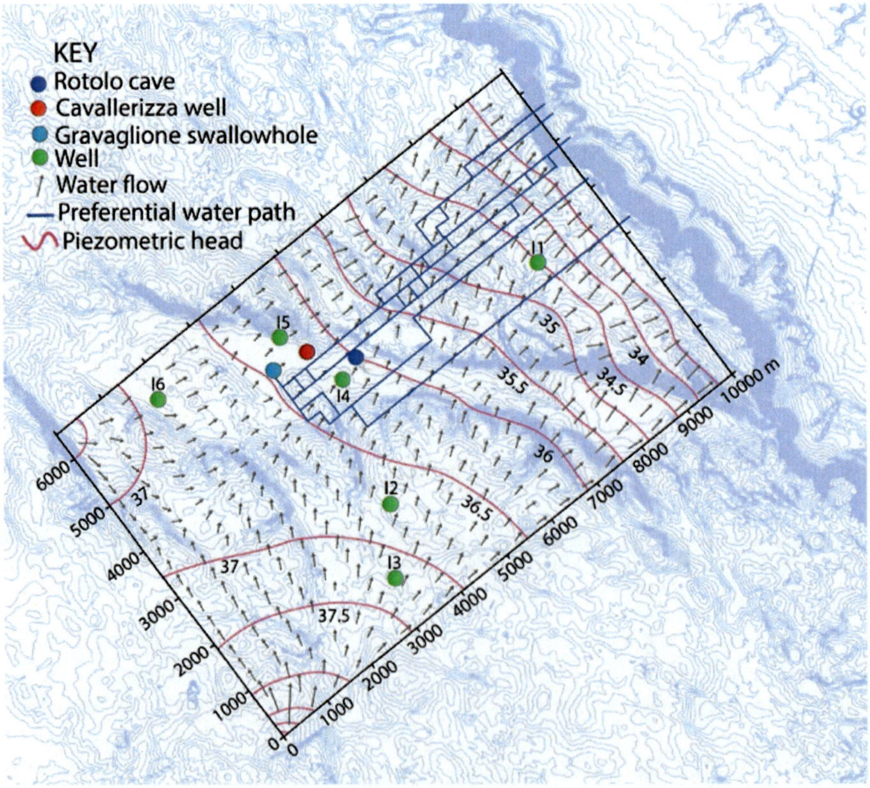


**Fig. 2** Inghiottitoio di Masseria Rotolo cave model output and 5 m elevation contour map, realized by University Aldo Moro of Bari and the CNR-IRSA

## 5 Discussion

The two study areas show highly different surface and underground karst features, in terms of lithologies, speleogenesis and hydrology. Although both caves open within topographically depressed areas, the articulated and complex development of the underground passages at Rotolo Cave does not allow direct connection between the surface and the aquifer, which, on the other hand, characterizes the Vora Bosco environment, with the consequent complete and immediate flooding of the cave, after intense or prolonged rainfall. This condition, together with the limited depth of the karst system, justifies the fact that the chemical and microbiological groundwater parameters recorded at Vora Bosco show a higher pollution, which seem to be absent at Rotolo Cave, at least based upon the data so far available.

The model indicates that there is only a very small storage function of the soil in the system, which results in generating recharge even after very small precipitation events.

The different behavior is further testified by the data trend recorded by the air micro-climatic sensors: these are about stable at Rotolo Cave, being not excessively influenced by the external weather conditions at all monitored sites, while they show a clear influence from the outside at Vora Bosco. The same considerations apply to the data recorded by the multi-parameter groundwater probe. In particular, the Rotolo water level is in agreement with the seasonal fluctuations of the Murge aquifer; occasionally, as a consequence of significant rainstorms, the water level rises, and this is likely due, at least partly, to the proximity with the Gravaglione swallow hole. The groundwater data series, recorded by the multi-parameter probe installed in Vora Bosco (time: 11 months) do not allow to make considerations on the seasonal groundwater trend. As regards water temperature, this is stable at Rotolo Cave, while at Vora Bosco it records sudden and immediate decreases in the groundwater temperature (and in electrical conductivity as well), as a result of intense rainfall events.

The two applied models have been selected in function of the above explained differences among the study areas. The Rotolo cave model (layered discrete fracture) allows to assess a preliminary value of underground water velocity and the principal groundwater flow direction. On the other hand, the Vora Bosco model (simple lumped) provides outputs on the hydrogeological parameters of the karst system, as soil and epikarst storage values. So far, both models are not able to represent precisely the fast hydraulic processes occurring in such karst settings. The models could be improved, including within the simulations some other karst processes and data, derived from surveys carried out both within the caves and at the surface. At this goal, it could be helpful to implement other monitoring actions as groundwater isotope studies, and/or tracer and infiltrometer tests in order to identify the main direction of groundwater flow, and be able to define the boundaries of the hydrogeological catchment and the water resident times.

**Acknowledgements** This research was performed by the first Author during her PhD at University Aldo Moro of Bari, and was partly carried out within the framework of *Protocollo d'intesa con Regione Puglia per l'attuazione dell'art. 45 "Interventi per esplorazione dei fenomeni carsici," comma 1 della L.R. n. 45 del 30/12/2013.*

## References

Bonacci O (2004) Poljes. In: Gunn J (ed), Encyclopedia of caves and karst science. Fitzroy Dearborn, pp 599–600

Bruno E, Calcaterra D, Parise M (2008) Development and morphometry of sinkholes in coastal plains of Apulia, southern Italy. Preliminary Sinkhole Susceptibility Assess Engng Geology 99:198–209

Castany G (1985) Idrogeologia. Principi e metodi. Dario Flaccovio (ed) 88–7758–036–4.

Chen Z, Auler AS, Bakalowicz M, Drew D, Griger F, Hartmann J, Jiang G, Moosdorf N, Richts A, Stevanović Z, Veni G, Goldscheider (2017a) The world karst aquifer mapping project: concept, mapping procedure and map of Europe. Hydrogeol J 25(3):771–785

Chen Z, Goldscheider N, Auler AS, Bakalowicz M, Broda S, Drew D, Hartmann J, Jiang G, Moosdorf N, Richts A, Stevanović Z, Veni G, Dumont A, Aureli A, Clos P, Krombholz M (2017b) World karst aquifer map (WHYMAP WOKAM). BGR, IAH, KIT, UNESCO

Cotecchia V (2014) Le acque sotterranee e l'intrusione marina in Puglia: dalla ricerca all'emergenza nella salvaguardia della risorsa. Mem Descr Carta Geol It 92:1–416

Cotecchia V (1955) Influenza dell'acqua marina sulle falde acquifere in zone costiere, con particolare riferimento alle ricerche d'acqua sotterranea in Puglia. Geotecnica 3

D'Angeli IM, De Waele J, Fiorucci A, Vigna B, Bernasconi SM, Florea LJ, Liso IS, Parise M (2021) Hydrogeology and geochemistry of the sulfur karst springs at Santa Cesarea Terme (Apulia, southern Italy). Hydrogeol J 29:481–498

Del Prete S, Iovine G, Parise M, Santo A (2010) Origin and distribution of different types of sinkholes in the plain areas of Southern Italy. Geodin Acta 23(1/3):113–127

Dogan U (2003) Sariot Polje, central Taurus (Turkey): a border polje developed at the contact of karstic and non-karstic lithologies. Cave and Karst Sci 30(3):117–124

Festa V, Fiore A, Parise M, Siniscalchi A (2012) Sinkhole evolution in the Apulian karst of southern Italy: a case study, with some considerations on sinkhole hazards. J Cave Karst St 74(2):137–147

Gams I (2005) Tectonic impacts on poljes and minor basins (case studies of Dinaric Karst). Acta Carsologica 34(1):25–41

Gams I (1978) The polje: the problem of definition. Zeitschrift fur Geomorphologie N.F. 22(2):170–181

Goldscheider N, Chen Z, Auler AS, Bakalowicz M, Broda S, Drew D, Hartmann J, Jiang G, Moosdorf N, Stevanović Z, Veni G (2020) Global distribution of carbonate rocks and karst water resources. Hydrogeol J 28:1661–1677

Gracia FJ, Gutierrez F, Gutierrez M (2003) The Jiloca karst polje—tectonic graben (Iberian Range, NE Spain). Geomorphol 52(3–4):215–231

Gunn J (2007) Contributory area definition for groundwater source protection and hazard mitigation in carbonate aquifer. In: Parise M and Gunn J (eds). Natural and anthropogenic hazards in karst areas: recognition, analyses and mitigation. Geological Society of London, London, sp. Publ. vol 279, pp 97–109
Hartmann A, Goldscheider N, Wagener T, Lange J, Weiler M (2014) Karst water resources in a changing world: review of hydrological modeling approaches. Rev Geoph 52:218–242
Kovacs A, Sauter M (2007) Modeling karst hydrodynamics. In: Goldscheider N, Drew D (eds) Methods in karst hydrogeology. International association of hydrogeologists, 26. Taylor and Francis group, London UK
Leins T, Liso IS, Parise M, Hartmann A (2022) Contrasting predicted changes of groundwater availability and prediction uncertainty in a data scarce karstic setting. In preparation
Liso IS, Parise M (2020) Apulian Karst Springs: A Review. J Environ Sci Eng Technol 8:63–83
Liso IS, Chieco M, Fiore A, Pisano L, Parise M (2020) Underground geosites and caving speleotourism: some considerations, from a case study in Southern Italy. Geoherit 12:13
Lopez Chicano M, Calvache ML, Martin-Rosales W, Gisbert J (2002) Conditioning factors in flooding of karstic poljes—the case of Zafarraya Polje (S Spain). CATENA 49:331–352
Maggiore M, Pagliarulo P (2004) Circolazione idrica ed equilibri idrogeologici negli acquiferi della Puglia. Geologi e Territorio 1: 13–35
Margiotta S, Negri S, Parise M, Quarta TAM (2016) Karst geosites at risk of collapse: the sinkholes at Nociglia (Apulia, SE Italy). Environ Earth Sci 75(1):1–10
Margiotta S, Marini G, Fay S, D'Onghia FM, Liso IS, Parise M, Pinna M (2021) Hydro-stratigraphic conditions and human activity leading to development of a sinkhole cluster in a Mediterranean water ecosystem. Hydrology 8:111
Margiotta S, Negri S (2004) Alla ricerca dell'acqua perduta. Congedo Editore, Galatina. vol 191
Milanovic P (2004) Dinaride poljes. In: Gunn J (ed) Encyclopedia of caves and karst science. Fitzroy Dearborn, New York, pp 291–293
Nicod J (2003) A little contribution to the karst terminology: special or aberrant cases of poljes? Acta Carsologica 32(2):29–39
Parise M (2006) Geomorphology of the Canale di Pirro karst polje (Apulia, southern Italy). Zeitschrift Für Geomorphol NF 147:143–158
Parise M (2011) Surface and subsurface karst geomorphology in the Murge (Apulia, southern Italy). Acta Carsologica 40(1):79–93
Parise M, Benedetto L, Chieco M, Fiore A, Lacarbonara M, Liso IS, Masciopinto C, Pisano L, Riccio A, Vurro M (2020) First outcomes of a project dedicated to monitoring groundwater resources in Apulia, southern Italy. In: Bertrand C, Denimal S, Steinmann M, Renard P (eds) Eurokarst 2018, Besançon, advances in the hydrogeology of karst and carbonate reservoirs. Springer, Advances in Karst Science, pp 243–249
Parise M, Benedetto L (2018) Surface landforms and speleological investigation for a better understanding of karst hydrogeological processes: a history of research in southeastern Italy. In: Parise M, Gabrovsek F, Kaufmann G, Ravbar N (eds) Advances in karst research: theory, fieldwork and applications. Geological Society, London, sp. publ. vol 466, pp 137–153
Parise M, Gabrovsek F, Kaufmann G, Ravbar N (2018) Recent advances in karst research: from theory to fieldwork and applications. In: Parise M, Gabrovsek F, Kaufmann G, Ravbar N (Eds.), Advances in Karst Research: Theory, Fieldwork and Applications. Geological Soc London, sp publ 466, pp. 1–24.
Pepe M, Parise M (2014) Structural control on development of karst landscape in the Salento Peninsula (Apulia, SE Italy). Acta Carsologica 43(1):101–114
Pieri P, Festa V, Moretti M, Tropeano M (1997) Quaternary tectonic activity of the Murge area (Apulian foreland, southern Italy). Ann Geofisica 40(5):1395–1404
Pisano L, Zumpano V, Liso IS, Parise M (2020) Geomorphological and structural characterization of the "Canale di Pirro" polje, Apulia (Southern Italy). J Maps 16(2):479–487
Sansò P, Margiotta S, Mastronuzzi G, Vitale A (2015) The geological heritage of Salento Leccese (Apulia, southern Italy). Geoheritage 7:85–101
Stevanović Z (2018) Global distribution and use of water from karst aquifers. In: Parise M, Gabrovsek F, Kaufmann G, Ravbar N (eds) Advances in karst research: theory, fieldwork and applications. Geological Soc London, sp publ. vol 466, pp 217–236
Zorzi L, Reina C (1955) Sulla necessità di controllare e disciplinare le utilizzazioni di acque sotterranee nella Penisola Salentina. In: 8th Conv. Naz. Ingegneri Italiani, Milano
Zumpano V, Pisano L, Parise M (2019) An integrated framework to identify and analyze karst sinkholes. Geomorphology 332:213–225

# Prediction of Future Interactions Between Karst and River Regarding to Climate Change Based on IPCC Scenarios: Application to a Mediterranean French River Basin (Cèze)

Y. Pascoletti, H. Chapuis, F. Paran, J. Jolivet, E. Van Den Broeck, and D. Graillot

**Abstract**

Groundwater in karst aquifers is a valuable resource in Mediterranean region which is particularly threatened by climate change. Climate change induces an increase in evapotranspiration and a decrease in rainfall available for groundwater resource. This impacts directly karst aquifer recharge and thus indirectly karst-river interactions. The study site is a karstic catchment located in the Mediterranean French river basin and known as the "Gorges of the Cèze River". This site has been studied through interdisciplinary research since 2014 to improve understanding on hydrogeological functioning of this catchment. The river at different points and each of its karstic springs have been equipped with probes and studied by multiple metrics (geochemistry, biology, radioactivity…). As a continuation of these researches, this paper presents an attempt to model the interactions between the whole karstic catchment of the Gorges and the Cèze river and then to predict their evolution regarding to climate change scenarios. The model has been developed using KarstMod software, which consists in an adjustable modelling platform. Firstly, the model has been calibrated on one hydrological cycle and then validated on the following cycle. After that, future climate dataset was simulated by the ALADIN model, according to Intergovernmental Panel on Climate Change (IPCC) Representative Concentration Pathway (RCP) scenario data, including the study area in the above-named model in order to predict karst and river interactions and water production for the period 2030–2100. The results of this work are presented in terms of volume of water produced for three simulations corresponding to three RCP scenarios (2.6, 4.5, and 8.5).

**Keywords**

Climate change • Groundwater resources • Mediterranean • Karst aquifers • Modeling • IPCC

## 1 Introduction

Groundwater in karst aquifers is a precious resource representing 25% of domestic water consumed in Mediterranean region (Margat 1998). Climate change is jeopardizing this resource by provoking more frequent and intense droughts (Cramer et al. 2018). Understanding groundwater flows, recharge, and discharge processes in karst aquifers is crucial to enable sustainable water management (Bakalowicz 2004). Groundwater flow modeling is recognized as a major tool for water management purpose. It can be achieved using lumped models which consist in a rather simple parameter-based approach to reproduce the hydrogeological functioning of an aquifer (Pascoletti 2018). The functioning of karstic springs has already been successfully simulated using lumped models (Fleury 2005). Nevertheless, modeling the water production of a whole and complex karstic catchment is still limited to few attempts. Furthermore, the hydrogeological modelling of a karstic catchment has still not been used to simulate the impacts of climate change. This paper presents an attempt to model the karstic water discharge along a whole section of the Cèze River and to predict its quantitative evolution according to IPCC's climate scenarios using KarstMod.

Y. Pascoletti (✉) · H. Chapuis · F. Paran · D. Graillot
Mines Saint Etienne—Procédés Pour L'Environnement Et Géoressources (PEG), Saint-Étienne, France
e-mail: yvan.pascoletti@hotmail.fr

Y. Pascoletti · H. Chapuis · F. Paran · D. Graillot
CNRS UMR Environnement Ville Société 5600, Lyon, France

J. Jolivet
CNRS UMR Espace 7300, Avignon, France

E. Van Den Broeck
GASOIL (Groupement Associatif de Spéléologues d'Orgnac, Issirac Et Labastide), Orgnac-l'Aven, France



B. Andreo et al. (eds.), *EuroKarst 2022, Málaga*, Advances in Karst Science,
https://doi.org/10.1007/978-3-031-16879-6_16

## 2 Material and Methods

### 2.1 Study Area

The study area (Fig. 1) belongs to a large karstic catchment located in the South of France. It consists in a calcareous plateau incised by the Cèze River, a tributary of the Rhône River, which length is 128.3 km and average flow 22 $m^3/s$. This site is, since 2014, the subject of a multidisciplinary research (Chapuis 2017) which aims to characterize exchanges between Cèze River and its aquifer (Pascoletti and Chapuis 2022) and to meet a territorial policy promoting sustainable water management in Mediterranean context. Geological studies suggest interactions between river and Lower Cretaceous, Barremian and Lower Aptian (so called Urgonian) formations which constitute a highly karstified calcareous geological unit. This aquifer is mainly recharged by rainfall in spite of some inflows from river losses. Rainfall is characterized by a dry period from June to September with an average of 33 mm/month and a wet period from October to December with an average of 88 mm/month. The discharge areas correspond to the karstic springs located along the river and contributing to its flow.

### 2.2 Data and Software

The model has been developed using KarstMod software which can reproduce the structure of conceptual lumped models of karst systems (Mazzilli et al. 2019). The hydrological data used to calibrate and verify the model presented below are provided by the French Ministry of Environment "Banque Hydro". Two stations, in Tharaux and in La Roque-sur-Cèze were chosen to encompass the studied catchment (Fig. 1). The average daily flow measured by these stations during two hydrological cycles (14/10/2013 to 13/10/2015) were thus retrieved. Meteorological data produced by the French weather service "Météo France" have been used as inputs to drive the model. The station located in Montclus has been chosen because of its central geographic position (Fig. 1). To simulate future karst hydrology, data produced by the National Centre for Meteorological Research (CNRM) according to IPCC works' (2014) were included in the model. These data proceed from meteorological simulations realized in 2014 from three RCP scenarios (2.6, 4.5 and 8.5). These data were produced at regional scale (12 km mesh). The coordinates of the chosen point on the grid (44.27090, 4.38973) corresponds precisely to our study site.

### 2.3 Model Calibration and Validation

From the above-described meteorological dataset, only the precipitation data has been injected into the model. The evapotranspiration served to determine a fitted sine function (regression coefficient: 0.84) to be used in the model (Fig. 2). Using a sine function rather than raw data seems to offer better results (Chapuis 2017).

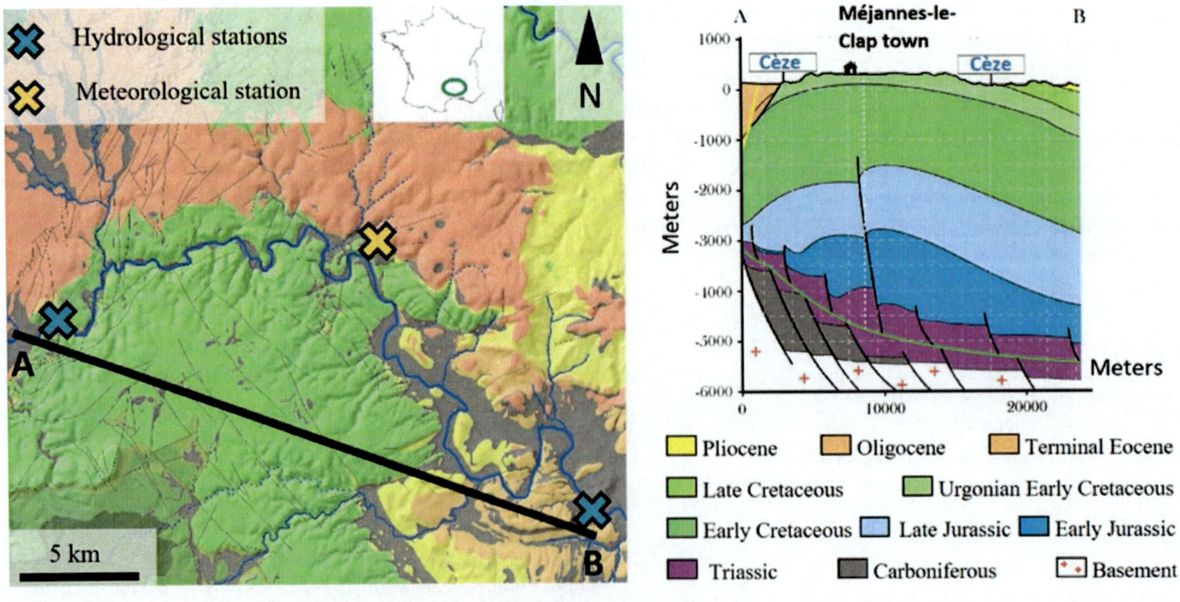


**Fig. 1** Location map of the study area (modified from Chapuis 2017)

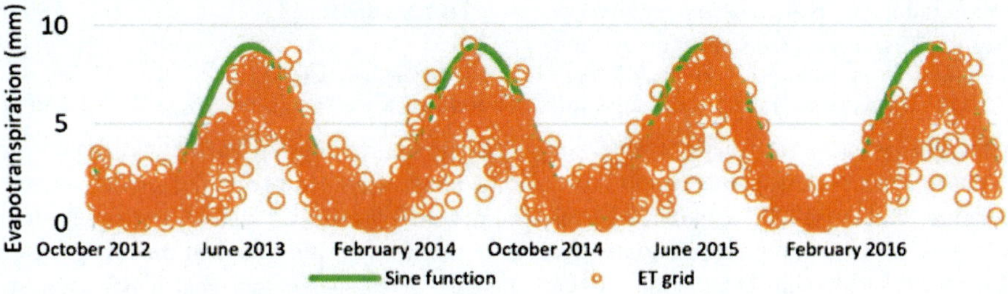


**Fig. 2** Fitted sine function for evapotranspiration

The hydrological data were used to calibrate the model which aims to simulate karst water net production, that is to say the total inflow from karst left in the river in the end of the studied section. The method used to evaluate it, for each day, consists in subtracting downstream flows (measured at Tharaux) from upstream flows, (measured at La Roque-sur-Cèze). Flows thus obtained comprehend a proportion of superficial water considered as minimal because Cèze tributaries in the studied catchment are mainly intermittent. These flows also include a few negative values due to river losses or measurement errors. Those were replaced by "NOINTERP" in the input file so that KarstMod does not take them into account. Moreover, the obtained data contained an odd value briskly reaching 163 $m^3/s$ on the September 13, 2015 while the flow of the previous day was −104 $m^3/s$. Both original flows (at Tharaux and at La-Roque-sur-Cèze) on that day were in fact estimated and not properly measured. Thus, this odd value has also been replaced by "NOINTERP".

The model structure (Fig. 3) includes three compartments and seven discharges. $Q_S$ is a sum of four outflows, two from the main compartment (*E*) and two others from two subcompartments (*M* and *C*). $Q_{EM}$ and $Q_{EC}$ are two different flows transferred from *E* to *M* and *C*, respectively. $Q_{loss}$ corresponds to lost outflows which not contribute to the final discharge.

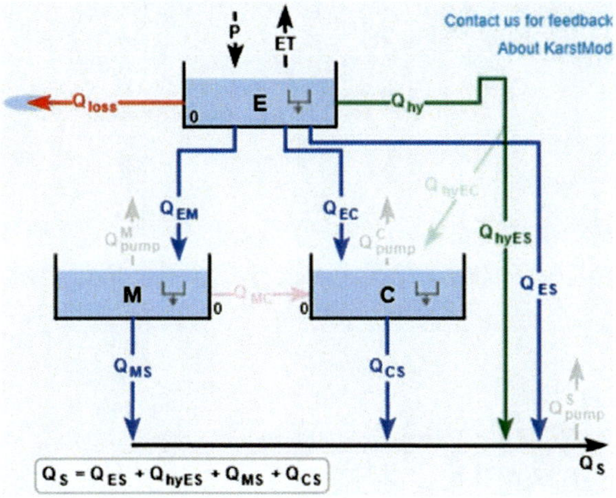


**Fig. 3** Model structure

The Nash–Sutcliffe Efficiency coefficient (NSE) has been set at 0.8 directly in the software, so that KarstMod only seeks for models above 0.8. The calibration period has been defined to span the first 365 days and the validation period the next 365 days. A maximal value of 600 $km^2$ for the recharge area has been defined and the other parameters have been left unsettled. KarstMod has then been launched and successfully determined a set of parameters allowing the model to meet the predefined performance objective.

## 2.4 Simulation of Future Karst-River Interactions

The calibrated and validated model has then been used to simulate net water production of the Gorges of the Cèze for 2030–2100 period considering three RCP scenarios. Three input files containing precipitation and evapotranspiration data for each scenario were constituted. While precipitation data (rain added up to snow) can be found directly in the CNRM dataset, evapotranspiration has to be calculated. It was done using the other available parameters and with Abtew (1996) method:

$$\mathrm{ET0} = \mathrm{K} \cdot \frac{R_s}{\lambda}$$

$R_s$ solar radiation expressed in $MJ \cdot m^{-2} \cdot d^{-1}$
$\lambda$ latent heat of vaporization equals to 2.45 $MJ \cdot kg^{-1}$
K dimensionless coefficient

In comparison with other methods, Abtew equation gives satisficing results (Xu and Singh 2000; Ndiaye et al. 2017). Besides, it only uses few meteorological parameters. Solar radiation ($R_s$) is also known as global radiation, which is the sum of direct shortwave radiation from the sun and diffuse radiation (Oudin 2004). Thus, the so called surface downwelling shortwave radiation in CNRM dataset can be used.

The latent heat of vaporization (λ) is the energy required to change water from liquid to vapor in a constant pressure and constant temperature process. The value of the latent heat varies as a function of temperature. As $\lambda$ varies only slightly over normal temperature ranges, a single value of 2.45 MJ $kg^{-1}$ can be used, which corresponds to the latent heat for an air temperature of about 20 °C. Finally, $K$ has to be determined in order to calibrate evapotranspiration to the data presented in Fig. 3. To make the calculation of the evapotranspiration closer to reality and to be coherent with CNRM simulations, solar radiation parameter proceeding from these has been used. Data from RCP 8.5 scenario have been chosen, because it seems to correspond to the late climate evolution (Schwalm et al. 2020). The calculated evapotranspiration fits the sine function when $K = 0.65$ (regression coefficient: 0.78).

# 3 Results

The model offers a rather close reproduction of karst net water production for two hydrological cycles (Fig. 4). It reaches an NSE value of 0.812 and 0.804 for calibration and validation periods and estimates a 593 $km^2$ recharge area.

Then, this model has been used to simulate karst net water production for the years to come. To analyze the data produced by the simulations, a graph comparing the cumulative volumes of water produced by the studied catchment according to the three different RCP scenarios has been drawn (Fig. 5).

# 4 Discussion

## 4.1 Interpretation

The model reproduces karstic net water production in a satisficing way. However, the model seems to have difficulties to emulate all peaks discharges, which is a very common flaw in hydrological modeling. As an example, in January 2014, when two peaks, both followed by a brisk recession, occurred, the model only simulates a concentration curve growing progressively. But in general, whilst simulated peaks are lower, the recession occurs more slowly. So, KarstMod smooths the curves and stretches the flows, probably to fit statistical laws. KarstMod also tends to give a regular aspect to the curves, whereas some curves' shapes proceeding from the original data seems odd, like the one in November 2014 which consists in a sudden growth and decrease.

Simulating future karstic hydrology with KarstMod produces results in terms of flows and volumes. For interpretation purposes, cumulative volumes are more relevant as they help to cast light on possible trends regarding to climate change. It seems that the curves presented in Fig. 5 are in correlation with RCP trends, because the more the scenario is pessimistic the less the karst produces water. Whilst water production curves for RCP 2.6 and RCP 4.5 scenarios are close, the one corresponding to RCP 8.5 scenario appears to be very distinct. RCP 8.5 is indeed a very pessimistic scenario for climate evolution and considering Fig. 5, it also seems to be very pessimistic for future karstic groundwater production and interactions between karst and river in terms

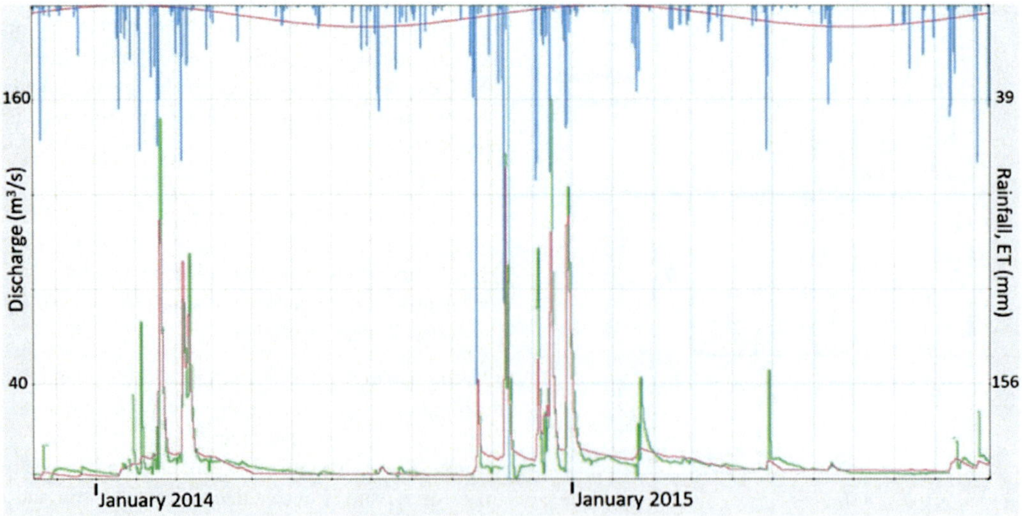


**Fig. 4** Observed flow in green versus simulated flow in pink

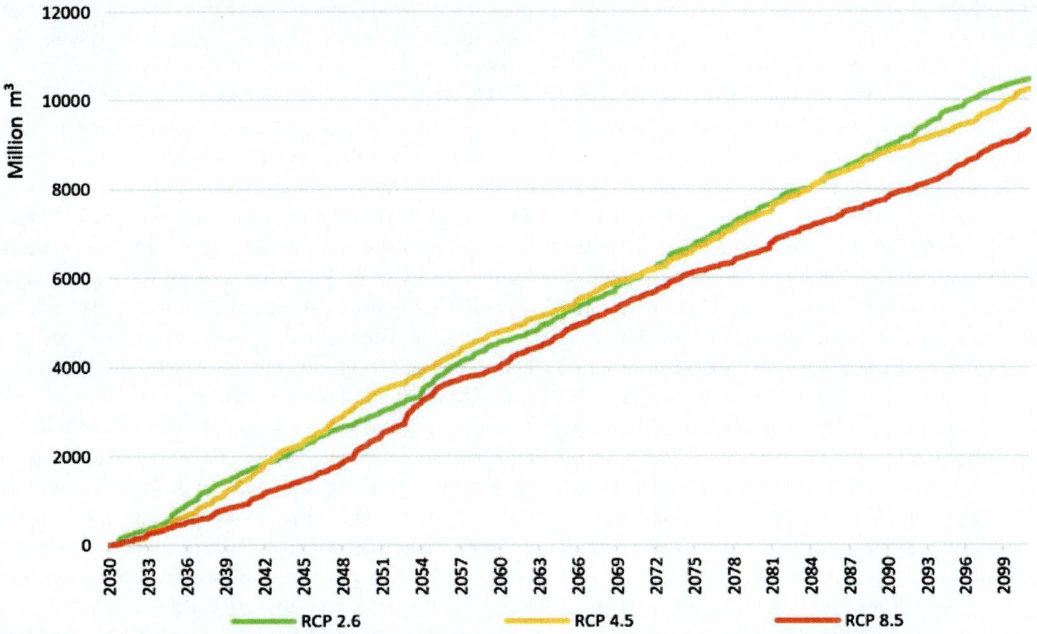


**Fig. 5** Cumulative volumes for each RCP scenarios

of flow. At the end of the simulation period, around 1 billion $m^3$ of water less has been produced by the catchment in the case of RCP 8.5 scenario compared to RCP 2.6 scenario. This is, on average, somewhat equivalent to an annual loss of about 14 million $m^3$ of water, which represents around 7% of the amount of water produced by the studied catchment during the hydrological cycle used for calibration step (2013–2014).

## 4.2 Limits

Such a modeling attempt necessarily implies some limits which have to be mentioned. Firstly, calibrating the model helps determine the best way to simulate the hydrological functioning of the studied catchment for a limited period of time, then, this functioning is used for the whole period 2030–2100, but really karst hydrological functioning may evolve because it's a dynamic process. Another common pitfall when modeling is the calculation of the recharge area. To meet the performance objectives, the calibration step may lead to an overestimated recharge area. Here it has been estimated at 593 $km^2$, a reasonable value although somewhat elevated. KarstMod determines this value during calibration step within a range of value set by the user. But, it is noticeable that the more the range is low, the less the performance objectives are easily met. Hydrological and meteorological data used to develop the model also encourage to be cautious because of uncertainty in measurement. Obviously, there are among hydrological data some surprising values that may be due to measurement errors.

Furthermore, some of them are not measured but estimated. The proposed methodology to determine karstic water production also implies limits, especially through the existence of negative flows, which can proceed from measurement errors or from important losses in the river that can occur "randomly". This makes the modeling approach even more complicated. The presented methodology also implies that karstic net water production contains a part of surface water, here considered as minimal but still existing. Concerning the meteorological data, evapotranspiration calculation is a frequent factor of uncertainty because it depends on a lot of other meteorological parameters. For the prediction part of this work, Abtew method has been selected because it seems trustworthy regarding to other works and can be applied even if few meteorological parameters are available for evapotranspiration calculation. But, this method is one among other existing methods. This work is also based on IPCC and CNRM climate models. But, it is noticeable that uncertainty also exists in climate projections, which clearly appears when comparing different models (Ly et al. 2019).

In sum, trying to predict future contribution of karstic flows remains an ambitious undertaking exposed to several uncertainties, even if the methodology presented here is coherent and even if the model proves to be reliable considering its performances.

### 4.3 Perspectives

The work presented here enables several perspectives to gain more in-depth knowledge of the subject. It would be interesting to observe how the karstic water production of the catchment evolves during the years to come to compare this evolution to the projections made in this paper and to adjust the model if necessary. It would also be interesting to improve the knowledge of the functioning of the studied catchment by extending the calibration period of the model. This would be an opportunity to understand karstic hydrological processes in a more dynamic way. For that purpose, it would be useful to evaluate the part of superficial water in the karstic net water production calculated here and to subtract it in order to gain a better appreciation of the exact groundwater outflows.

Furthermore, the methodology that has been followed throughout the study can easily be applied to another karstic catchment. As an example, it could easily be used for a study site like the Ardèche River and its karstic catchment that shares the same climate regime and the same geological context, even if its hydrological scale is larger.

Finally, the model constructed for the research presented here rest upon a triplet of models: RCP/GCM/RCM (respectively: Representative Concentration Pathway, General Circulation Model, and Regional Circulation Model). But different types of GCM and RCM models are available. As an example, simulations obtained from climate models developed by the Pierre-Simon-Laplace Institute (IPSL) or by the Swedish Meteorological and Hydrological Institute are also available. Using these simulations as input data and comparing the results with those presented here could be interesting. Moreover, climate models are constantly updated. New simulations, provided by French laboratories of climate simulations are available since 2020. The latest IPCC report is also available and will probably be translated soon in new climate simulations adapted to French regional scales.

## 5 Conclusion

The model developed with KarstMod is reproducing karstic water net production of the studied area, which is the sum of flows provided by the karst to the Cèze River and remaining in it at the end of the studied section. The structure of the model is rather simple, because it comprehends only a few compartments and flow exchanges. But its performance is satisficing as a NSE value of 0.812 and 0.804 is reached for calibration and validation periods each spanning one hydrological cycles. Then the model has been used to simulate the future karstic water net production by using CNRM meteorological simulations at regional scale which rest upon three RCP scenarios from the IPCC. It appears that the hydrological trends presented in this paper are in correlation with RCP trends. In fact, the more the scenario is pessimistic, climate wise, the less the karst produces water. Whilst water production evolution for RCP 2.6 and RCP 4.5 scenarios looks similar, the one corresponding to RCP 8.5 scenario is very distinct. RCP 8.5 appears to be very pessimistic for future karstic groundwater production and interactions between karst and river in terms of flow. RCP 8.5 scenario represents around 1 billion $m^3$ of water less in comparison with RCP 2.6 scenario for the whole simulation period. That being said, one should focus on trends rather than such an estimation because trying to number the evolution of karstic production of water remains very challenging and submitted to uncertainties.

Still, the knowledge acquired through this study aroused the interest of water stakeholder. Thereby, the model developed is also expected to be a pedagogical tool to help water management policy, considering that the model presented here has the advantage of focusing on a regional scale. Climate projections and observation of hydrological trends could be a way to anticipate and adapt to change, which could be decisive in a sensitive area like the Mediterranean region.

## References

Abtew W (1996) Evapotranspiration measurement and modeling for three wetland systems in S. Florida. Water Resour Bull 32(3):465–473

Bakalowicz M (2004) Karst groundwater: a challenge for new resources. Hydrogeol J 13:148–160. https://doi.org/10.1007/s10040-004-0402-9

Chapuis H (2017) Caractérisation, évaluation, modélisation des échanges entre aquifères karstiques et rivières: application à la Cèze (Gard, France). Autre. Université de Lyon. Français. ⟨NNT: 2017LYSEM026⟩

Cramer W, Guiot J, Fader M et al (2018) Climate change and interconnected risks to sustainable development in the Mediterranean. Nature Clim Change 8:972–980. https://doi.org/10.1038/s41558-018-0299-2

Fleury P (2005) Sources sous-marines et aquifères karstiques côtiers méditerranéens. Fonctionnement et caractérisation. Sciences de l'environnement. Université Pierre et Marie Curie—Paris VI. Français

IPCC (2014) Climate change 2014: synthesis report. Contribution of working groups I, II and III to the fifth assessment report of the intergovernmental panel on climate change [Core writing team, Pachauri RK, Meyer LA (eds)]. IPCC, Geneva, Switzerland, p 151

Ly M, Segnon AC, D'Haen S, Totin E, Noblet M, Camara I, Pfleidere P (2019) Comprendre et interpréter les sorties des modèles climatiques pour la conduite des études de vulnérabilité: guide à l'attention des praticiens. Climate analytics gGmbH, Berlin

Margat J (1998) Les eaux souterraines dans le bassin méditerranéen. Ressources et utilisations. Documents BRGM 282. BRGM, Orleans, France

Mazzilli N, Guinot V, Jourde H, Lecoq N, Labat D et al (2019) KarstMod: a modelling platform for rainfall—discharge analysis and modelling dedicated to karst systems. Environ Model Softw 122

Ndiaye PM, Bodian A, Diop L, Djaman K (2017) Évaluation de vingt méthodes d'estimation de l'évapotranspiration journalière de référence au Burkina Faso. Physio-Géo 11(1):129–146

Oudin L (2004) Recherche d'un modèle d'évapotranspiration potentielle pertinent comme entrée d'un modèle pluie-débit global. Sciences of the Universe [physics]. ENGREF (AgroParisTech)

Pascoletti Y (2018) Modélisation du fonctionnement hydrologique de la tourbière du Col du Luitel (Isère): recherches menées dans le cadre de l'étude du rôle hydrologique et hydrogéologique des zones humides de tête de bassin versant dans le soutien d'étiage (projet ZABR/AE-RMC). Génie des procédés

Pascoletti Y, Chapuis H (2022) Connaître et quantifier les échanges hydrauliques karst/rivière, Recommandations méthodologiques, Retour d'expérience d'études menées sur les gorges de la Cèze (30)

Schwalm CR, Glendon S, Duffy PB (2020) RCP 8.5 tracks cumulative $CO_2$ emissions. Proc Natl Acad Sci USA

Xu CY, Singh VP (2000) Evaluation and generalization of radiation-based methods for calculating evaporation. Hydrol Process 14:339–349

# A Smart Analytical and Numerical Interpretation of Injection Tests in Unsaturated, Fractured and Karstified Carbonate Reservoirs

C. Danquigny, J. Coqueret, G. Massonnat, P. Léonide, M. Barbier, L. Dal Soglio, and J. L. Lesueur

**Abstract**

The hydraulic conductivity of karst medium largely varies through space and with the observation scale. Here, we investigate the petrophysical properties of carbonates from centimetric to metric scale. Several boreholes a few meters apart were cored and exhaustively characterized. Petrophysical measurements were performed on rock samples, while packer injection tests investigated some metric intervals of the unsaturated medium surrounding the wells. These hydraulic tests are interpreted using both analytical solutions and numerical models. Results show that discontinuities unsampled at centimetric scale induce a dual or even multi-media behavior at metric scale. The tests allow characterizing the permeability of both media.

**Keywords**

Karst • Permeability • Scale • Matrix • Fracture • Hydraulic tests

## 1 Introduction

In karst systems more than in other reservoirs, the hydraulic conductivity largely depends on the observation scale and can largely vary from one place to another (e.g., Jazayeri Noushabadi et al. 2011). Fractures and karst conduits strongly and very locally augment the medium heterogeneity, giving the karst its specificities. In most cases, these elements are highly permeable and increase the hydraulic conductivity as both the scale of observation and their relative volume increase. Their network structure and connectivity can also considerably increase the size or even question the existence of a representative elementary volume, i.e., a scale at which equivalent homogeneous properties as permeability can be defined.

These issues often hamper simulation of flow physics in these media. Physically based gridded flow models, which are useful management tools usually applied to other aquifers, are rarely applied to karst aquifers (Hartmann et al. 2014). Indeed, such models raise the delicate question in karst environment of the homogenization of properties at the mesh scale and that of taking into account the networks of fractures and karst conduits and their flow physics. Addressing such questions require multi-scale characterization of the permeability field (e.g., Galvao et al. 2016).

The mesh size of these models, generally between metric and hectometric, is also the scale at which few measurements are available. While laboratory measurements are sub-metric, well tests or springs hydrographs investigate large areas of several thousand squared meters or even several squared kilometers. When hydraulic tests nevertheless target small volumes, potentially smaller than the representative elementary volume, they are tricky to interpret: the geometry and volume of the area investigated are uncertain, the assumptions associated with the analytical solutions used for interpretation may not be met, and in general, the results depend largely on the location of the measurement relative to the highly permeable features (Jazayeri Noushabadi et al. 2011; Maréchal et al. 2014). The scarcity of data and the lack of multi-scale measurements to address multi-scale heterogeneity often leads to erroneous extrapolations of the properties measured at wells.

C. Danquigny · J. Coqueret · G. Massonnat · J. L. Lesueur
TotalEnergies, Centre Scientifique et Technique Jean Féger (CSTJF), 64000 Pau, France

C. Danquigny (✉)
Avignon Université, UMR EMMAH, 84916 Avignon, France
e-mail: charles.danquigny@univ-avignon.fr

P. Léonide
Aix-Marseille Université, CNRS, IRD, CEREGE UM 34, 13331 Marseille, France

M. Barbier ·
L.Dal Soglio
Modis, 4 rue Jules Ferry, 64000 Pau, France


B. Andreo et al. (eds.), *EuroKarst 2022, Málaga*, Advances in Karst Science,
https://doi.org/10.1007/978-3-031-16879-6_17

This study is part of a wider research project (e.g., Danquigny et al. 2019b) dedicated to the understanding and modeling of fluid dynamics at various scales of a vast karstified carbonate aquifer with respect to its multiphase geological history. Here, we are interested in the characterization of the permeability of different carbonate rock types (facies, texture, structural features) from centimetric to metric scale and the relation with their geological characteristics. Several boreholes were cored and exhaustively characterized, which revealed the complex heterogeneity of the rock at various scales, the result of an equally complex, multiphase geological history involving sedimentological, diagenetic and tectonic processes (Cochard et al. 2020, 2021). Petrophysical measurements were performed on rock samples taken from the cores, while water injection tests between packers investigated some metric intervals of the medium surrounding the wells.

In previous work, after establishing a robust facies model for the geological formation of interest, namely the Urgonian of southeastern France (Tendil et al. 2018; Lanteaume et al. 2020), we quantified the high intra- and inter-facies variability of petrophysical properties at small scales and compared some of these measurements with larger-scale measurements (Danquigny et al., 2019a). In this paper, we rely on a more detailed description of the tested intervals and a better interpretation of these hydraulic tests to further characterize these relationships.

Due to the location and the low matrix permeability of the targeted facies, considered hydraulic tests have two particularities: (i) they are "constant-head tests" or "constant pressure injection tests", i.e., water is injected at a constant pressure, while the decline of the injection rate is studied, (ii) they were performed in the unsaturated zone of the aquifer, thus the initial saturation of the medium surrounding the well is poorly known, varies in space and time and is affected by the experiment. This configuration raises questions concerning both the interpretation methodology and the possibility of being able to evaluate the properties of the medium relative to a variable saturation. Here, in the absence of an analytical solution adapted to the unsaturated medium, we have tested and confronted both several analytical solutions assuming the saturated medium and a numerical approach assuming the unsaturated medium. Results are faced to the small-scale measurements and to the geological description of the tested intervals.

## 2 Material and Methods

### 2.1 The Experimental Site

In the south of the Fontaine-de-Vaucluse catchment area (SE France), the LSBB is an underground research laboratory giving access to the carbonate rock along a 3.8 km long and almost horizontal tunnel, which reaches a depth of 519 m below the top of the surrounding mountain (Fig. 1). The laboratory offers a unique, highly interdisciplinary experimental setting (e.g., Danquigny et al. 2021).

There five boreholes were cored at a 280 m depth, in the so-called GAS area. Called from P1 to P5, they are spaced by a few meters only and have an average depth of 21 m. The cores were precisely described to establish lithostratigraphic logs. These logs show two main facies belonging to the stratigraphic unit Urgonian 2 (U2): porous rudist facies and fine peloid-foraminiferal facies. Porosity and permeability profiles were established thanks to numerous porosity and permeability plug measurements. The values indicate low permeability ranging from 0.01 mD to less than 10 mD and porosity varying from less than 5% to more than 25%. Well imagery allowed a comprehensive description and reporting of the discontinuities affecting the reservoir (Cochard 2018). Two intervals seem to be intensely affected by fractures, karst or stylolites. On the well P2 for instance, these intervals are located at 4–6 m depth, with poorly opened stylolithes and joints, and at 16–19 m depth, with evidence of karstification and very opened discontinuities.

### 2.2 Description of the Tested Intervals

Two main campaigns of hydraulic testing were led on the GAS boreholes. During the first one, eight intervals of 1 m thickness were tested in the P2 borehole. Numerical simulations of these tests provided some estimations of permeability (Jeanne et al. 2013). These are considered but not reinterpreted in this study.

The second campaigns targeted ten intervals of 2.48 m thickness, three intervals in the P5 borehole numbered from 1 to 3, and seven intervals in the P2 numbered from 5 to 11 (Fig. 2). Several types of tests were performed in a row on each interval, including slug tests, injection by pressure steps, and constant-head tests. Cochard (2018) provided a first estimation of hydraulic conductivity from these tests.

We synthetized these rough estimates in a previous paper, highlighting the differences between the measurements at different scales (Danquigny et al., 2019a). Here, constant-head tests and good quality constant pressure steps from the second campaign are further interpreted and simulated.

### 2.3 Hydraulic Tests

#### 2.3.1 Measurements Protocol

Constant-head or constant pressure tests consist in keeping pressure constant while injecting or pumping water into the aquifer and recording flowrate. Performed less frequently than the other methods (e.g., pumping and slug tests), these

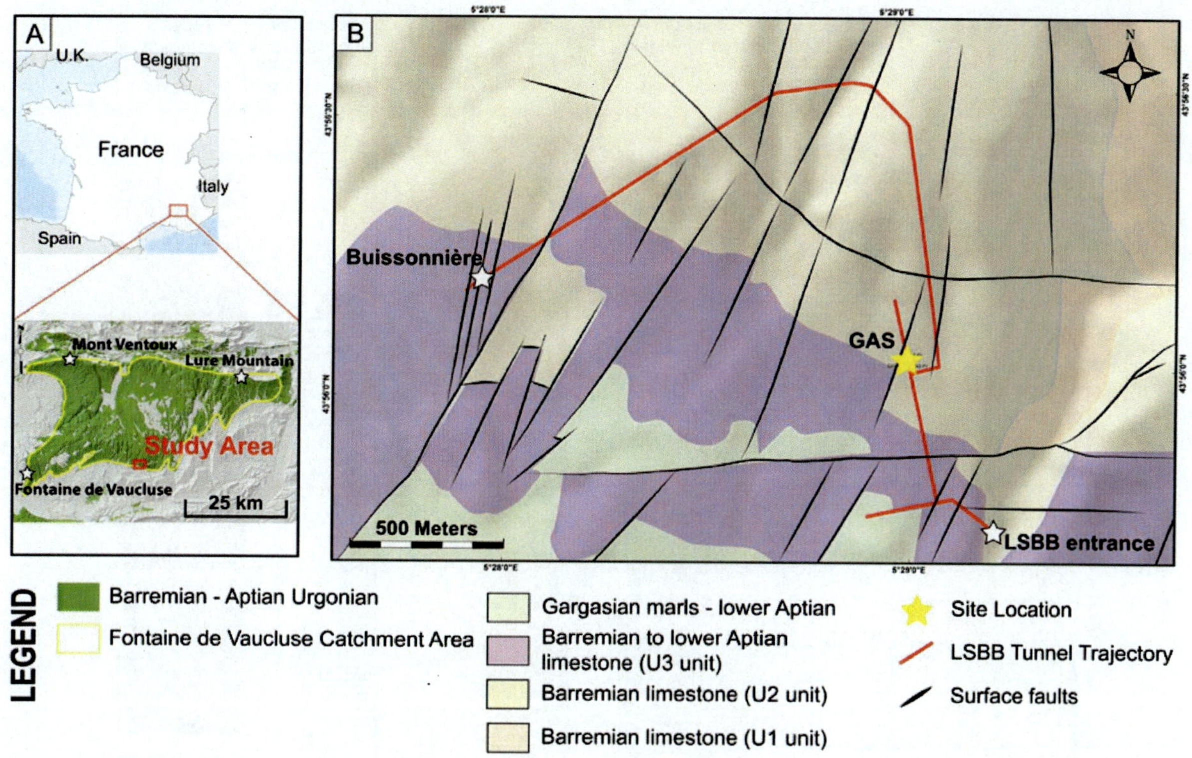


**Fig. 1** Location and geological context of the study site

tests are more suitable to low permeability formations (Mejias et al. 2009).

In the case of this study, injection tests at constant pressure were performed between inflatable packers. This system allows the isolation of sealed-off targeted sections. A probe, located between the packers, measures the pressure into the isolated interval. A pump, located at surface, controls the injection of water in the chamber and an electromagnetic flowmeter measures the flowrate at surface. The pump and the flowmeter are directly linked to an acquisition station, which allows the regulation of the flowrates (or the pressure) thanks to a control gauge.

### 2.3.2 Analytical Interpretations

Similarly to the other hydraulic testing methods, constant pressure injection tests can be interpreted using type-curve matching. However, the pressure being constant, classical analytical models cannot be applied. The first transient rate analysis for constant pressure injection tests was developed by Jacob and Lohman (1952). This model has been widely used and accepted in the literature. However, it makes serious assumptions and might oversimplify the flow in the formation. The generalized radial flow (GRF) model and its extension to dual-porosity media have generalized the interpretation of hydraulic tests to a $n$ dimension, i.e., non radial, flow around the well in a fully saturated, homogenous and isotropic reservoir (Barker 1988). Doe (1991) presented a generalized solution for constant pressure tests of the GRF model.

Other analytical solutions exist, more dedicated to particular cases and less commonly applied. However, we have not identified any solution adapted to tests in unsaturated medium. We will apply the solutions of Jacob and Lohman (1952) and Barker (1988) to the tests and evaluate their applicability to unsaturated media using numerical modeling. The detailed equations for each solution can be found in the related literature. They have been implemented in different software. For this study, we used the AQTESOLV software (Duffield 2007).

### 2.3.3 Numerical Simulations

The relevancy of analytically estimated properties is checked through a numerical approach. 3D flow models centered on the well interval tested are built in FEFLOW software (Diersch 2014). Due to the scale considered, the well radius cannot be neglected and is represented within the grid, which is locally refined. Pressure values from the tests data are used as boundary condition.

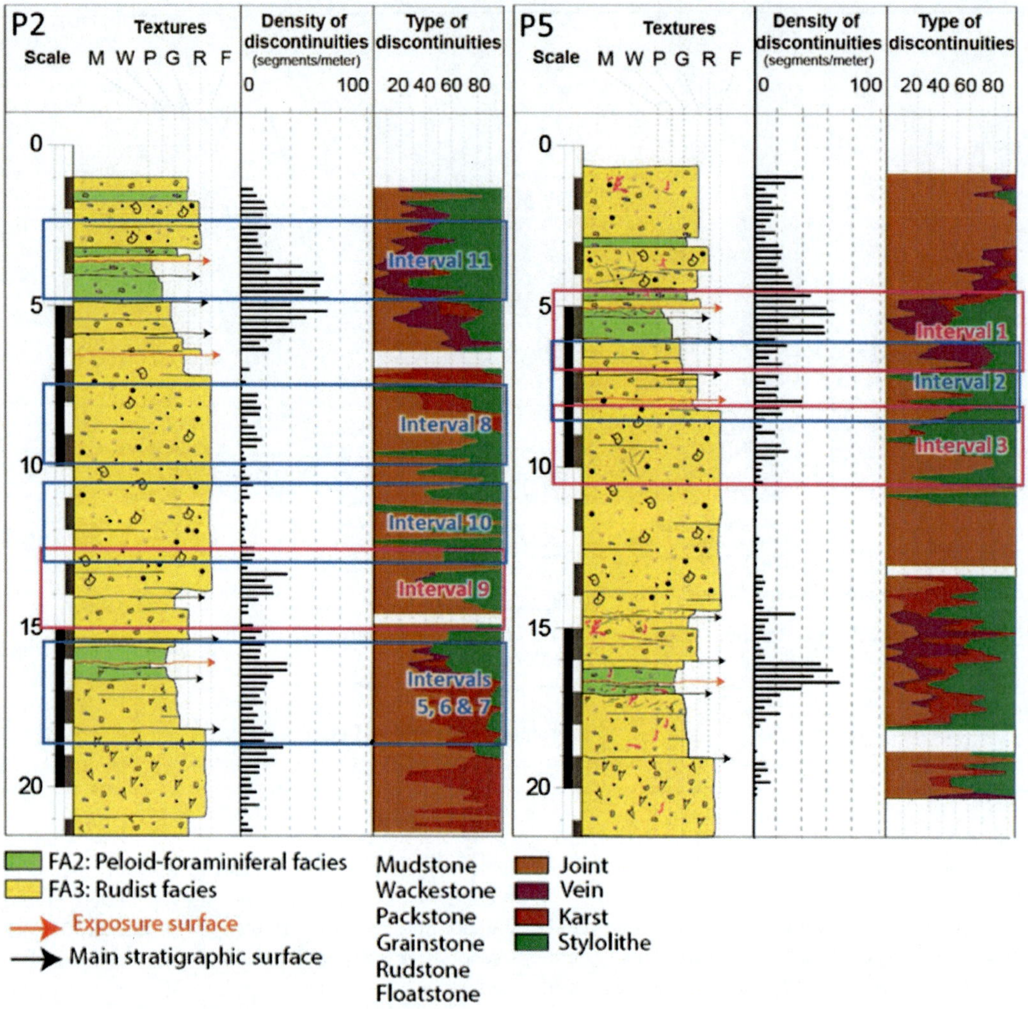


**Fig. 2** Description of the tested intervals (after Cochard et al. (2020), modified)

These models aim at reproducing the injection process, considering the unsaturated nature of the reservoir, to estimate properties numerically and compare with analytical results. To define the relationship between water saturation, relative permeability and pressure, the empirical model of Van Genuchten (1980) is applied. The two fitting parameters needed ($\alpha$ ($m^{-1}$) and $n$ (unitless)) have been rarely characterized for carbonate rock (Andriani et al. 2021). Here, we performed a sensitivity analysis around values from the literature (e.g., Kordilla et al. 2012).

In most of the intervals tested, several tests were performed successively. Thus, injected volume before the studied test is injected in the model before the simulation, to define a consistent initially saturated volume around the well. Before any prior injection, initial saturation in the reservoir is defined at 0.75, as no measurements are available. This rather high value can be justified by the depth of the well (280 m below the surface) and by numerous drips observed in the tunnel, indicating a high saturation. A sensitivity analysis around this value has been performed.

Porosity is input from plug measurements. Single-medium permeability and storage capacity are investigated parameters by trial-and-error process.

## 3 Results

### 3.1 A Systematic Dual-Medium Behavior

All tests show dual-medium type behavior on the log–log plot of the derivative. Thus, Barker's analytical solution for dual-porosity reservoirs provides the best data match. For

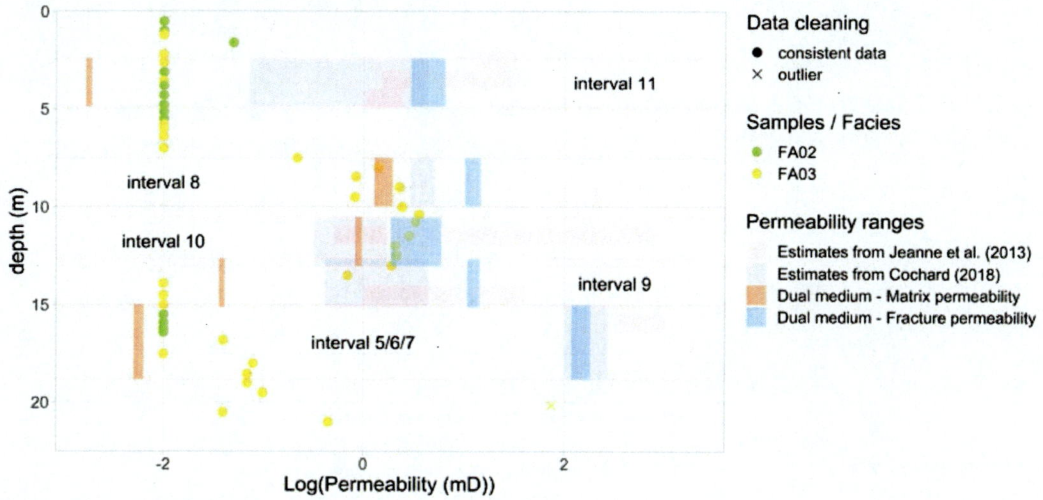


**Fig. 3** Comparison of the different permeability values in the P2 borehole. The points, colored according to the facies, represent small-scale measurements (limit of quantification = $10^{-2}$ mD). The rectangles represent the interpretations of the injection tests (the vertical extension corresponds to the thickness of the tested interval, and the horizontal extension corresponds to the permeability range deduced from the interpretation of the different tests)

the sake of usage and simplicity, the two media characterized are named matrix and fractures regardless of their geological characteristics. We will distinguish them from plug measurements. For all the tests, the matrix permeability varies between $2 \times 10^{-3}$ mD and 2 mD, the fracture permeability, between 2 and 300 mD. Figure 3 compares the permeability ranges for the set of tests processed per interval of well P2 to the previous estimates and the small-scale measurements.

In general, the dual permeability ranges are less spread out than the single-medium rough values estimated in previous studies. These latter estimated an average equivalent permeability of the whole medium. This average permeability appears to be essentially controlled by fracture permeability when matrix permeability is low, as shown in Fig. 3. However, interval 8 (P2), whose matrix permeability is both relatively high and not very different from the fracture permeability, illustrates different weights in the contribution of each medium. It is, therefore, difficult to relate geological characteristics to a single-medium permeability value estimated for a medium with multi-media behavior.

Most of the matrix permeability values are consistent with the plug measurements, validating in this case the often questioned representativeness of small-scale measurements. The interpretation of the tests nevertheless highlights permeability values of the order of $10^{-3}$ mD, lower than the limit of quantification ($10^{-2}$ mD) of the laboratory measurements, reached for intervals 1 (well P5), 5, 6, 7 and 11 (well P2). But at this level of permeability, the impact is probably very small.

Note the difference between the respective interpretations of P2 intervals 8 and 10: while the results of interval 8 show a matrix permeability of the same order of magnitude as the sample measurements and a fracture permeability of an order of magnitude higher, the results of interval 10 show a fracture permeability of the same order of magnitude as the sample measurements and a lower matrix permeability. Recall that, in the case of a dual-medium interpretation, the notions of matrix and fractures are, respectively, akin to two media of different capacity and conductivity, in most cases but not necessarily a matrix continuum of low permeability and more conductive fractures, hence the abuse of language. In the present case, no discontinuities were observed on interval 10, but multi-centimetric cemented fossils of rudists are present and purposely avoided during sampling. These cemented impermeable elements could induce this dual-medium response with a lower matrix permeability. The presence of fine discontinuities in interval 8 masks this behavior. Indeed, the fractured and karstified carbonate medium investigated is a complex multi-medium with a multi-scale heterogeneity whose dual media interpretation only identifies the two main end-members.

## 3.2 Multi-scale Petrophysical Properties According to the Geological Features

The intervals 1 (P5), 5, 6, 7 and 11 (P2), which exhibit the lowest matrix permeability ($<6 \times 10^{-3}$ mD), are characterized by a strong representation of the FA02 facies and a

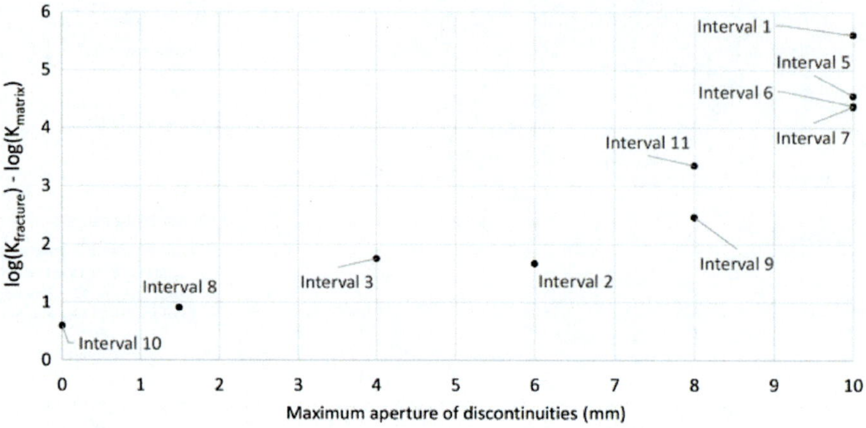


**Fig. 4** Dual-medium contrast of permeability as a function of discontinuities maximum aperture

predominantly mudstone-wackestone texture. These intervals are also those with the highest fracture permeability, up to 300 mD for interval 1 (P5), in agreement with the high density and large aperture of discontinuities observed.

On the other hand, intervals 3 (P5), 8 and 10 (P2) are characterized by rudist facies (FA03) and grainstone-rudstone textures with very few, poorly open discontinuities. They have relatively high matrix permeability, of a few mD according to sample measurements and between 0.2 and 2 mD according to test interpretation. The fracture permeability is not very contrasting compared to the matrix permeability's, ranging from 2 to 16 mD.

The permeability contrast between the two media appears to be correlated to the maximum aperture of the discontinuities quantified on well imagery (Fig. 4).

### 3.3 Insights from Numerical Simulations

Figure 5 shows the comparison between the injection rates from the simulation results and the data, for interval 6 and interval 10. Interval 6 is mainly constituted of wackestone facies FA02 and marked by several stratigraphic and tectonic discontinuities, some with centimetric aperture due to karstification. Conversely, interval 10 is made up of grainstone-rudstone rudist facies (FA03) without any discontinuities observed. A rather good match is obtained for both intervals.

For interval 10, the best match is obtained for a permeability of 3 mD, which is of the same order of magnitude as the fracture permeability values deduced from the analytical interpretation, between 2 and 6 mD. For interval 6, a permeability increasing with time, from 92 to 109 mD, provides the best match. This evolution with time is a proxy for the evolution of the equivalent permeability of the increasing saturated volume investigated by the test. This evolution is small and the values are of the same order of magnitude as the permeability of the fracture according to the analytical interpretation (126 mD). Nevertheless, it highlights the variability in permeability that may be due to the increase in saturation of the medium and/or an increase in permeability away from the well.

All sensitivity tests revealed low sensitivity of the simulations to the Van Genuchten model parameters and initial saturation. The most important parameter for the calibration of these tests is the permeability, and the values obtained analytically by applying solutions adapted to the saturated medium are consistent with the calibration values of a simulation in unsaturated medium.

## 4 Conclusions

The purpose of this study was to investigate the evolution of permeability from centimetric to metric scale for different carbonate facies. It confirms the interest of injection tests between packers to characterize the petrophysical properties of the rock at a scale closer to that of the numerical models.

While the results show the representativeness of small-scale measurements for a homogeneous matrix, they also confirm the inability of these measurements to characterize the effect on the permeability of larger discontinuities, therefore not sampled, such as cemented levels or dissolved fractures.

Moreover, the tests show that at the metric scale and for the different rock types investigated, these discontinuities induce a behavior at least dual medium that must be taken into account in large-scale models.

Finally, the numerical simulations showed that these tests were not very sensitive to the unsaturated character of the surrounding medium, perhaps due to a relatively high initial saturation. Thus, the applied analytical solutions give consistent results. On the other hand, it seems difficult to characterize the parameters of the water retention model with this

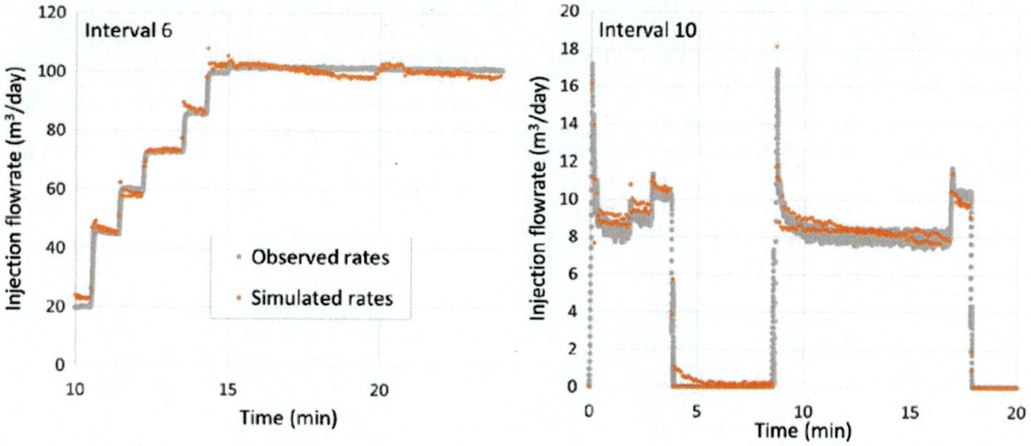


**Fig. 5** Simulation results compared to data for successive tests in interval 6 and interval 10

kind of tests. This aspect would merit further simulations in both double and unsaturated media.

**Acknowledgements** The authors thank TotalEnergies for funding this R&D project and giving permission to publish this paper.

## References


Andriani GF, Pastore N, Giasi CI, Parise M (2021) Hydraulic properties of unsaturated calcarenites by means of a new integrated approach. J Hydrol 602(126730):15. https://doi.org/10.1016/j.jhydrol.2021.126730

Barker JA (1988) A generalized radial flow model for hydraulic tests in fractured rock. Water Resour Res 24:1796–1804

Cochard J (2018) Analyses des propriétés réservoirs d'une série carbonatée microporeuse fracturée: approches multi-échelle sédimentologiques, diagénétiques etmécaniques intégrées. PhD Thesis, Aix-Marseille Univ., France

Cochard J, Léonide P, Borgomano J, Guglielmi Y, Massonnat G, Rolando JP, Marié L, Pasquier A (2020) Reservoir properties of Barremian-Aptian Urgonian limestones, SE France, part 1: influence of structural history on porosity-permeability variations. J Pet Geol 43:75–94

Cochard J, Leonide P, Borgomano J, Guglielmi Y, Massonnat G, Rolando JP, Marie L, Pasquier A (2021) Reservoir properties of Barremian-Aptian Urgonian limestones, SE France, part 2: influence of diagenesis and fracturing. J Pet Geol 44:97–113

Danquigny C, Massonnat G, Mermoud C, Rolando JP (2019a) Intra- and inter-facies variability of multi-physics data in carbonates. New insights from database of ALBION R&D project. In: Abu Dhabi international petroleum exhibition & conference 2019a, Abu Dhabi, UAE, SPE-197836-MS. https://doi.org/10.2118/197836-MS

Danquigny C, Massonnat G, Tendil A, Dal Soglio L, Cochard J, Dumont I, Mazzilli N, Emblanch C, Chalikakis K, Léonide P, Borgomano J, Fournier F, Sénéchal G, Rousset D, Garambois S, Voisin C, Brito D, Rebelle M, Rolando JP (2019b) Interdisciplinary characterisation of carbonate reservoirs. ALBION, a multi-scales dynamic analogue. In: 46th IAH congress, 22–27 September 2019b, Malaga, Spain, p 422

Danquigny C, Massonnat G, Barbier M, Bouxin P, Dal Soglio L, Lesueur JL (2021) Contribution of the ALBION dynamic analogue in understanding the diversity of fluid flows in fractured carbonate reservoirs. The example of the LSBB instrumented site. In: Abu Dhabi international petroleum exhibition & conference 2021, Abu Dhabi, UAE, SPE-207664-MS. https://doi.org/10.2118/207664-ms

Diersch HJ (2014) FEFLOW. Finite element modeling of flow, mass and heat transport in porous and fractured media. Berlin Heidelberg, Springer-Verlag

Doe TW (1991) Fractional dimension analysis of constant-pressure well tests. In: SPE annual technical conference and exhibition 1991. SPE-22702-MS

Duffield GM (2007) AQTESOLV for windows version 4.5 user's guide. Reston, Virginia, USA

Galvao P, Halihan T, Hirata R (2016) The karst permeability scale effect of Sete Lagoas, MG, Brazil. J Hydrol 532:149–162

Hartmann A, Goldscheider N, Wagener T, Lange J, Weiler M (2014) Karst water resources in a changing world: review of hydrological modeling approaches. Rev Geophys 52:218–242

Jacob CE, Lohman SW (1952) Nonsteady flow to a well of constant drawdown in an extensive aquifer. EOS Trans Am Geophys Union 33:559–569

Jazayeri Noushabadi MR, Jourde H, Massonnat G (2011) Influence of the observation scale on permeability estimation at local and regional scales through well tests in a fractured and karstic aquifer (Lez aquifer, Southern France). J Hydrol 403:321–336

Jeanne P, Guglielmi Y, Cappa F (2013) Dissimilar properties within a carbonate-reservoir's small fault zone, and their impact on the pressurization and leakage associated with $CO_2$ injection. J Struct Geol 47:25–35

Kordilla J, Sauter M, Reimann T, Geyer T (2012) Simulation of saturated and unsaturated flow in karst systems at catchment scale using a double continuum approach. Hydrol Earth Syst Sci 16:3909–3923

Lanteaume C, Massonnat G, Samson P, Léonide P, Borgomano J, Rebelle M, Michel J, Danquigny C (2020) Carbonate facies models, fake or real? Comparison of the Urgonian Formation South-East France with the Kharaib-Shuaiba formations middle east—insights from the ALBION R&D project. In: Abu Dhabi international petroleum exhibition & conference 2020. Abu Dhabi, UAE, SPE-203106-MS. https://doi.org/10.2118/203106-MS

Maréchal JC, Ladouche B, Dewandel B, Fleury P, Dörfliger N (2014) Diagnostic plots applied to well-tests in karst systems. In: Mudry J, Zwahlen F, Bertrand C, Lamoreaux JW (eds) H2Karst research in limestone hydrogeology. Springer International Publishing

Mejias M, Renard P, Glenz D (2009) Hydraulic testing of low-permeability formations—a case study in the granite of Cadalso de los Vidrios, Spain. Eng Geol 107:88–97

Tendil AJB, Frau C, Leonide P, Fournier F, Borgomano JR, Lanteaume C, Masse JP, Massonnat G, Rolando JP (2018) Platform-to-basin anatomy of a Barremian-Aptian Tethyan carbonate system: new insights into the regional to global factors controlling the stratigraphic architecture of the Urgonian Provence platform (southeast France). Cretac Res 91:382–411

Van Genuchten MT (1980) A closed-form equation for predicting the hydraulic conductivity of unsaturated soils. Soil Sci Soc Am J 44:892–898

# On the Choice of a Performance Metric for Model Calibration Scheme Using Discharge-Age Information

K. Ö. Çallı, D. Bittner, and A. Hartmann

**Abstract**

The robustness of a hydrological model is mainly dependent on how well the simulations resemble a quantity of interest, typically discharge, which is mainly achieved by successful model calibration. Performance metrics are used to quantify to what degree the system of interest is represented by a model. However, as there is not a general procedure and a rule for the selection of a performance metric, it is ultimately the modeller's subjective decision. This study explores the contribution of discharge young water fractions—derived from stable water isotopes—to the selection of a hydrologically appropriate performance metric for the model calibration. For our analysis, we examined different metric combinations by a multi-variable calibration scheme on the Pareto-optimal frontier. By searching for a trade-off between the model performance on discharge and the model performance on discharge young water age, we not only satisfy the minimization of model simulation bias but also bring the process-based discharge-age information content into the model parametrization. To test our hypothesis, we applied our approach to one of the karstic water sources in Austria. Our finding indicates that the information content of young water age supports the proper choice of a performance metric for the model calibration scheme while reducing the modeller's subjectivity on performance metric selection, thereby ensuring physically more plausible parameter sets.

**Keywords**

Hydrological modelling • Water age • Young water fraction • Model confinement • Performance metric

## 1 Introduction

Model parameterization is one of the main challenges in hydrological modelling studies as most of the model parameters are estimated by model calibration. This stage is particularly essential for modelling studies in karst aquifers since a large number of model parameters are often required to capture the complex hydrological functioning (Bittner et al. 2018; Chen et al. 2017) and to resemble the heterogeneities with prior knowledge (Frank et al. 2021). One key decision in the model calibration is the selection of a performance metric (hereinafter referred to as metric) (Mizukami et al. 2019) as it renders an indication of how well and to what degree the model represents the system reality while providing a degree of belief in model simulations. For that reason, the robustness of model simulations, including an identified model parameter set, is predominantly dependent on which performance metric is formulated for the model calibration.

The successful model calibration is mainly attributed to the value of the performance metric that verifies the simulation accuracy with its targeted value. In this regard, the overall goal of the model calibration is to get a 'targeted value' of the selected metric, while decreasing the discrepancies between simulated and observed states. However, as there is not a general procedure or a rule as to how to choose a performance metric for the model calibration, the selection is ultimately the modeller's subjective decision (Krause et al. 2005).

Different metrics consider different aspects of the model simulations due to the differences of their mathematical/statistical index formulations. To a certain extent, while

K. Ö. Çallı (✉) · D. Bittner · A. Hartmann
TU Dresden, Institute of Groundwater Management, Dresden, Germany
e-mail: kuebra.oezdemir-calli@tu-dresden.de

D. Bittner
Department of River Basin Management, Erftverband, Bergheim, Germany


B. Andreo et al. (eds.), *EuroKarst 2022, Málaga*, Advances in Karst Science,
https://doi.org/10.1007/978-3-031-16879-6_18

calibration metrics in the time domain mainly capture the discrete hydrological dynamics (e.g. high flows, peak flow, baseflow) (Reusser et al. 2009), hydrological signatures (e.g. flow duration curve, aridity index, autocorrelation) reveal time-invariant system properties (Dubois et al. 2020). For this reason, to compensate the inadequacy of a (selected) metric, the model parameterization is intentionally carried out by either using different metric combinations (multi-criteria calibration approach) or by considering multiple data sources (multi-variable calibration approach). However, despite the considerable literature delivering a metric plethora, the discussion on which metric would be the reasonable choice for the model calibration is still open to the modelling community (Jackson et al. 2019). Therefore, it is fair to question as to how to decide which metric would be a more reasonable choice in the intended modelling framework.

Water age is commonly served as an auxiliary information source for model parameterization as it mainly provides process-based knowledge that reveals how catchments store and release water (Berghuijs and Kirchner 2017). Therefore, this information content can be served as a critical descriptor to improve the model simulation accuracy (Sanford 2011), to constrain the flow or transport model parameters (Solomon et al. 2010), and to identify the different model compartments as a model flux indicator (Zhang et al. 2020).

In this study, the terminology water 'age' refers to the 'apparent age' obtained by tracer tests, whereas 'old'/'young' water specifies the water parcel that is older/younger than a certain threshold defined by stable isotopes (such as oxygen-18). In this context, since the reliable quantification of 'mean transit time' is often more difficult (Sprenger et al. 2019) and even misleading (Woolfenden and Ginn 2009), young water age has been a subject of high interest in the modelling studies as an additional information source (Jing et al. 2019; Maxwell et al. 2019; Rusjan et al. 2019), especially in capturing time-variant characteristics of heterogeneous hydrological systems (Kirchner 2016a, b; Zhang et al. 2020).

The knowledge of discharge young water age is an important benchmark to reveal and compare the system-specific dynamics which are mainly dominated by preferential flow paths (e.g. fractures, conduits) as in fractured and karst aquifers. For that reason, this knowledge could be used as a model diagnostic to characterize system hydrological functioning as well as to improve model reality. But, beyond that, it could be served as an unbiased calibration constraint to identify hydrologically reasonable metric either by falsifying or confirming the mathematical/statistical adequacy of the selected performance metric.

The study explores the contribution of discharge young water fractions ($F_{\text{yw}}$)—derived from stable water isotopes—to the performance metric selection for the model parameterization. We hypothesized that $F_{\text{yw}}$ can be used as a model assessment criterion that inherently allows us to select a hydrologically—more—adequate performance metric. To test our hypothesis, we applied our approach to one of the karstic drinking water sources in Schneealp, Austria.

## 2 Case Area

The case area lies in the Northern Calcareous Alps between Styria and lower Austria (Fig. 1). As one of the sub-catchments in the Schneealpe karst massive, the area is characterized by the highly karstified Triassic-age limestone and dolomite (Rank et al. 2006), and thus approximately 75% of precipitation quickly infiltrates into the karstic system mainly via sinkholes (Hilberg 2016). Of the numerous springs of the karst massive, Wasseralm spring is an important one as one of the drinking water supplies for Vienna (Austria).

For our experimental design, we used daily meteorological forcing data including precipitation and air temperature (01/01/1975–31/12/2016), which were obtained in Nasswald (Wasseralm) (774 m) and Mürzzuschlag (758 m) weather stations, respectively (Table 1). Daily discharge (01/01/1995–31/12/2016) was downloaded from https://ehyd.gv.at/#, while the monthly stable isotope concentration data for the precipitation ($\delta^{18}O_P$) and spring discharge ($\delta^{18}O_Q$) was digitized from Maloszewski et al. (2002). Potential evapotranspiration was estimated by the Hargreaves equation, and snow accumulation and snowmelt were calculated by the degree-day concept.

## 3 Model Set-Up and Experimental Design

To simulate Wasseralm spring discharge, we used the VarKarst model developed by Hartmann et al. (2013). For the simulation of the $\delta^{18}O$ concentration in spring discharge ($\delta^{18}O_Q$) and corresponding discharge young water fractions ($F_{\text{yw}}$), we used StorAge Selection function (SAS) approach (hereinafter referred to as the SAS model) by Benettin and Bertuzzo (2018). After coupling the SAS model with VarKarst, we run the coupled model such that the SAS model was fed by VarKarst outputs including effective recharge and discharge. That way, we simultaneously simulated spring discharge, $\delta^{18}O_Q$, and $F_{\text{yw}}$.

The SAS model is a solute transport model that identifies water age relationships between an input and output water parcel. The solute concentration is simulated based on the 'age-ranked storage' concept defined by the volume of water in storage that is younger (or equal to) than a certain age. To simulate the concentration of conservative tracer (for our case $\delta^{18}O$), the probability density function (PDF) of rank

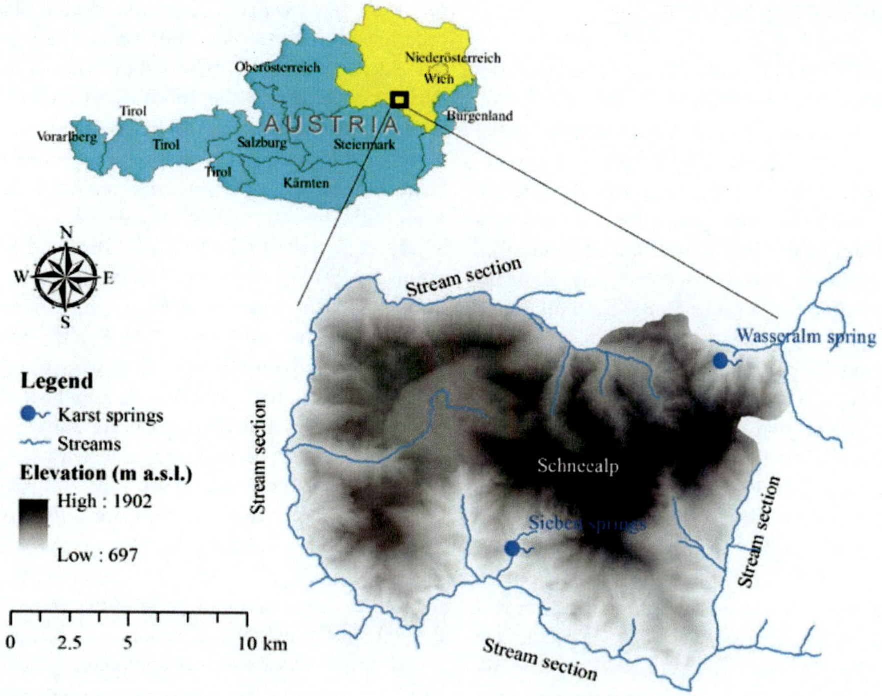


**Fig. 1** Map demonstrating the location of Schneealp karst massive and two karstic springs (Wasseralm and Sieben springs). The catchment area is delineated based on the stream sections

**Table 1** Data availability, data properties, and sources for the experimental site

| Variable | Unit | Time resolution | Time period | Source |
|---|---|---|---|---|
| Spring discharge | [l s$^{-1}$] | Daily | 1995–2016 | https://ehyd.gv.at/# |
| Precipitation | [mm] | Daily | 1975–2016 | https://ehyd.gv.at/# |
| Air temperature | [°C] | Daily | 1975–2016 | ZAMG[a] (Histalp) http://www.zamg.ac.at/histalp |
| Snow depth | [cm] | Daily | 1994–2016 | ZAMG |
| Fresh snow depth | [cm] | Daily | 1994–2016 | ZAMG |
| $\delta^{18}O_P$ and $\delta^{18}O_Q$ | [‰] | Monthly | 1975–1990 | Maloszewski et al. (2002) |

[a]ZAMG: *Central Institution for Meteorology and Geodynamics in Austria*

storage, $\omega_Q(S_T, t)$, is considered. Therefore, to obtain discharge $\delta^{18}O$ concentration, $C_Q(t)$, each water parcel in reservoir storage is characterized by the inflow $\delta^{18}O$ concentrations in precipitation input, $C_j(S_T, t)$.

$$C_Q(t) = \int_{T=0}^{\infty} C_j(S_T t)\omega_Q(S_T, t)\mathrm{d}S_T \quad (1)$$

Here, $\omega_Q(S_T, t)$ is solved by a suitable functional form for the SAS model (e.g. power law, beta function, and gamma function). For our study, we used the beta function with two shape parameters ($\alpha$ and $\beta$). For each simulation of $C_Q(t)$, we simultaneously obtained $F_{\mathrm{yw}}$ using the following formula:

$$F_{\mathrm{yw}} = \int_{\tau=0}^{\tau_{\mathrm{yw}}} \tau^{\alpha-1}(1-\tau)^{\beta-1}\mathrm{d}\tau \quad (2)$$

where $\tau$ is the travel time and $\tau_{\mathrm{yw}}$ is the threshold age of discharge young water. The mismatch between the observed $F_{\mathrm{yw}}$ ($F_{\mathrm{yw},o}$) and simulated $F_{\mathrm{yw}}$ ($F_{\mathrm{yw},s}$) was then estimated by

$$\mathrm{misfit}(F_{\mathrm{yw}}) = 1 - \sum_{n=1}^{N} \frac{\left(F_{\mathrm{yw},o} - F_{\mathrm{yw},s}\right)^2}{F_{\mathrm{yw},s}} \quad (3)$$

Here, $\mathrm{misfit}(F_{\mathrm{yw}})$ is normalized to be varied between 0 and 1 in such a way that the values approaching to 1 indicate the better model performance on $F_{\mathrm{yw}}$.

## 4 Model Conditioning Scheme

We carried out a multi-variable calibration scheme to search the trade-off between model performance on spring discharge and model performance on $F_{yw}$. To sample model parameters, we used Latin hypercube sampling (LHS) scheme based on the uniform parameter distribution assumption. The feasible parameter range for VarKarst was obtained by Hartmann et al. (2016) and Maloszewski et al. (2002) (Table 2), while the shape parameters ($\alpha$ and $\beta$) of the beta function were defined by a set of model calibration and set to 0.85 [-] and 4.56 [-], respectively.

The coupled model was run by $3 \times 10^4$ times from 01/01/1975 to 31/12/2016. Since the data periods for $\delta^{18}O$ concentration ($\delta^{18}O_P$ and $\delta^{18}O_Q$) and spring discharge do not overlap each other (see Table 1), the coupled model was conditioned by different periods. The period of 01/01/1975–31/12/1990 was used to simulate $\delta^{18}O_Q$ concentration and corresponding $F_{yw}$, whereas the discharge was calibrated for 01/01/1995–31/12/2016. A 2-year warm-up was considered in each model conditioning. Since our main goal is to reveal hydrologically more relevant performance metrics, we did not perform model validation on each data set. Instead, we assigned the whole data series only for calibration, and thus 15 years for $\delta^{18}O$ concentration and 22 years for spring discharge were used for the parameterization.

We used 16 performance metrics—12 deterministic metrics and 4 hydrological signatures—to evaluate the model performance on spring discharge. For the exploration of the hydrologically more relevant metric from the trade-off between performance metric and $F_{yw}$, we applied bivariate kernel density estimation (KDE2D) method in R-Studio. In doing that, we examined different metric combinations on the Pareto-optimal frontier by considering the model performance over the multi-variables. Therefore, on this frontier, we explored the optimal combinations by the model performance on discharge and discharge young water age. After selecting hydrologically appropriate metrics, we constrained the initial model ensemble ($3 \times 10^4$ realizations) and extracted the behavioural 100 parameter sets for each performance metric. The parameter uncertainty was then assessed based on the percentile approach (25th and 75th percentiles).

## 5 Results and Discussion

### 5.1 Exploration of Hydrologically Appropriate Performance Metric

The model output constrained by the model performance on $F_{yw}$ and model performance on discharge is demonstrated by the surface plots in Fig. 2. Here, each plot indicates two-dimensional distribution for two different variables at one location in which $F_{yw}$ was used as a model competing criterion to choose an appropriate performance metric. Considering this, we explored the Pareto-optimal frontier on each surface where the model performance on $F_{yw}$ and model performance on discharge simultaneously provide high performance. This Pareto-optimal frontier is, therefore, characterized by a high-density region with a more condensed and narrower 'point cloud'.

Figure 2 reveals that $NSE_Q$, $MSE_Q$, and $KGE_Q$ are more prone to convey the process-based information content throughout the model output space, whereas $NSE_Q$ and $MSE_Q$ make the best couple with $F_{yw}$ by demonstrating the central tendency on the frontier. This particular finding can be associated with the fact that NSE and MSE better capture the hydrological functioning of the karst aquifer. Further, they would better resemble the dual characteristics of aquifer including recharge (concentrated and diffusive) and storage (matrix and conduit) as compared to metrics such as $PBIAS_Q$, $VE_Q$, and $RSR_Q$. For that reason, it is fair to imply that when $NSE_Q$ (or $MSE_Q$) is selected as a performance metric, it could convey more informative model realizations throughout the model calibration.

As for the hydrological signatures ($ACF_Q$, $CCF_Q$, and $FDC_Q$),[1] they do not necessarily indicate high-density regions in which the model performance on $F_{yw}$ cannot be further improved without the value of corresponding performance metric worsening (Fig. 2). Here, the only exception is $SVI_Q$[2] in that it captures an informative model cluster in the region of $0.75 < F_{yw} < 1$ (indicated by the high-density region over the Pareto-frontier). One possible explanation would be the index formulation of SVI since it mainly reflects hydrological variability in spring discharge considering the maximum, minimum, and median flow characteristics. Conversely, $ACF_Q$ and $CCF_Q$ focus on the certain hydrodynamics in the system such as system memory.

### 5.2 The Physical Realism of Model Parameters Considering Aquifer Physical Properties

The posterior distributions of the model parameters are provided in Fig. 3. Overall, the behavioural parameter sets obtained by $KGE_Q$, $NSE_Q$, and $MSE_Q$ are represented with physically reasonable parameter ranges, mainly considering the typical characteristics of the karst aquifer. For instance, $\alpha_f$[-] characterizes the conduit-dominated flow characteristics,

[1] denotes hydrological signatures: auto-correlation function, cross-correlation function, flow duration curve (formulated for the median flow), respectively.

[2] refers to spring variability index: the formulation considers the maximum, minimum, and median flow characteristics of the spring discharge.

**Table 2** VarKarst model parameters, definitions, and ranges

| Parameter | Description | Unit | Parameter range | |
|---|---|---|---|---|
| | | | Lower | Upper |
| $V_S$ | Mean soil storage capacity | [mm] | 0 | 1500 |
| $V_E$ | Mean epikarst storage capacity | [mm] | 0 | 1500 |
| $\alpha_{SE}$ | Soil/epikarst depth variability constant | [-] | 0 | 2 |
| $K_E$ | Epikarst mean storage coefficient | [d] | 1 | 50 |
| $\alpha_f$ | Recharge separation variability constant | [-] | 0 | 2 |
| $K_C$ | Conduit storage constant | [d] | 1 | 10 |
| $\alpha_{GW}$ | Groundwater variability constant | [-] | 0 | 2 |
| $A$ | Recharge contributing area | [km$^2$] | 20 | 50 |

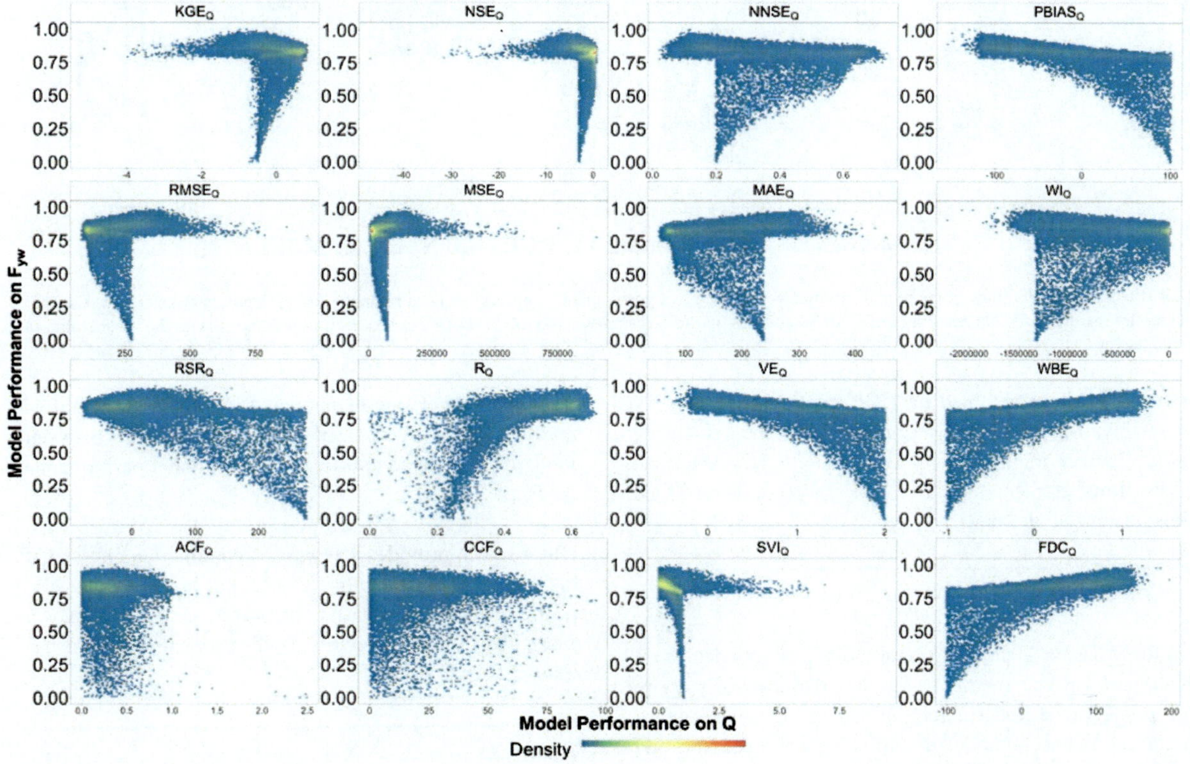


**Fig. 2** Pareto-optimal model output space constrained by the model performance on young water fractions ($F_{yw}$) and model performance on spring discharge ($Q$) represented by the two-dimensional kernel density estimates. The surface plots demonstrate the two-dimensional distribution (for two different variables at one location). Here, the dots demonstrate $3 \times 10^4$ realizations, while red indicates the high-density region

thereby indicating the karstification degree in the system. Here, $\alpha_f$ is represented by the mean values of $NSE_Q$ (0.36), $MSE_Q$ (0.33), and $KGE_Q$ (0.34). Further, $\alpha_f$ is strongly constrained by each metric, thus leading to a substantial reduction in the parameter uncertainty. Similarly, $KGE_Q$ confirms the further parameter confinement in the range for the parameter $V_S$ [mm] while providing physically more plausible values with a mean value of 690.57 [mm].

The hydrological functioning of preferential pathways in epikarst is represented by $K_E$ [d] in the VarKarst model structure. Therefore, obtained range for $K_E$ by $NSE_Q$ is more than acceptable by the values ranging between 5 [d] and 13.23 [d]. Here, $NSE_Q$ ensures a narrower interquartile range than those obtained by $KGE_Q$ and $MSE_Q$, thereby allowing a reduction in the parameter uncertainty. Moreover, $NSE_Q$ allows better confinements for $A$ [km$^2$] and $a_{GW}$ [-] as

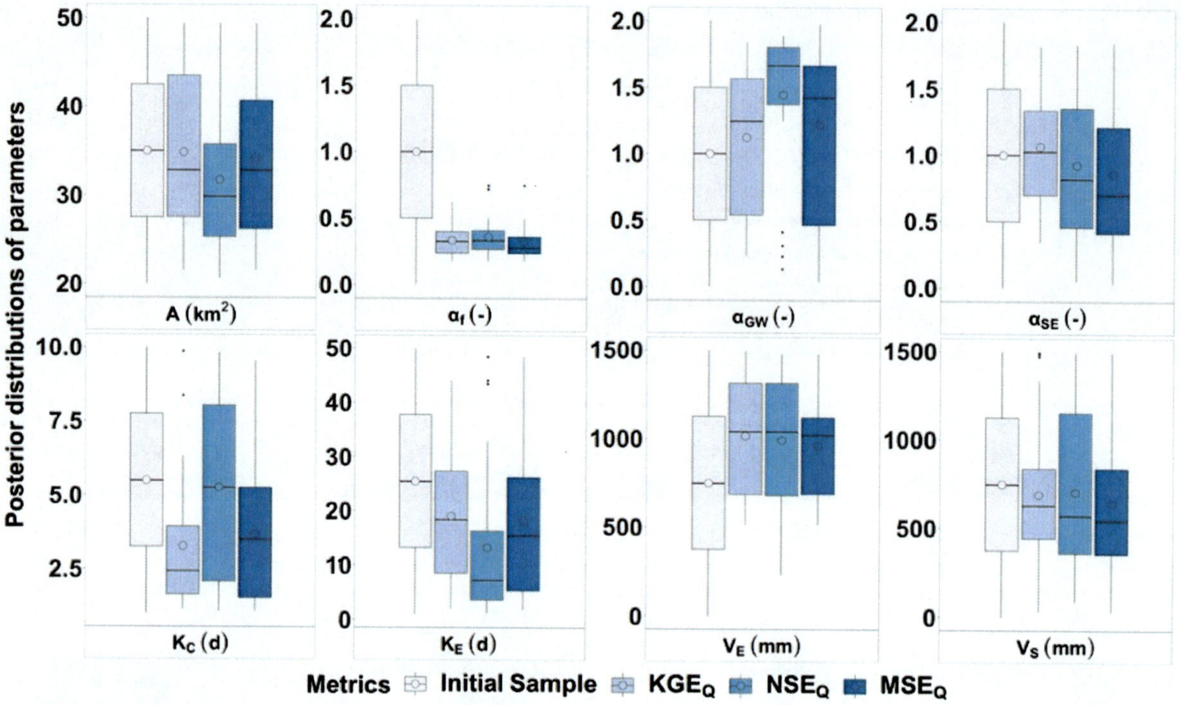


**Fig. 3** Posterior distributions of the model parameters obtained by hydrologically appropriate performance metrics. Here, parameter uncertainty is assessed by the distance between 25th and 75th percentiles on the box-and-whisker plots. The circle on each plot demonstrates the mean value of the parameter

compared to those obtained by $KGE_Q$ and $MSE_Q$ (Fig. 3). Conversely, $K_C$ [d] is well resembled by $KGE_Q$ since the karstic channel in the system is essentially characterized by its low storage capacity and short transfer time (Maloszewski et al. 2002).

## 6 Conclusions

This study has made the first attempt to objectively decide on which performance metric would be hydrologically more appropriate for the model calibration scheme. By searching for a trade-off between model performance on discharge and model performance on discharge young water age ($F_{yw}$), we not only satisfy the minimization of model simulation bias yet also bring the process-based discharge-age information content into the model parametrization. The major findings are as follows:

- The proposed metric selection procedure could improve the model realism either by reducing the subjectivity in the metric selection or by ensuring physically more plausible parameter sets.
- $F_{yw}$ is a potentially unbiased calibration metric to identify hydrologically more informative model realization(s) from the model output in that the process-based information is conveyed throughout the trade-off between model performance on discharge and model performance on $F_{yw}$.

Our research provides a new research direction on to what extent the model reliability could be improved by the inclusion of $F_{yw}$ information content into the calibration as a hydrological system-specific and time-variant process diagnostic metric.

## References

Benettin P, Bertuzzo E (2018) Tran-SAS v1. 0: a numerical model to compute catchment-scale hydrologic transport using StorAge selection functions. Geosci Model Dev 11(4):1627–1639

Berghuijs WR, Kirchner JW (2017) The relationship between contrasting ages of groundwater and streamflow. Geophys Res Lett 44 (17):8925–8935

Bittner D, Narany TS, Kohl B, Disse M, Chiogna G (2018) Modeling the hydrological impact of land use change in a dolomite-dominated karst system. J Hydrol 567:267–279

Chen Z, Hartmann A, Goldscheider N (2017) A new approach to evaluate spatiotemporal dynamics of controlling parameters in distributed environmental models. Environ Modell Softw 87: 1–16

Dubois E, Doummar J, Pistre S, Larocque M (2020) Calibration of a lumped karst system model and application to the Qachqouch karst spring (Lebanon) under climate change conditions. Hydrol Earth Syst Sci 24(9):4275–4290

Frank S, Goeppert, N, Goldscheider N (2021) Improved understanding of dynamic water and mass budgets of high-alpine karst systems obtained from studying a well-defined catchment area. Hydrol Process 35(4):e14033

Hartmann A, Barberá JA, Lange J, Andreo B, Weiler M (2013) Progress in the hydrologic simulation of time variant recharge areas of karst systems–Exemplified at a karst spring in Southern Spain. Adv Water Resour 54:149–160

Hartmann A, Kobler J, Kralik M, Dirnböck T, Humer F, Weiler M (2016) Model-aided quantification of dissolved carbon and nitrogen release after windthrow disturbance in an Austrian karst system. Biogeosciences 13(1):159–174

Hilberg S (2016) Natural tracers in fractured hard-rock aquifers in the Austrian part of the Eastern Alps–previous approaches and future perspectives for hydrogeology in mountain regions. Hydrogeol J 24 (5):1091–1105

Jackson EK, Roberts W, Nelsen B, Williams GP, Nelson EJ, Ames DP (2019) Introductory overview: error metrics for hydrologic modelling—a review of common practices and an open-source library to facilitate use and adoption. Environ Modell Softw 119:32–48

Jing M, Heße F, Kumar R, Kolditz O, Kalbacher T, Attinger S (2019) Influence of input and parameter uncertainty on the prediction of catchment-scale groundwater travel time distributions. Hydrol Earth Syst Sci 23(1):171–190

Krause P, Boyle DP, Bäse F (2005) Comparison of different efficiency criteria for hydrological model assessment. Adv Geosci 5:89–97

Kirchner JW (2016a) Aggregation in environmental systems—part 1: seasonal tracer cycles quantify young water fractions, but not mean transit times, in spatially heterogeneous catchments. Hydrol Earth Syst Sci 20(1):279–297

Kirchner JW (2016b) Aggregation in environmental systems—part 2: catchment mean transit times and young water fractions under hydrologic nonstationarity. Hydrol Earth Syst Sci 20(1):299–328

Maloszewski P, Stichler W, Zuber A, Rank D (2002) Identifying the flow systems in a karstic-fissured-porous aquifer, the Schneealpe, Austria, by modelling of environmental 18O and 3H isotopes. J Hydrol 256(1–2):48–59

Maxwell M, Condon LE, Danesh-Yazdi M, Bearup LA (2019) Exploring source water mixing and transient residence time distributions of outflow and evapotranspiration with an integrated hydrologic model and Lagrangian particle tracking approach. Ecohydrology 12(1):e2042

Mizukami N, Rakovec O, Newman AJ, Clark MP, Wood AW, Gupta HV, Kumar R (2019) On the choice of calibration metrics for "high-flow" estimation using hydrologic models. Hydrol Earth Syst Sci 23(6):2601–2614

Rank D, Wieselthaler F, Papesch W, Kuschnig G (2006) Separation of discharge components at a karst spring on the basis of event investigations (Wasseralmquelle, Austria). Austrian J Earth Sci 99:18–23

Reusser DE, Blume T, Schaefli B, Zehe E (2009) Analysing the temporal dynamics of model performance for hydrological models. Hydrol Earth Syst Sci 13(7):999–1018

Rusjan S, Sapač K, Petrič M, Lojen S, Bezak N (2019) Identifying the hydrological behavior of a complex karst system using stable isotopes. J Hydrol 577:123956

Sanford W (2011) Calibration of models using groundwater age. Hydrogeol J 19(1):13–16

Solomon DK, Genereux DP, Plummer LN, Busenberg E (2010) Testing mixing models of old and young groundwater in a tropical lowland rain forest with environmental tracers. Water Resour Res 46(4)

Sprenger M, Stumpp C, Weiler M, Aeschbach W, Allen ST, Benettin P, Dubbert M, Hartmann A, Hrachowitz M, Kirchner JW, McDonnell JJ (2019) The demographics of water: a review of water ages in the critical zone. Rev Geophys 57(3):800–834

Woolfenden LR, Ginn TR (2009) Modeled ground water age distributions. Groundwater 47(4):547–557

Zhang Z, Chen X, Cheng Q, Soulsby C (2020) Characterizing the variability of transit time distributions and young water fractions in karst catchments using flux tracking. Hydrol Process 34(15):3156–3174

# Understanding Water Table Fluctuations in a Karstic Semiarid Mediterranean Aquifer Through Numerical Modeling: The Case of Almudaina-Segaria Aquifer

M. C. Ruiz, C. Pla, J. Valdés-Abellán, M. Fernández-Mejuto, J. A. Hernández-Bravo, and D. Benavente

**Abstract**

The exploitation of underground water has an important role in the province of Alicante, Southeast Spain, where it is mostly used for agriculture and the supply of populations. The Almudaina-Segaria hydrogeological domain, in northern Alicante, constitutes the water supply for La Vall d'Ebo, Planes and Gorga villages. This domain includes the Almudaina-Segaria system, a 167 $km^2$ carbonated karstic aquifer with 700 m of permeable thickness made of limestones, dolomites, and marls. In this project the aquifer's water flow was modeled using MODFLOW and ModelMuse for the interval 1980–2019, aiming to understand its dynamics. The water input is provided by the infiltration of precipitation (approximately 730 mm and 68 $hm^3$ of annual precipitation and recharge, respectively), while the output is conducted through direct pumping (up to 2.31 $hm^3$/yr), natural drainage to the Vergel and Pego aquifers and, mainly, discharge through the Finca Rincón del Rosario drainage and the Racons River (54.44 $hm^3$/yr). The groundwater flow occurs in a SW–NE direction toward the discharge zone. The aquifer modeling is a useful tool for an appropriate management of water resources in the north of the Alicante province, given its importance as a main support for the populations and activities that take place there.

**Keywords**

Alicante • Groundwater • Karst • Modeling • MODFLOW

## 1 Introduction

Throughout history, groundwater has been a major source of water for sustaining human life. Because it is buffered from short-term variability in weather patterns, groundwater has often been considered a stable and reliable resource (Fienen and Arshad 2016). Karst aquifers provide potable water sources around the world, with the total population served by karst sources estimated to be between 9 and 20% of the world's population according to different studies, and about 13% of the total water withdrawal (Ford and Williams 2013; Stevanović 2019). Although they are complex and highly heterogeneous, assessing the flow pathways and dynamics in karstic systems is an essential issue considering current concerns such as water management in a scenario of climate change.

The southeast of Spain is also sensitive to this scenario, since a rise of the average temperature and a reduction of total annual rainfall, together with the intensification of extreme events, rank among the most likely effects of climate change in the Mediterranean (IPCC 2014). Groundwater resources are vulnerable to such changes because decreasing precipitation and increasing evapotranspiration can lead to a reduction of recharge rates. This could trigger unwanted effects in those areas where aquifers provide the main source of water supply (Pulido-Velazquez et al. 2018).

Despite the above scenario, climate change is highly unpredictable. Hence, there is a need to endow water

M. C. Ruiz (✉) · D. Benavente
Department of Earth and Environmental Sciences, University of Alicante, 03690 Alicante, Spain
e-mail: mcandela.ruiz@ua.es

D. Benavente
e-mail: david.benavente@ua.es

C. Pla · J. Valdés-Abellán
Department of Civil Engineering, University of Alicante, 03690 Alicante, Spain
e-mail: c.pla@ua.es

J. Valdés-Abellán
e-mail: javier.valdes@ua.es

M. Fernández-Mejuto · J. A. Hernández-Bravo
Área de Ciclo Hídrico, Diputación Provincial de Alicante, 03006 Alicante, Spain


B. Andreo et al. (eds.), *EuroKarst 2022, Málaga*, Advances in Karst Science,
https://doi.org/10.1007/978-3-031-16879-6_19

managers with tools that are sufficiently simple to use on a regular basis, yet accurate enough to underpin decisions (Aslam et al. 2018). To understand better the functioning of karst aquifers, a range of models have been used in the literature, including reservoir models, black box or transfer function models, machine learning models, and physically-based distributed models. Each of these approaches have their advantages and drawbacks depending on the application. When it comes to simulate in detail what is physically happening in the different compartments of a karst system, the distributed models can represent the complex flow and transport processes across an entire catchment. These numerical models can be used to predict different scenarios (changes in pumping or climate) and hence impacts on resources, to test previous conceptual models or to estimate hydraulic parameters (Duran and Gill 2021).

This research aims to describe and parameterize the behavior of the Almudaina-Segaria aquifer, modeling for this purpose the water inflows and outflows from the system, its hydraulic properties, and the variation in its piezometric heads. To achieve this goal, we built and calibrated a numerical model using the MODFLOW code, obtaining an accurate simulation of the underground dynamics for the last 40 years.

## 2 Study Area Description

The Almudaina-Segaria aquifer (also called Almudaina-Alfaro-Segaria) is in southeastern Spain, in the extreme northeast of the Alicante province (Fig. 1a). The simulated aquifer belongs to the also called Almudaina Segaria hydrogeological domain, along with 8 other smaller aquifers.

As seen in Fig. 1b, the Almudaina-Segaria aquifer has an elongated shape of about 35 × 5 km, except in the Segaria (eastern) sector where it narrows approximately to the half. It has an elongated morphology in SW–NE direction and occupies a surface area of 167 $km^2$, of which about 145 $km^2$ correspond to permeable materials. The coordinates of the rectangle surrounding the aquifer are 38° 42′ 30.31″ N, 0° 23′ 53.57″ W for its south-western vertex, and 38° 51′ 27.89″ N, 0°0′ 11.53″ W for its north–eastern vertex. It is in an area of steep slopes, as a component of the (from west to east) Almudaina, Alfaro, La Carrasca, Mediodía and Segaria ranges.

The permeable layer is made of carbonate formations with an average thickness of about 800 m in the western sector, and 300 m in the eastern sector. The main aquifer section is determined by 400 m of materials with Aptian-Albian marly intercalations, Eocene limestone and sandstones, conglomerates, and lower Miocene limestones, present the latter only in reduced sectors. This section, with a high permeability, overlies a less permeable layer of Lower Cretaceous bioclastic limestones. The aquifer impermeable bottom is determined by Barremian marl and marl limestones with an estimated thickness of 400 m (IGME-DPA 2015).

All the Upper Cretaceous permeable sections are in hydraulic connection, constituting a single aquifer except in its southern sector where the Upper Aptian marly horizons isolate permeable sections defining various hanging aquifers

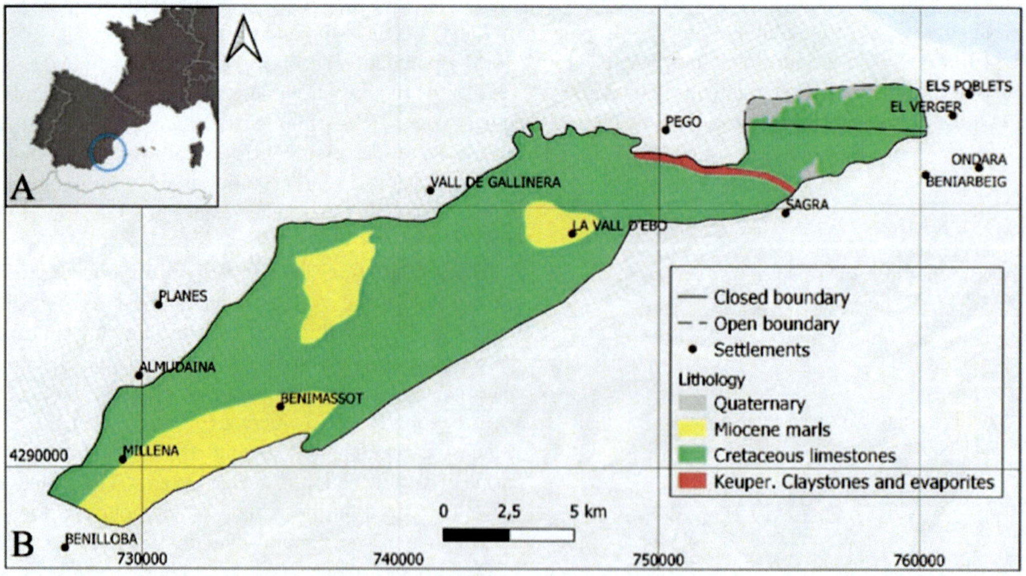


**Fig. 1** **a** Almudaina-Segaria aquifer location in Alicante, Spain. **b** Almudaina-Segaria aquifer

that are disconnected from the main aquifer (IGME-DPA 2015). In the central east sector of the aquifer there are outcrops of Keuper facies, which constitute a hydraulic barrier that causes a sudden drop in the piezometric level.

The western and northern boundaries are constituted by Tap facies (marlstone), which, along the reverse fault that extends to the city of Pego, seal the permeable materials of the aquifer. It is, therefore, a closed boundary, except at its easternmost end (east of Pego), through which it establishes a connection with the detrital materials of the Vergel aquifer. The eastern limit is closed in the southern sector due to contact with the marly Miocene. The southern limit is constituted, in the western zone, by the Tap facies as observable in the Ceta Valley. In the central sector, it is assumed that there is a disconnection with the Mediodía aquifer caused by tectonic elevation of the base impermeable layer (IGME-DPA 2015).

In the Almudaina sector, the level of the piezometric surface is only known at very specific points and varies between about 320 m.a.s.l. and the 150 m.a.s.l. in the western area, which implies depths of more than 530 m, and values close to 50–70 m.a.s.l. SW of Pego. At the structural flow threshold that separates the Almudaina and Segaria sectors there is a significant drop that lowers the levels below 5 m.a.s.l., even reaching negative values in the eastern end of the Segaria sector.

The recharge of the system is produced via the infiltration of rainfall, while the discharge takes place by pumping for urban, agricultural, and industrial uses, as well as by lateral drainage, and superficial drainage to Finca Rincón del Rosario and the Racons River (IGME-DPA 2015).

## 3 Methodological Approach

The aquifer was simulated using the MODFLOW-2005 code (Harbaugh et al. 2017), with ModelMuse (Winston 2020) as the Graphic User Interface. The model was built for reconstructing and actualizing the existing previous simulations. For this project, first a steady-state simulation was built and, using this model as initial condition, the transient-state was developed and calibrated. For the transient simulation almost 40 years of daily data were processed, from 01/01/1980 until 31/10/2019.

### 3.1 Conceptual Model and Simulation Building

The aquifer was discretized in 300 $\times$ 300 m finite cells resulting an array of 123 columns and 28 rows within a single permeable layer, being active cells those intersected and enclosed by the aquifer's contour. The upper and lower boundaries were defined by using digital elevation models, with the terrain's surface as Model Top, while the aquifer bottom was built by integrating the existing information (drillings and previous studies).

As stated, the underlying layer is impermeable and, thus, there is no vertical flow through the aquifer bottom. The aquifer's permeable thickness is in between 800 m in the western sector, and 300 m in its easternmost portion.

The boundary conditions are defined based on the hydrogeological setting. The existence of TAP sequences and marl facies results in no-flow boundaries throughout the perimeter of the aquifer except for the eastern sector, where the discharges of the system occur. Groundwater discharge takes place through outflow towards the Racons River, Finca Rincón del Rosario artificial drainage, two sectors of lateral drain and extraction of wells, located as seen in Fig. 2. The Racons River and Finca Rincón del Rosario were assigned to the River Package with a bed conductance of 3540 and 5170 $m^2/d$, respectively. In the sectors named Pego and Vergel, there is hydraulic connection with the adjacent detrital aquifer and, thus, there is a flow boundary assigned to the drain package with conductances of 320 and 74 $m^2/d$, respectively. For each well, the time series of extractions were loaded into MODFLOW.

The horizontal hydraulic conductivity ($K$) was initially assigned by sectors with values between 0.005 and 30 m/d, while the storage coefficient was between 0.001 and 0.05.

The Almudaina-Segaria aquifer is recharged exclusively by the infiltration of rainfall through the permeable unsaturated zone, which is the entire surface of the aquifer except for the Millena portion, to the southwest. To build the input for the Recharge package, nearly 40 years of daily

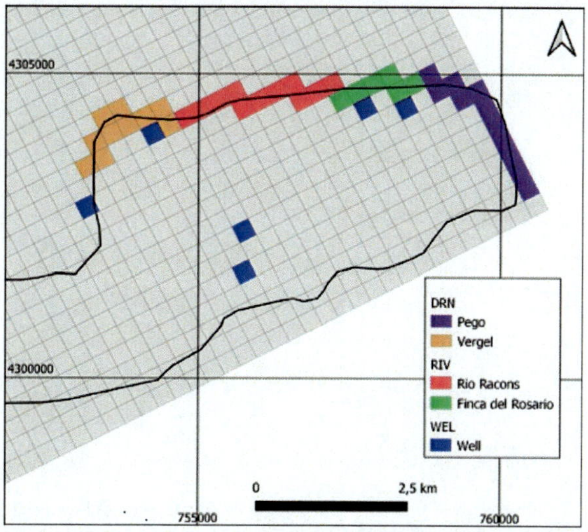


**Fig. 2** System water outputs grouped by MODFLOW Package used. DRN: drain package, RIV: river package, WEL: well package

precipitation data were transformed into recharge considering evapotranspiration, runoff threshold, land slope, soil type and its uses, depth of roots, and soil parameters such as initial moisture, field capacity, and wilting point. Considering a mean annual rainfall of 730 mm, the effective infiltration and thus aquifer recharge is around 55% of precipitation.

### 3.2 Model Calibration

The calibration was carried out using the Head-Observation Package and the piezometric heads time series from 5 wells which are representative of the different sections of the aquifer. Each of these calibration wells has daily measurements corresponding to different time periods, with a total of more than 5000 individual measurements subjected to comparison with simulated heads. To adjust the simulation, the parameters modified by trial and error were hydraulic conductivity and specific storage. The calibration's accuracy and thus the fit between the observed ($h_{obs}$) and simulated ($h_{sim}$) water heads was quantified by calculating the correlation coefficient, root mean square error (RMSE) and normalized root mean square error (NRMSE), following Eqs. (1) and (2).

$$\mathrm{RMSE} = \sqrt{\frac{\sum(h_{obs} - h_{sim})^2}{2}} \quad (1)$$

$$\mathrm{NRMSE} = \frac{\mathrm{RMSE}}{(h_{max} - h_{min})} \quad (2)$$

## 4 Results and Discussion

### 4.1 Model Calibration

The comparison between the observed and simulated groundwater heads shows a significant correlation (Fig. 3), which indicates that the model performs an accurate representation of the aquifer dynamics. The calibrated hydraulic conductivities vary between 0.07 and 25.3 m/d, and the storage coefficients are between 0.006 and 0.01, being the highest values for both parameters in the easternmost sector. The RMSE is 5.3 m, the NRMSE is 0.014% and $R^2$ = 0.996, showing a high correlation between observed and simulated values.

### 4.2 Groundwater Flow and Variations Through Time

As expected, the groundwater flow occurs in a SW–NE direction, toward the discharge zone and the Mediterranean Sea. The results show an important temporal variation in the piezometric levels in the western portion of the aquifer, with variations over 50 m throughout the evaluated time series, in accordance with periods of greater and lower rainfall. In this sector, the piezometric surface is between 350 and 250 m.a.s.l., and it descends in the flow direction reaching values even below the sea level in the far east. In this last sector, the level remains relatively stable through time.

The groundwater budget shows that the total of the water inputs to the system is through the infiltration of precipitation, with an annual average of 68.12 $hm^3$. The system's water outputs are through the Rio Racons and the Finca Rincón del

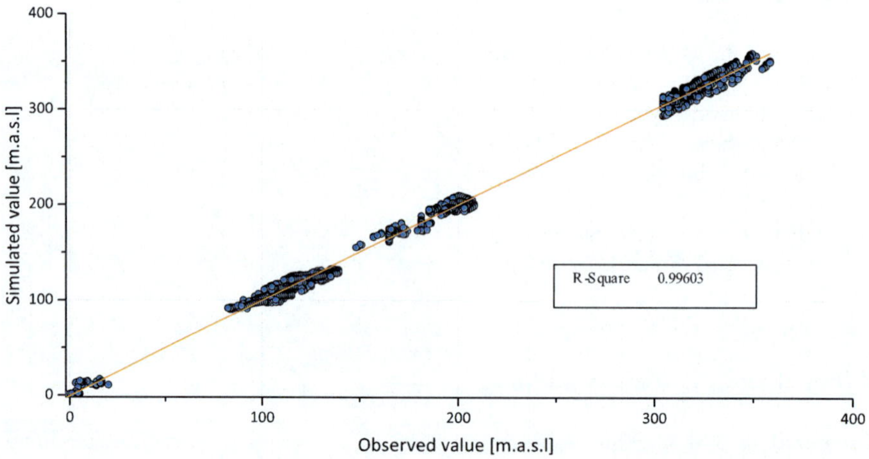


**Fig. 3** Observed and simulated piezometric heads of calibration wells. The simulated and observed data are distributed close to the 1:1 line with a correlation coefficient of 0.996. The ranges without points correspond to the areas where there are no wells available, and thus no measuring of piezometric heads

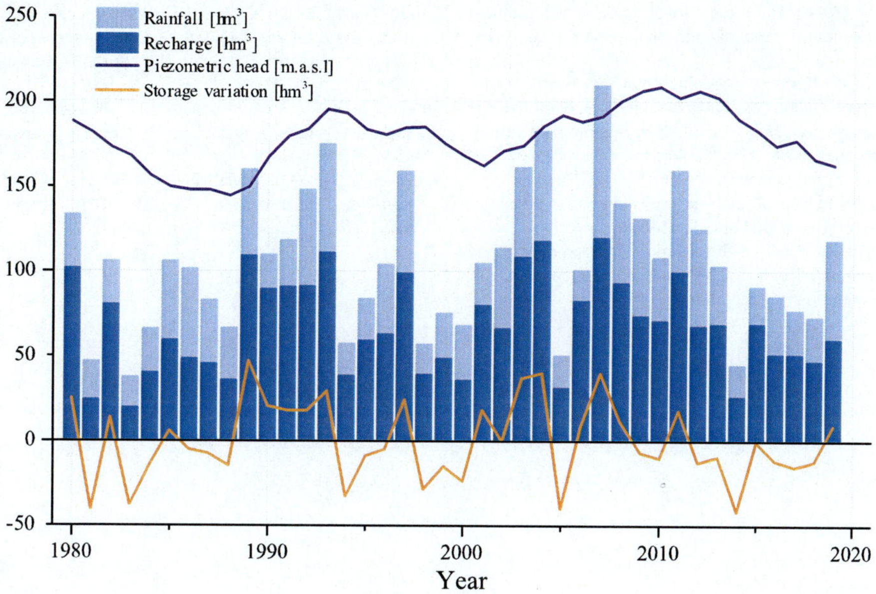


**Fig. 4** Rainfall, recharge, piezometric heads and storage variation comparison chart. There is an evident recharge response to rainfall, which determines the variation in heads and storage

Rosario, with a joint annual average of 54.44 $hm^3$, the Pego and Vergel drains with 11.36 $hm^3$/yr, and extractions through wells, with an annual average of 2.31 $hm^3$.

As seen in Fig. 4, precipitation and system recharge have varied through the period considered, with periods of greater and lesser precipitation being identifiable. Recharge is shown as a variable always proportional to precipitation. The simulated piezometric levels for a point in the western sector of the aquifer, near Benimasot, respond to variations in recharge. The orange line shows the variation in aquifer storage: values below zero are indicators that the aquifer is losing water volume, while positive values indicate that the aquifer is being recharged. The line of variation in storage and recharge have a similar behavior, showing that the aquifer is in equilibrium with the recharge through the studied period. It is then observed, for example, that in years with significantly lower rainfall (and, thus, recharge) the variation in storage is negative, and therefore the aquifer is losing water. In some periods, depending on the conditions in previous years, this is also accompanied by a decrease in piezometric levels. The dynamic is reversed in periods of higher rainfall.

## 5 Conclusions

Based on the hydrogeological information available on this sector of northern Alicante, we designed a conceptual model of groundwater flow for the studied aquifer. Thereafter, we built, calibrated, and validated a numerical model based on the conceptual model. The results show an acceptable representation of the real dynamics considering the absence of data referring to structures, secondary porosities, and the lack of depth drillings in the area, considering that the version of MODFLOW used for the construction of the model is, in theory, not ideal to represent the complex and heterogeneous nature of karstic aquifer systems.

The Almudaina-Segaria aquifer is a valuable resource for the northern sector of the province of Alicante, as it is the only source of drinking water for different towns and settlements, and as a source of support for local agriculture. Its natural dynamics depends mainly on rainfall and responds to wet and dry periods. For the studied period, the aquifer seems to be in equilibrium with recharge. The aquifer modeling constitutes the first stage in the study of the regional hydrology and the responsible management of different aquifers in the area, considering the occurrence of future scenarios in which the relevance of the groundwater resources becomes transcendental.

**Acknowledgements** This work was supported by the Department of Innovation, Universities, Science and Digital Society of the Generalitat Valenciana (GRISOLIAP/2020/124).

## References

Aslam RA, Shrestha S, Pandey VP (2018) Groundwater vulnerability to climate change: a review of the assessment methodology. Sci Total Environ 612:853–875. https://doi.org/10.1016/j.scitotenv.2017.08.237

Duran L, Gill L (2021) Modeling spring flow of an Irish karst catchment using Modflow-USG with CLN. J Hydrol 597. https://doi.org/10.1016/j.jhydrol.2021.125971

Fienen MN, Arshad M (2016) The international scale of the groundwater issue. In: Integrated groundwater management: concepts, approaches and challenges

Ford D, Williams P (2013) Karst hydrogeology and geomorphology

Harbaugh AW, Langevin CD, Hughes JD, Niswonger RN, Konikow LF (2017) MODFLOW-2005 version 1.12.00, the US geological survey modular groundwater model. US Geological Survey Software Release, 3 Feb 2017. https://doi.org/10.5066/F7RF5S7G

IGME-DPA (2015) Atlas hidrogeológico de la Provincia de Alicante. Instituto Geológico y Minero de España, Madrid

IPCC (2014) Climate change 2014: synthesis report. In: Contribution of working groups I, II and III to the fifth assessment report of the intergovernmental panel on climate change

Pulido-Velazquez D, Collados-Lara AJ, Alcalá FJ (2018) Assessing impacts of future potential climate change scenarios on aquifer recharge in continental Spain. J Hydrol 567. https://doi.org/10.1016/j.jhydrol.2017.10.077

Stevanović Z (2019) Karst waters in potable water supply: a global scale overview. Environ Earth Sci 78. https://doi.org/10.1007/s12665-019-8670-9

Winston RB (2020) ModelMuse version 4.3. US Geological Survey Software Release, 16 August 2020. https://doi.org/10.5066/P9XMX92F

# Karst Caves, Geomorphology, Landscape and Natural Heritage

# Understanding Karst Conduit Size Distribution by Numerical Speleogenesis Modeling

A. Maqueda, P. Renard, and M. Filipponi

**Abstract**

Large karstic conduits are rare within a rock formation, and therefore, the probability of encountering them during the construction of a tunnel is relatively low. However, the consequences can be dramatic for tunnel worker safety and/or the economic aspects of the tunnel realization, therefore, the risk must be accounted for. One of the critical aspects when evaluating the karst-related hazards is to estimate the statistics of the size of the conduits, as well as the connectivity of the karst network. This information is fundamental for a reliable risk assessment. Statistical analysis can be carried out from data collected by speleologists, but a large portion of the karst conduit network is not directly accessible, and therefore, the resulting statistics are incomplete. An alternative to assess the inaccessible areas of a karst conduit network is the use of numerical simulations of the speleogenesis processes. The premise is that the numerical simulation of groundwater flow on a connected fracture network, including mineral dissolution, will produce a conduit network that shares properties of real networks. We simulated the enlargement of fracture networks and conducted a statistical analysis of the results to obtain possible statistical distributions of conduit sizes and spatial distribution of the conduit network. These findings will better constrain hazards concerning the construction of tunnels. Furthermore, our results confirm numerically a conceptual model of staged cave development previously introduced.

A. Maqueda (✉)
AFRY Schweiz AG, Herostrasse 12, CH-8048 Zürich, Switzerland
e-mail: axayacatl.maqueda@afry.com

P. Renard
Centre d'hydrogéologie (CHYN), Rue Emile-Argand 11, CH-2000 Neuchâtel, Switzerland

M. Filipponi
National Cooperative for Disposal of Radioactive Waste (Nagra), Hardstrasse 73, CH-5430 Wettingen, Switzerland

**Keywords**

Karst • Mineral dissolution • Speleogenesis • Conduit statistics

## 1 Introduction

Karst aquifers develop when carbonate rocks are locally dissolved by groundwater. The global behavior of these aquifers is dominated by fast groundwater flow within the karst conduits. When tunnels are constructed in such environments, they face the risk of encountering a conduit, and when these accidents occur, the consequences can be dramatic because the tunnel can rapidly be flooded. Usually, large karst conduits are rare within a rock formation, and therefore, the probability of encountering them is relatively low, but the consequences being potentially very severe the risk must be accounted for. Casagrande et al. (2005), Filipponi et al. (2012) and Jeannin et al. (2015) report several examples of such events and discuss methods to evaluate the risks. One of the critical aspects when evaluating the karst-related hazards is to estimate the statistics of the dimensions of the conduits, as well as the connectivity of the karst network. The work presented in this paper aims to generate possible statistical distributions of the conduit size and spatial distribution of the karst conduit network to better constrain hazards concerning the construction of tunnels. For that purpose, we simulated speleogenesis processes and assessed how the distribution of the conduit size evolves before and after breakthrough flow in fractures.

## 2 Speleogenesis Conceptual Model and Software Development

We developed the FEFLOW 7 plug-in **K**arstification **S**imulation **P**lug-in (KSP) to simulate the enlargement of fractures by mineral dissolution. FEFLOW is a finite element software



B. Andreo et al. (eds.), *EuroKarst 2022, Málaga*, Advances in Karst Science,
https://doi.org/10.1007/978-3-031-16879-6_20

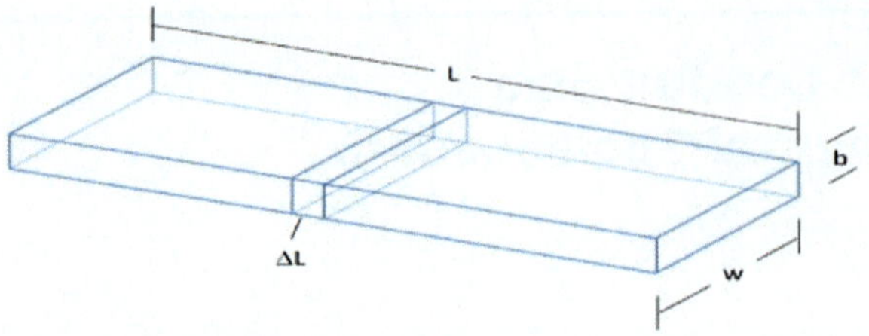


**Fig. 1** Fracture enlargement conceptual model

to simulate groundwater flow and reactive transport in porous and fractured media. The current development is based on the research work and past software development at the University of Neuchâtel (Maqueda 2017). The modeling approach is based on the conceptual model presented by Dreybrodt et al. (2005). The model represents the feedback loop between mineral dissolution and the flow of reactive water through fractures. The model computes the growth of fracture opening due to mineral dissolution and then the flow field is updated. The basic model assumes water with a known solute concentration flows through a planar fracture (Fig. 1).

The conceptual model relies on two assumptions. The first assumption is the fracture walls are assumed to be pure calcite rock ($CaCO_3$) therefore soluble by reactive water. The second assumption is the dissolution reaction occurs at the rock surface only, and the effect of calcite dissolution is a retreat of the fracture wall. FEFLOW can simulate both laminar and turbulent flow in fractures. KSP accounts in addition for the transition between laminar and turbulent flow and can estimate the wall retreat for both flow regimes. Water flow in fractures is described by the Hagen-Poiseuille equation for laminar flow and Manning–Strickler for turbulent flow. Mass transport is described by the advective-diffusion equation applied to only one dissolved species: calcium ion ($Ca^{2+}$). The simulation of calcite dissolution is based on the kinetics chemistry model of Dreybrodt et al. (2005) which was developed from laboratory experiments of pure calcite dissolution under laminar and turbulent flow conditions. Mineral dissolution is the only driver of fracture growth with KSP. Other processes known to contribute to the growth of karst conduits in nature are not accounted for: erosion of walls by suspended solids in water, a local increase of water acidity (e.g., pyrite weathering), or rock detachment along the conduit walls due to mechanical stress.

## 3 KSP Benchmark Test

We first applied KSP to replicate a simulation of the enlargement of a single fracture published by Dreybrodt et al. (2005) to test the KSP code. The available data include the results of a numerical simulation and an analytical approximation for the transition from laminar to turbulent flow. There are two fundamental differences between the original paper and our implementation. In the benchmark, flow occurs only in fractures and hydrodynamic dispersion of mass is not accounted for. In our simulation, fractures are embedded in a porous rock matrix with low hydraulic conductivity ($<1 \times 10^{-6}$ m/s), where flow and solute transport still occur at a minimal rate, and we use an advective-dispersion transport model that considers hydrodynamic dispersion. Hydrodynamic dispersion accounts for the heterogeneity in water flow velocity in fractures.

The benchmark model consists of a rectangular fracture having a width of 1 m and an initial aperture of 0.0002 m (0.2 mm) and a length of 1000 m. The hydraulic boundary condition is a constant hydraulic head of 50 m at the inlet point and 0 m at the outlet. The solute boundary condition is water fully unsaturated with calcite at the inlet. The calcite equilibrium concentration is 2 mmol/l. The linear and fourth-order reaction kinetics constant values are $4 \times 10^{-7}$ and $4 \times 10^{-4}$, respectively.

Figure 2 (top) presents the evolution of fracture aperture in both the KSP simulation (dashed lines) and the benchmark (solid lines). At simulation times of 13,100 and 17,800 yrs., the fracture aperture in both simulations is nearly identical. The transition to turbulent flow occurs at 18,850 yrs., and the fracture apertures in the KSP simulation are slightly smaller than the benchmark (blue lines). The greatest difference in fracture aperture is observed at a simulation time of 19,032 yrs. The deviations can be explained by differences due to the presence of porous media in KSP and the different transport equations. At a simulation time of 19,152 yrs., the difference in fracture aperture is reduced after flow becomes turbulent. Figure 2 (bottom) presents the evolution of flow rate for both KSP simulation and benchmark. Simulated flow rates with KSP are very similar before and after the transition to laminar flow (nearly vertical increase in flow rate). Only a small difference is observed by the end of the simulation.

We conclude from this test that KSP reproduces the main trends of fracture aperture and flow rate evolution of the benchmark. In addition, it can model the transition between laminar and turbulent flow. Finally, we consider that this test shows that KSP is capable to simulate reasonably well fracture growth from $10^{-4}$ m to 10 m. In the next section, we apply KSP to fracture networks.

## 4 Speleogenesis Simulation in Synthetic Fracture Networks

We apply KSP to investigate the evolution of synthetic fracture networks under conditions that we consider closer to aquifers in nature. The groundwater flow rate is limited by precipitation and subsoil recharge in the vadose zone, see the conceptual model in Fig. 3. The top boundary of the model

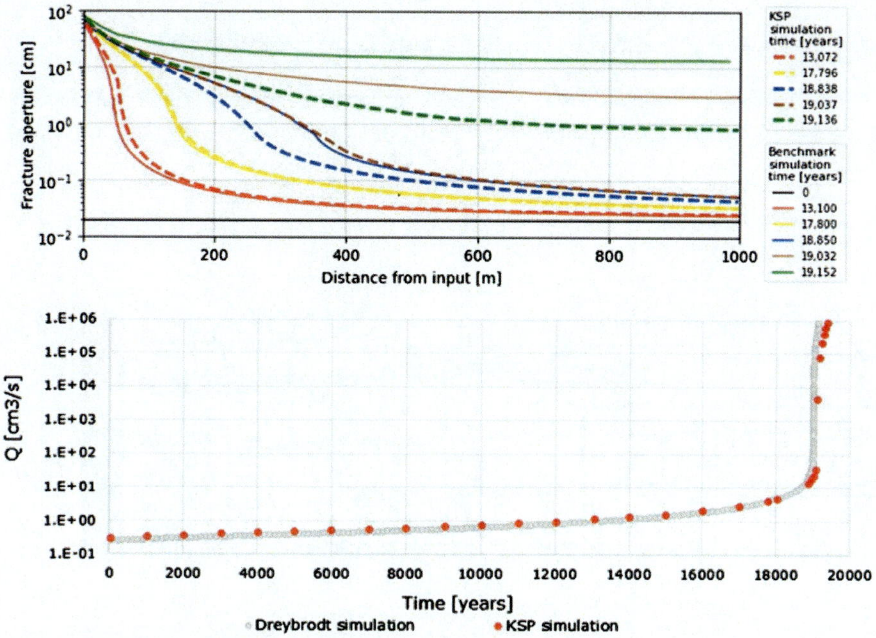


**Fig. 2** Evolution of fracture aperture and flow rate in Dreybrodt benchmark and KSP simulation

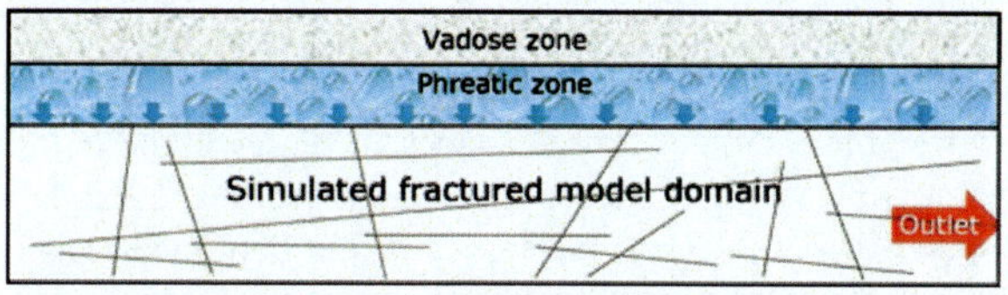


**Fig. 3** Conceptual model of a fracture network under phreatic conditions in a carbonate aquifer

is assumed to be the bottom of a shallow unconfined aquifer. Water inflow is not constant because rainfall is occasional. The vadose zone and an aquifer on top of the model domain act as buffers for intermittent precipitation. The aquifer provides a nearly constant hydraulic head on top of the simulated domain. When fracture apertures are small (mm), the hydraulic gradient is high because small fractures have a great resistance to flow. When fracture apertures grow to the size of karst conduits, groundwater recharge is the limiting factor. Therefore, we assumed a hydraulic boundary condition of high initial constant head and maximum flow rate.

The model domain for a fracture network has a size of 2000 m horizontally and 500 m vertically. The initial fracture network was generated using an object-based model implemented in Python code. We defined 4 families of discontinuities that correspond to a simplified realistic situation where a rock massif is in extension. The horizontal family represents bedding planes or horizontal discontinuities. The vertical one represents sub-vertical tension cracks, and in addition, there are two conjugate families of extension fractures. All the fractures are simulated independently. Their position is generated following a Poisson random point process with a density that is different for every fracture family. The distribution of the length of the fractures follows a truncated power-law distribution, with an exponent that has been kept constant. The orientations follow a von Mises distribution for each fracture family. All the parameters of those statistical distributions are provided in Table 1. In total, 6257 fractures were generated, and the total accumulated fracture network length is 57,627.83 m.

In numerous geological environments, it has been shown that fracture apertures are variable in space. Their apertures are spatially correlated at multiple scales (Tatone and Grasselli 2012). To generate a simple but plausible initial distribution of fracture aperture, we modeled it using a random multi-Gaussian field. The method is described in detail in Chilès and Delfiner (2009). The generation of the apertures is conducted in two steps. First, we simulated a correlated random multi-Gaussian field $Z(x, y)$ which has a Gaussian marginal distribution. The field is generated using the Sequential Gaussian Simulation (SGS) algorithm within the Ar2GEMS software. The parameters used for the simulation are: (i) a Gaussian variogram model with a range equal to 150 m in the horizontal direction and 50 m in the vertical direction, and (ii) a mean equal to 0 and a variance equal to 1. The resulting distribution of apertures for the 2000 m by 500 m model domain has a log-normal distribution with

**Table 1** Input parameters to generate initial fracture network

| Discontinuity family | Sub-horizontal fractures | Sub-vertical fractures | Conjugate fractures 1 | Conjugate fractures 1 |
|---|---|---|---|---|
| Min orientation [°] | 88 | −2 | 20 | −40 |
| Max orientation [°] | 92 | 2 | 40 | −20 |
| Min length [m] | 50 | 50 | 20 | 20 |
| Max length [m] | 500 | 300 | 300 | 300 |
| Density | $2.5 \times 10^{-4}$ | $2.5 \times 10^{-4}$ | $1.4 \times 10^{-4}$ | $1.4 \times 10^{-4}$ |

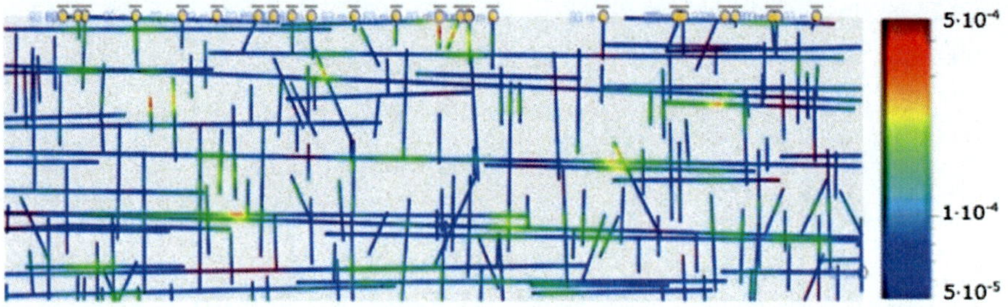


**Fig. 4** Hydraulic boundary conditions and initial fracture apertures

aperture values from $5.0 \times 10^{-5}$ m to $5.0 \times 10^{-4}$ m. Figure 4 presents the fracture network with the initial fracture aperture values. Fractures are discretized into smaller elements which have a unique aperture value. The hydraulic boundary conditions are 46 recharge nodes on the top model boundary and a single outlet node in the right boundary near the bottom. The hydraulic boundary condition is a constant hydraulic head of 33 m limited to a maximum flow rate of 1 l/s.

Figure 5 presents the evolution of flow rate before stabilization for scenario A, where we observe 4 sudden increases

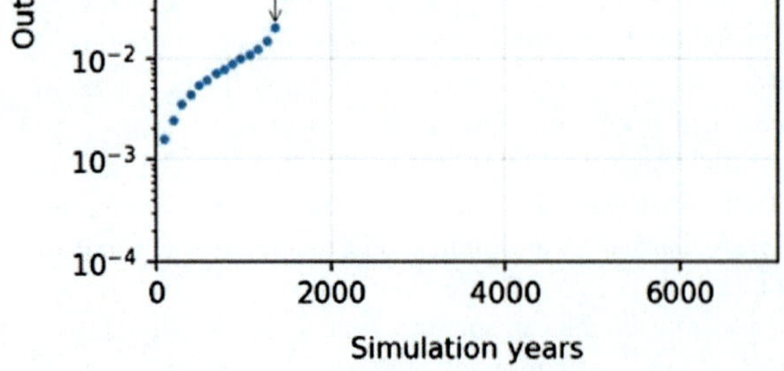


**Fig. 5** Breakthrough times (1, 2, 3, 4) observed in flow rate evolution

in flow rates. Every increase in the outflow is related to a breakthrough of enlarged fractures connecting the inlet nodes with the outlet (see Fig. 6). The subnetwork of initial fractures that suddenly connects inlets and the outlet are identified and are highlighted with a dashed rectangle. Fracture growth in these regions caused sudden increases in the outflow rate. Increase 1 is the consequence of 1st breakthrough presented in Fig. 6 (top). Increase 2 in the outflow rate is caused by the second breakthrough (Fig. 6 second from top), and so on until the last increase in outflow rate caused by the 4th and final breakthrough as presented in Fig. 6 (bottom). At the outflow stabilization time, the sum of the length of fractures that transitioned to turbulent flow is 7730 m out of a total fracture length of 57,627 m (13%).

The breakthrough of enlarged fractures influences the fracture aperture distribution as shown on the histograms of fracture aperture (see Fig. 7). At time $T = 0$ yrs., the fracture aperture has a log-normal distribution with a mode or peak 0 (histogram on the top left). At $T = 1{,}560$ yrs., the 1st breakthrough causes an aperture mode (peak 1) of ~ 0.1 m. At $T = 1{,}855$ yrs., the 2nd breakthrough causes a 2nd aperture mode (peak 2) of ~ 0.1 m. By then, the fractures of the 1st breakthrough have a new mode of ~ 0.3 m (peak 1). At $T = 3{,}618$ yrs., the third breakthrough causes another aperture mode (peak 3) of ~ 0.1 m. At this time, peaks 1 and 2 of mode have merged into a single peak of ~ 1 m. At $T = 5{,}667$ yrs., the 4th and final breakthrough occur and yet another mode (peak 4) of ~ 0.2 m emerges. By then, peaks 1, 2 and 3 have almost converged to an aperture between 1 and 3 m and can be regarded as karst conduits. At $T = 5{,}667$ yrs., inflow rate stabilizes (see Fig. 5), and the simulation is

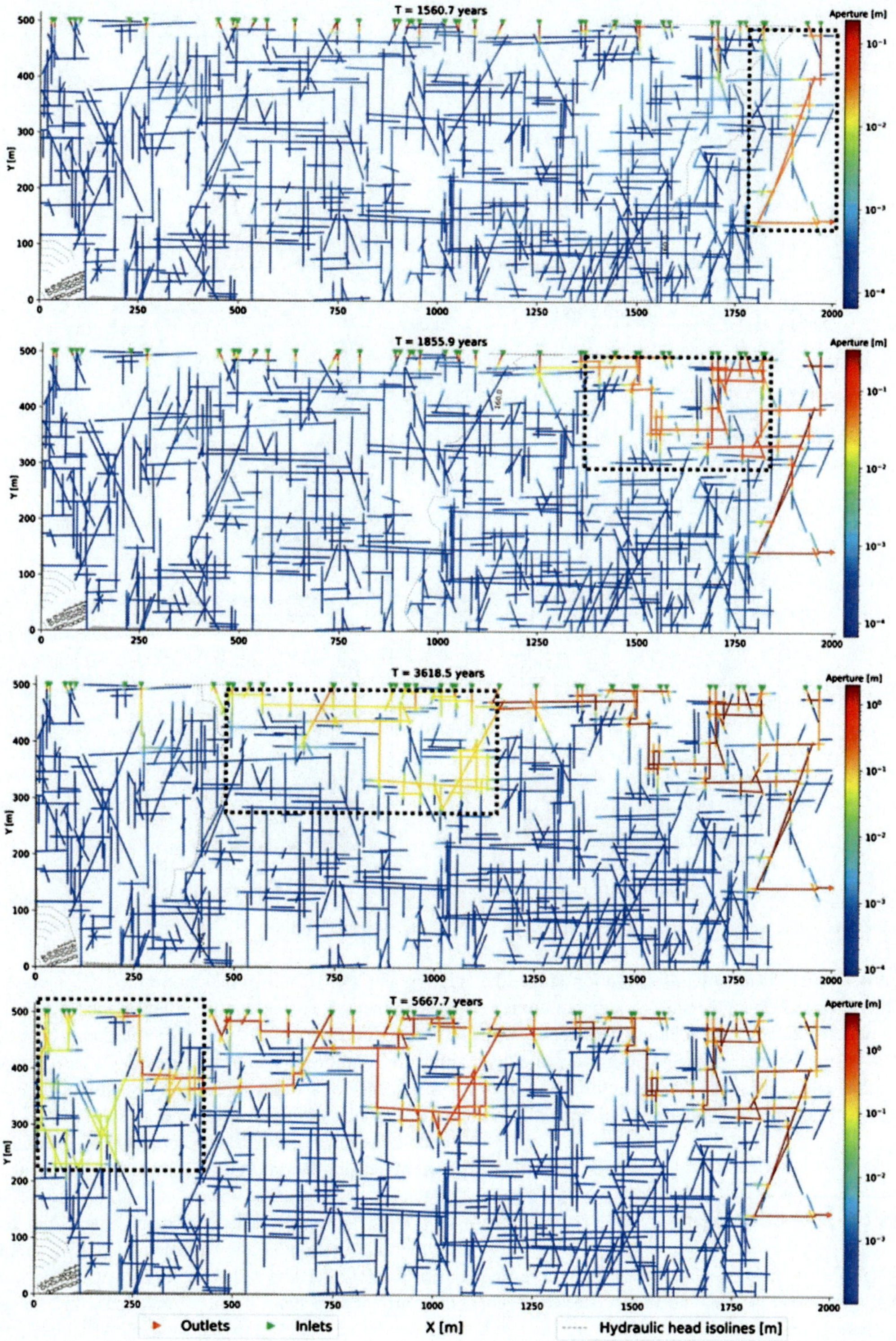


**Fig. 6** 1st (1,561 yrs.), 2nd (1,857 yrs.), 3rd (3,618 yrs.), and 4th breakthrough (5,677 yrs.) and stabilization of flow rate to 1 l/s at every inlet fracture

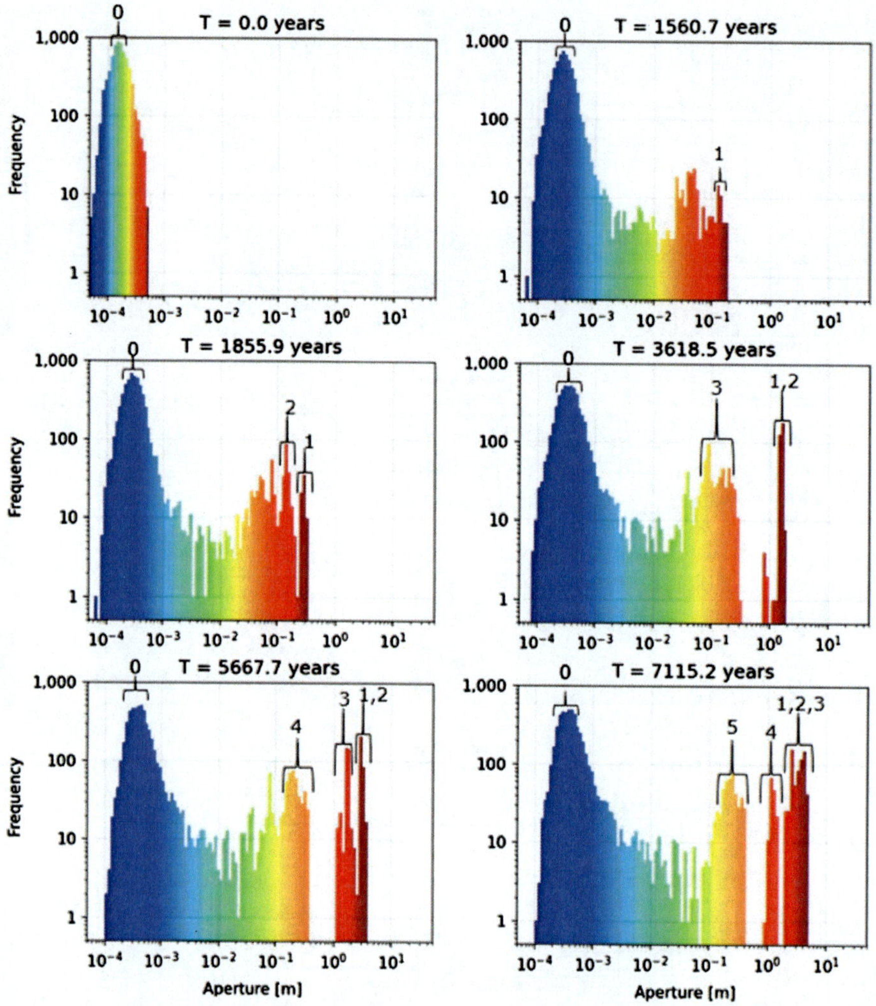


**Fig. 7** Histograms of fracture aperture at 1st, 2nd, 3rd, 4th breakthrough and final simulation time

left to run until $T = 7{,}115$ yrs. At this evolution stage, the fractures of the 4th breakthrough (peak 4) almost converge with previous peaks 1, 2 and 3, and a 5th mode emerges with an aperture $\sim$ 0.2 m emerges. Since the inflow rate has stabilized, the flow in these fractures does not transition from laminar to turbulent flow and it is not expected that fractures of peak 5 grow into karst conduits.

## 5 Discussion and Conclusions

One striking feature of our results is that we observe and describe quantitatively that the karstification process does not occur homogeneously in a basin, but instead, it proceeds in a series of multiple breakthroughs corresponding to the different stages of karstification (Fig. 8) as it was proposed earlier by Filipponi (2009). The original model included 3 phases in cave development: (i) inception, (ii) gestation and (iii) development. The inception phase is the start of dissolution in fractures under phreatic conditions. The outlet or spring is assumed to be the consequence of a valley incision in a soluble rock massif. The outlet organizes flow in the aquifer and at time 1 a cave gestation zone emerges. This is comparable to the initial conditions of our simulation. At time 2, breakthrough occurs, and the new karst conduit network acts as a spring for the upstream section of the model, which is comparable to the 1st breakthrough in our simulation. The karst conduits (cave development phase) offer less resistance to flow, thus the water table drops (hydraulic head in our simulation), and the gestation and inception zones move upstream. At time 3, the cave development keeps advancing upstream, which is comparable to the 2nd, 3rd and 4th breakthroughs in our simulation.

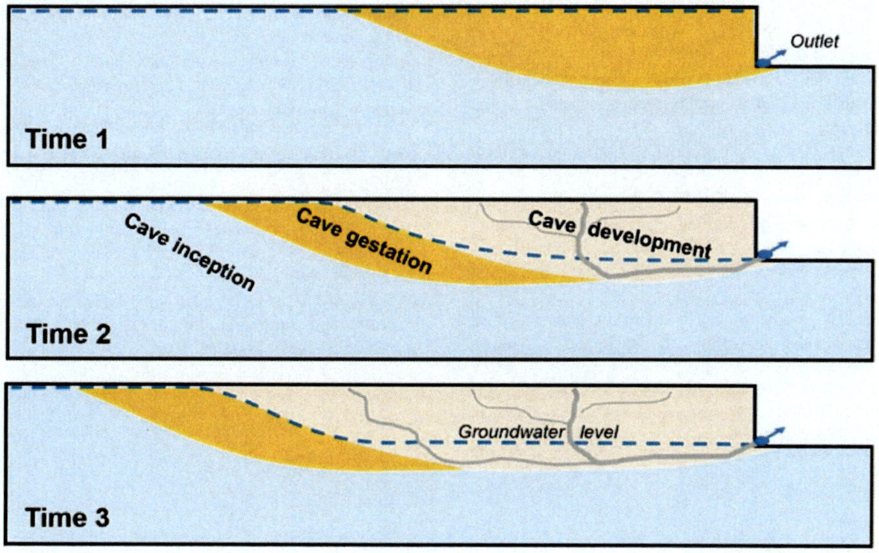


**Fig. 8** Conceptual model for the special development of karst conduit network

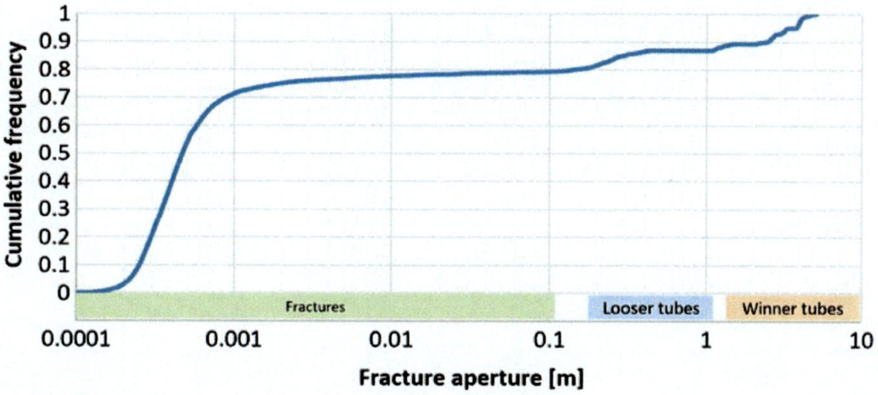


**Fig. 9** Cumulative fracture aperture at simulation time $T = 7115$ yrs.

The growth process is heterogeneous, only 13% of the initial fracture length transitioned from laminar to turbulent flow. Fracture paths with greater resistance to flow stay at the initial size, and this is observed qualitatively in Fig. 6 and quantitatively in the histograms of Fig. 7. Figure 9 presents the cumulative frequency of fracture aperture at the end of the simulation. The distribution is trimodal with a first local maximum at undeveloped or slightly enlarged fractures smaller than 1 mm. A second mode at a size between 10 and 40 cm represents karst conduits that developed until the inflow rate stabilized. A third mode represents the karst conduits that continue growing while capturing most of the inflow after flow rate stabilization. By observing the evolution of the geometry of the fracture network and the aperture distribution, we classify them as fractures, looser tubes and winner tubes.

Finally, an important result is the statistical distribution of conduit sizes. In this framework, we show that it tends to develop into a multimodal log-normal distribution even if the setup is very simple. The magnitude of the larger conduits, or winner tubes and their statistical distribution seems to be reasonable as compared to field observations in caves as presented in histograms of conduit size in Frantz et al. (2021) and Maqueda (2017).

## References

Casagrande G, Cucchi F, Zini L (2005) Hazard connected to railway tunnel construction in karstic area: applied geomorphological and hydrogeological surveys. Nat Hazards Earth Syst Sci 5:243–250

Chiles JP, Delfiner P (2009) Geostatistics: modeling spatial uncertainty. Wiley, New York

Dreybrodt W, Gabrovšek F, Romanov D (2005) Process of speleogenesis: a modelling approach. ZRC Publishing, Postojna-Ljubijana

Frantz Y, Collon P, Renard P, Viseur P (2021) Analysis and stochastic simulation of geometrical properties of conduits in karstic networks. Geomorphology 377

Filipponi M (2009) Spatial analysis of karst conduit networks and determination of parameters controlling the speleogenesis along preferential lithostratigraphic horizons (Doctoral dissertation). EPFL, Laussane

Filipponi M, Parriaux A, Schmassmann S, Jeannin P-Y (2012) KarstALEA: Wegleitung zur Prognose von karstspezifischen Gefahren im Untertagbau. Forschung im Strassenwesen Bericht 1395. UVEK, ASTRA, Bern

Jeannin YP, Malard A, Rickerl D, Weber E (2015) Assessing karst hydraulic hazards in tunneling the Brunnmühle spring system Bernese Jura, Switzerland. Environ Earth Sci 74(12):7655–7670

Maqueda A (2017) Process based modeling of the karstification processes in the region of Tulum, Mexico. PhD thesis. University of Neuchâtel, Switzerland

Tatone BS, Grasselli G (2012) Quantitative measurements of fracture aperture and directional roughness from rock cores. Rock Mech Rock Eng 45(4):619–629

# Unroofed Cave—An Underground Form on the Karst Surface

M. Knez and T. Slabe

**Abstract**

According to origin, karst caves are, of course, an underground phenomenon, but in the development of the karst, they frequently appear as surface phenomena that we call "unroofed caves". Due to the denudation, dissolution and disintegration of carbonate rock and the subsequent lowering of the karst surface, which can be many dozen metres in a million years, underground karst caves are uncovered and with further dissolution can gradually disappear completely. Here we show the frequency, significance and characteristic forms of unroofed caves and place them correctly in the development scheme of the karst cycle. Many years of research of the Classical Karst in Slovenia have shown that unroofed caves are a relatively frequent surface karst form, certainly more frequent that we imagined before expressway construction works in Slovenia's karst regions uncovered the karst surface. The first attempts to typify their characteristic forms and to explain their formation were born. Our results show the development from old caves to unroofed caves due to the lowering and dissection of the karst surface. Old caves are preserved by sediments, primarily alluvium and flowstone. We present forms of unroofed caves and determine their significance in studying the cavernosity of karst aquifers.

**Keywords**

Unroofed cave • Classical Karst • Slovenia

Caves are one of the most typical karst forms. Water flows through carbonate rock and through dissolution and mechanical erosion carves caves during various stages in the development of karst aquifers. Denudation also lowers the karst surface up to many dozen metres in a million years (Gams 1965) and uncovers caves that we call "unroofed caves" (Figs. 1 and 2d, e). Many years of research of the Classical Karst in Slovenia (Kranjc 1997) have shown that unroofed caves are a relatively frequent surface karst form, more frequent that we imagined before recent expressway construction works in Slovenia's karst regions uncovered the karst surface (Knez and Slabe 2001, 2002, 2004a, b, 2005, 2006, 2007; Knez et al. 1994, 2004, 2008; Kogovšek 1995; Kogovšek and Petrič 2004; Kogovšek et al. 1997; Mihevc 1999, 2001; Slabe 1996, 1997, 1998; Šebela and Mihevc 1995). These are karst phenomena with an entirely different origin than those that develop as surface phenomena due to various other factors, but their proportion among forms found on the karst surface is considerable and significant (Knez and Slabe 1999, 2004a, 2005, 2007; Mihevc et al. 1998). Knowledge of their origin and various forms is therefore an essential condition for understanding the formation and development of the three-dimensional karst landscape that occupies about twenty per cent of the world's surface (Ford and Williams 2007).

Only caves that were completely filled with fine-grained sediments by flooding waters were preserved in their original form after the lowering of the water table and the denudation of the karst surface. The ratio between the speed of the washing away of sediment from the exposed caves and of the lowering of the surrounding surface as well as the form of the caves and of the lowering surface that cuts them determine the distinctiveness and form of unroofed caves as a surface phenomenon.

M. Knez (✉) · T. Slabe
Research Centre of the Slovenian Academy of Sciences and Arts, Karst Research Institute, Titov trg 2, 6230 Postojna, Slovenia
e-mail: knez@zrc-sazu.si

e-mail: slabe@zrc-sazu.si

UNESCO Chair On Karst Education, University of Nova Gorica, Glavni trg 8, 5271 Vipava, Slovenia

International Joint Research Center for Karstology, Yunnan University, Xueyun rd. 5, Kunming, CN-650223, China


B. Andreo et al. (eds.), *EuroKarst 2022, Málaga*, Advances in Karst Science,
https://doi.org/10.1007/978-3-031-16879-6_21

**Fig. 1** During the motorway construction, a softer sediment was discovered among limestones at the level of the future roadway. To preserve static stability, the sediment was removed, which revealed the bottom part of a cave passage—an unroofed cave

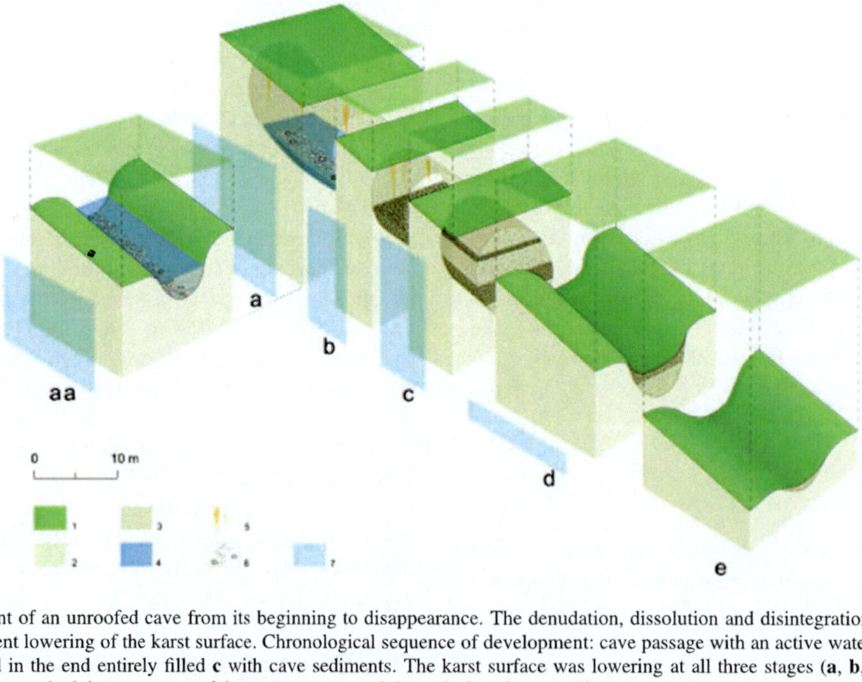


**Fig. 2** Development of an unroofed cave from its beginning to disappearance. The denudation, dissolution and disintegration of carbonate rock caused the subsequent lowering of the karst surface. Chronological sequence of development: cave passage with an active water flow **a** which was later partially **b** and in the end entirely filled **c** with cave sediments. The karst surface was lowering at all three stages (**a**, **b**, **c**). At stage d, the lowered karst surface reached the upper part of the cave passage and the majority of cave sediments was removed. Today, the last remnants of cave sediments can be seen on the surface. Legend: 1 current karst surface; 2 former karst surface; 3 circumference of passage; 4 water; 5 flowstone; 6 sand and pebbles; 7 flooded zone. Drawn by Tamara Korošec

If the sediment from unroofed caves is washed away slowly, they appear to be just part of the karst surface with unique soil and vegetation or we can possibly find just cave sediment and flowstone lying on the surface. With the more rapid washing away of sediment, unroofed caves acquire more distinctive forms. When the surface cuts the vertical part of a cave network, forms similar to dolines develop. Dolines, large funnel- and bowl-shaped hollows that along with karst poljes, collapse dolines and karren are among the most common karst phenomena, develop with the vertical percolation of rainwater through the karst surface. The diameter of unroofed caves similar to dolines can reach many metres. They are like half-dolines when a steep slope due to denudation cuts a gently sloping or horizontal passage. Unroofed caves similar to dolines can also develop in the middle of old passages that have opened in several places. They frequently reveal the direction and size of more extensive caves. Oblong depressions develop from old passages without roofs that run parallel to the surface (Figs. 1 and 2d, e). They can be several hundred metres long and uniform, interrupted by preserved roofs or cut by dolines.

Under the fine-grained sediments in unroofed caves, there are frequently very well preserved cave walls with rock relief forms, the traces of the last dominant factors of their formation and various types of flowstone. Paleomagnetic studies of the fine-grained sediment and datation of flowstone reveal that the sediments found in the unroofed caves that occur in the Classical Karst are several million years old (Bosák et al. 2007; Zupan Hajna et al. 2008). The cave sediments from profiles near Kozina are older than the Brunhes/Matuyama border (0.78 million years). From the distribution of individual magnetozones, it is possible to confirm that the sediment is older than the end of the Olduvai Chron (1.77 million years) since the magnetostratigraphic profile ends with a magnetozone of reverse polarity and contains two zones of normal polarity. The fossilization of the cave system and consequently the preservation of certain cave passages entirely filled with cave sediments was connected with the rapid tectonic uplift of the area when around 5.3 million years ago the Strait of Gibraltar opened and the Mediterranean Basin filled again. Further fossilization was the result of changes in the regional water level and hydrological conditions that occurred with the gradual development of the surface and neotectonics of this part of the Classical Karst as well as changes in the sea level in the Mediterranean Basin.

With the uncovering of larger cave networks, various forms of unroofed caves formed one beside the other. The forms of unroofed caves are often also interwoven with other distinct forms of the formation of the karst surface, dolines and karren. One such network, which was uncovered during the construction of the expressway on the Classical Karst, was visible on the surface as a network of more or less distinct oblong depressions and shallow dolines. In addition to sections without roofs, the cave network also included a smaller hollow passage known before the construction work and passages with roofs. Most of the cave was filled with fine-grained sediment and flowstone (Fig. 3). Sharp-edged gravel, which in some places covered the surface or filled the caves, is the result of the disintegration of the rock surface during the cold Pleistocene period. We found it frequently above old sediments or filling spaces below overhangs at the beginning of passages. The bottoms of larger dolines formed on the edge of passages of the old cave network were covered with several-metre-thick layers of brown and red soil. On their walls were rock forms that distinctly showed the percolating of water downward and at their bottoms were the mouths of narrow shafts (Knez and Slabe 1999).

A special form of unroofed caves originated from glacial activity, one of the most important factors in the dissection of mountain karst. With mechanical erosion, which lowered the surface and carved valleys, the glaciers uncovered and opened parts of cave networks from the side (Klimchouk et al. 2006).

The karst surface can lower more rapidly than the level of underground water and only in places due to the removed roof does the water become a surface feature, either as a water course or as standing water. One such example occurs in the remarkable Rakov Škocjan area near Postojna (Slovenia) where cave passages, natural bridges and forms similar to river valleys (Fig. 2a) that are actually unroofed passages alternate (Habič and Gospodarič 1989). In the event of a rapid local removal of the roof of an underground space, a collapse doline can form (Fig. 4). If the collapse includes a cave space, we can classify such a collapse doline as an unroofed cave.

We have learned that unroofed caves are relatively frequent karst surface forms, certainly more frequent than we imagined. During the fruitful karstological monitoring of a 70 km-long and on average 25-m-wide expressway right of way, 350 caves were discovered, of which 90 were unroofed caves filled with various sediments. They are ever more clearly readable phenomena of the karst surface. Our good knowledge enables us to recognize them through a thorough study of the surface and accurate topographical maps without uncovering the surface.

**Fig. 3** When making the road cut, the contractors cut through a cave passage just below the surface, which was entirely filled with cave sediments

Unroofed caves are part of the polygenetic formation of the karst surface and epikarst and vadose zones, the upper part of karst aquifers created by water percolating from the surface. They originated in the period of the development of this part of the aquifers in the flooded zone or at the water table before the lowering of the level of underground waters, as the sediment in them confirms. Until the underground water level dropped, water streams flowed through the caves in the Classical Karst, the region of Slovenia that gave its name to this unique landscape formed from carbonate rock and where karstology began to develop (Kranjc 1997). The aquifer was still highly enclosed and covered by flysch. In one of the last periods of the development of the caves, before which they were already dry and flowstone had been deposited in them, flooding waters filled them with fine-grained sediment. The sediment filled the fissures and was preserved during the long period of the aquifer's development when the surface lowered by several dozen metres. The sediment found on the karst surface was generally explained as traces of a former surface stream when the level of underground water was close to the surface, but now we know that it is largely cave sediment.

From the beginning of their formation to their disappearance (Fig. 2a–e), we can trace in the caves the development of the karst aquifer in different geological conditions and changing geomorphological, hydrological and climate conditions through an important part of the long-term karst cycle.

**Fig. 4** Due to the denudation of the karst surface, the ceiling of an empty cave passage thinned and collapsed, creating consecutive collapse dolines; the Škocjan Caves (Slovenia), UNESCO heritage site. *Photo* Škocjan Caves Park

**Acknowledgements** The authors acknowledge the financial support from the Slovenian Research Agency (research core funding No. P6-0119). Research was supported and included in the framework of projects »Development of research infrastructure for the international competitiveness of the Slovenian RRI Space—RI-SI-EPOS« and »Development of research infrastructure for the international competitiveness of the Slovenian RRI Space—RI-SI-LifeWatch«, both operations are co-financed by the Republic of Slovenia, Ministry of Education, Science and Sport and the European Union from the European Regional Development Fund, the UNESCO IGCP project No. 661 and UNESCO Chair on Karst Education.

## References

Bosák P, Knez M, Pruner P, Sasowsky I, Slabe T, Šebela S (2007) Palaeomagnetic research into unroofed caves opened up during the highway construction at Kozina, SW Slovenia. Annales 17(2):249–260

Ford D, Williams P (2007) Karst hydrogeology and geomorphology. Wiley, Chichester, p 562

Gams I (1965) H kvartarni geomorfogenezi ozemlja med Postojnskim, Planinskim in Cerkniškim poljem (summary: On the Quaternary geomorphogenesis of the area among the karst poljes of Postojna, Planina and Cerknica). Geografski Vestn 37:61–101

Habič P, Gospodarič R (1989) Rakov Škocjan karst valley. Int J Speleol 18(1–2):43–48

Klimchouk A, Bayari S, Nazik L, Törk K (2006) Glacial destruction of cave systems in high mountains, with special reference to the Aladaglar massif, Central Taurus, Turkey. Acta Carsolog 35(2):111–121

Knez M, Slabe T (1999) Unroofed caves and recognising them in karst relief (discoveries during motorway construction at Kozina, South Slovenia). Acta Carsolog 28(2):103–112

Knez M, Slabe T (2002) Unroofed caves are an important feature of karst surfaces: examples from the classical karst. Z Geomorphol 46 (2):181–191

Knez M, Slabe T (2004a) Karstology and the opening of caves during motorway construction in the karst region of Slovenia. Int J Speleol 31(1):159–168

Knez M, Slabe T (2004b) Highways on karst. In: Gunn J (ed) Encyclopedia of caves and karst science. Fitzroy Dearborn, New York, pp 419–420

Knez M, Slabe T (2006) Dolenjska subsoil stone forests and other karst phenomena discovered during the construction of the Hrastje-Lešnica motorway section, Slovenia. Acta Carsolog 35(2): 103–109

Knez M, Slabe T (2007) Krasoslovna spremljava gradnje, raziskave ter načrtovanje avtocest prek slovenskega krasa. Kraški pojavi, razkriti med gradnjo slovenskih avtocest (summary: Karst features discovered during motorway construction in Slovenia). ZRC Publishing, Postojna-Ljubljana, pp 9–22

Knez M, Kranjc A, Otoničar B, Slabe T, Svetličič S (1994) Posledice izlitja nafte pri Kozini (The effect of oil split near Kozina). Ujma 9:74–80
Knez M, Slabe T, Šebela S (2004) Karstification of the aquifer discovered during the construction of the expressway between Klanec and Črni Kal, Classical Karst. Acta Carsolog 33(1):205–217
Knez M, Slabe T, Šebela S, Gabrovšek F (2008) The largest karst caves discovered in a tunnel during motorway construction in Slovenia's Classical Karst (Kras). Environ Geol 54(4):711–718
Knez M, Slabe T (2001) Karstology and expressway construction. In: Proceedings of the 14th IRF road world congress (on CD-ROM), Paris
Knez M, Slabe T (2005) Caves and sinkholes in motorway construction, Slovenia. In: Waltham T, Bell F, Culshaw M (eds) Sinkholes and subsidence. Karst and cavernous rocks in engineering and construction. Springer, Praxis, Chichester, pp 283–288
Kogovšek J (1995) Podrobno spremljanje kvalitete vode, odtekajoče z avtoceste in njen vpliv na kraško vodo (Detailed monitoring of the quality of the water that runs off the motorway and its impact on karst water). Annales 5(7):149–154
Kogovšek J, Petrič M (2004) Advantages of longer-term tracing—three case studies from Slovenia. Environ Geol 47(1):76–83
Kogovšek J, Slabe T, Šebela S (1997) Motorways in Karst (Slovenia). In: Byerly DW (ed) Proceedings and a fieldtrip excursion guide of the 48th Highway geology symposium. University of Tennesee, Department of Geological Studies, Knoxwille, pp 49–55
Kranjc A (ed) (1997) Kras: Slovene classical karst. ZRC Publishing, Ljubljana, p 254
Mihevc A, Slabe T, Šebela S (1998) Denuded caves–an inherited element in the karst morphology: the case from Kras. Acta Carsolog 27(1):165–174
Mihevc A (1999) The caves and the karst surface–case study from Kras, Slovenia. In: Karst 99: Colloque européen Des paysages du karst au géosystème karstique: dinamiques, structures, et enregistrement karstiques. Études de Géographie Physique, suppl 28, Aix-en-Provence, pp 141–144
Mihevc A (2001) Geomorfološko kartiranje na delu trase AC Razdrto–Vipava (Rebernice), ki poteka v območju krajinskega parka (Geomorphological Mapping on Motorway Section Razdrto–Vipava). Karst Research Institute, Postojna, p 49
Šebela S, Mihevc A (1995) The problems of construction on karst–the examples from Slovenia. In: Beck BF, Pearson FM (eds) Karst Geohazards, engineering and environmental problems in karst terrain. In: Proceedings of the 5th multidisciplinary conference on dolines and the engineering and environmental impacts on karst. A. A. Balkema, Rotterdam, pp 475–479
Slabe T (1996) Karst features in the motorway section between Čebulovica and Dane. Acta Carsolog 25:221–240
Slabe T (1997) Karst features discovered during motorway construction in Slovenia. Environ Geol 32(3):186–190
Slabe T (1998) Karst features discovered during motorway construction between Divača and Kozina. Acta Carsolog 27(2):105–113
Zupan Hajna N, Mihevc A, Pruner P, Bosák P (2008) Palaeomagnetism and Magnetostratigraphy of Karst Sediments in Slovenia. Carsologica 8, Založba ZRC, Ljubljana, p 266

# Geophysical Researches to Detect Karst Caves in the Main Polje of Apulia

G. Romano, M. De Girolamo, and M. Parise

**Abstract**

Canale di Pirro, the main polje in Apulia, extends for 12 km in the Low Murge karst area. It hosts the deepest cave in the region, reaching the depth of −324 m. Geophysical analyses have been performed in correspondence of two sites of interest. The first is located about 0.5 km upstream of the cave and is the main swallow hole in the polje. During rainstorms, most of the water accumulated in this sector is drained at this site, with time of absorption ranging from a few hours to some days, in function of the saturation condition of the sediments. The survey was addressed to verify the existence of caves, and the possibility to open a passage in the cover deposits. By adopting a multi-array and multi-scale strategy, based on the acquisition and joint inversion of geoelectrical data collected by using different arrays and with electrode spacing ranging from 5 to 20 m, a high-resolution model of the electrical subsurface structure was obtained. With an investigation depth up to ~150 m below the ground and high surficial resolution, the model shows features compatible with the presence of large caves, providing useful information on their possible extension and the thickness of the cover deposits. The second site is a small sinkhole (3 m depth, 7 m diameter) within a vineyard in the eastern sector of the polje. The survey was addressed to detect likely ways of water infiltration at the site and possible presence of underground voids.

**Keywords**

Geophysics • Karst • Cave • Sinkholes • Apulia

## 1 Introduction

Canale di Pirro is the main polje in the Apulian karst of SE Italy, extending for more than 12 km in W–E direction in the Murge area (central sector of the region; Parise 2006, 2011; Pisano et al. 2020). Besides being a remarkable example of surface karst landscape, it hosts the deepest cave in the region *(Inghiottitoio della Masseria Rotolo)*, one of the two where cavers directly reach the water table at depth of −264 m below the ground surface (Parise and Benedetto 2018; Liso et al. 2020). Speleodiving explorations added further 60 m, so that the total depth of the system is −324 m.

Geologically, the bedrock in the area consists of Cretaceous limestones belonging to the formations of Bari Limestone and Altamura Limestone, with an overall sub-horizontal bedding. Pleistocene–Holocene residual deposits fill the flat bottom of the polje, reaching maximum exposed thickness of some 4–5 m at its borders. Low permeability of these deposits is at the origin of the periodic flooding registered in the polje after major rainstorms, as usual in the Apulian karst (Parise 2003; Mossa 2007; Cotecchia and Scuro 2010; Martinotti et al. 2017).

Following the recent explorations at *Inghiottitoio della Masseria Rotolo* (Parise and Benedetto 2018) and looking at other possible caves below the polje, it was decided to perform geophysical surveys in selected sites within the Canale di Pirro polje, where evidence of sinkhole and swallow holes were present. Actually, sinkholes of different origin (Gutierrez et al. 2014; Parise et al. 2018) represent the main surface landforms recognizable in wide sectors of Apulia (Zumpano et al. 2019), and they have been object of interdisciplinary studies, including geophysical research, at many sites of the region (Del Prete et al. 2010; Festa et al. 2012; Margiotta et al. 2012).

G. Romano · M. Parise (✉)
Department of Earth and Environmental Sciences, University Aldo Moro, Bari, Italy
e-mail: mario.parise@uniba.it

G. Romano
e-mail: gerardo.romano@uniba.it

M. De Girolamo
National Research Council, IMAA, Tito Scalo (Potenza), Italy
e-mail: michele.degirolamo@imaa.cnr.it


B. Andreo et al. (eds.), *EuroKarst 2022, Málaga*, Advances in Karst Science,
https://doi.org/10.1007/978-3-031-16879-6_22

In detail, the two selected sites (Fig. 1) are represented by a small swallow hole in the eastern sector of the polje, at the mouth of one of the tributary valleys, and by the Gravaglione. This latter is the main swallow holes in the polje, and is located in its central sector: it is an elongated depression which collects most of the surface runoff from the western sector of the catchment.

## 2 Data Collection and Analysis

The Gravaglione is located about 0.5 km upstream of the cave before mentioned, and is the main swallow hole in the polje. It is an elongated opening in the ground, where 4 m of terra rossa deposits are exposed. During rainstorms, most of the water accumulated in this sector of the polje is drained at this site, with a time of absorption ranging from a few hours to some days, also in function of the pre-storm saturation condition of the sediments. The survey was addressed to verify the existence of caves large enough to be explored by man, and therefore the possibility to open a passage in the cover deposits. The geophysical survey was aimed to obtain hints to confirm the presence of a cavity, and to provide information on its possible extension (both laterally than in depth).

The size, position, and depth of a subsurface cavity may be recognized through the implementation of several geophysical methods such as Georadar, refraction seismic, and geoelectric (Mochales et al. 2008; Cardarelli et al. 2010; Kaufmann et al. 2011; Kaufmann 2014; Das and Mohanty 2016; Leucci et al. 2016). Among these, in the operative framework here presented, we chose to apply the geoelectrical method in consideration of its investigation depth, capability of investigating large areas in a reduced time, and easy application. First, a multi-scale approach was adopted to assess the presence of the cave and it's in depth extension; secondly, investigations were carried to define its lateral continuity.

Several studies (Martinez-Lopez et al. 2013) highlighted the difficulties related to the clear detection of a cavity by using the geoelectrical method. Quality of cavity detection and characterization depends on the inter-electrodic spacing, as well as on the geoelectrical array. As regards the former factor, the cavity detection is optimal when the distance between electrodes is comparable to the diameter of the cavity. If this condition is not fulfilled, the geoelectrical investigation may only show vague traces of the cavity. As for the latter factor, Dipole–Dipole and Wenner–Schlumberger arrays seem to lead to the best results even if the final choice of the array (arrays) to use should be decided based upon the type of local geology.

To overcome these problems, the multi-scale approach consisted in the acquisition of three geoelectrical profiles (ERT) with inter-electrode spacing of 5 m (ERT1), 10 m (ERT2), and 20 m (ERT3), centered on the study area, oriented W-E and overlapping each other's (Fig. 2). Data were acquired by using Dipole–Dipole, Dipole–Dipole reverse, Wenner–Schlumberger, and Wenner configurations. In this contribution, the results of the joint inversion of the Dipole–Dipole and Dipole–Dipole data will be presented only.

All ERT images show a resistivity distribution which is generally compatible with the geological environment of the area. ERT1, having the highest surficial resolution, shows the presence of a shallow conductive layer ($\rho < 30\ \Omega$*m) at the topsoil, extending along the whole profile, down to a maximum depth of few meters from the ground surface. Below this layer, the subsurface is characterized in all ERTs by lateral and in depth resistivity variability. This feature could be indicative of a bedrock composed of Calcare di Altamura Fm. in a not uniform condition in terms of

**Fig. 1** Two sites object of this study: to the left, Gravaglione; to the right, the sinkhole at the eastern sector of the polje

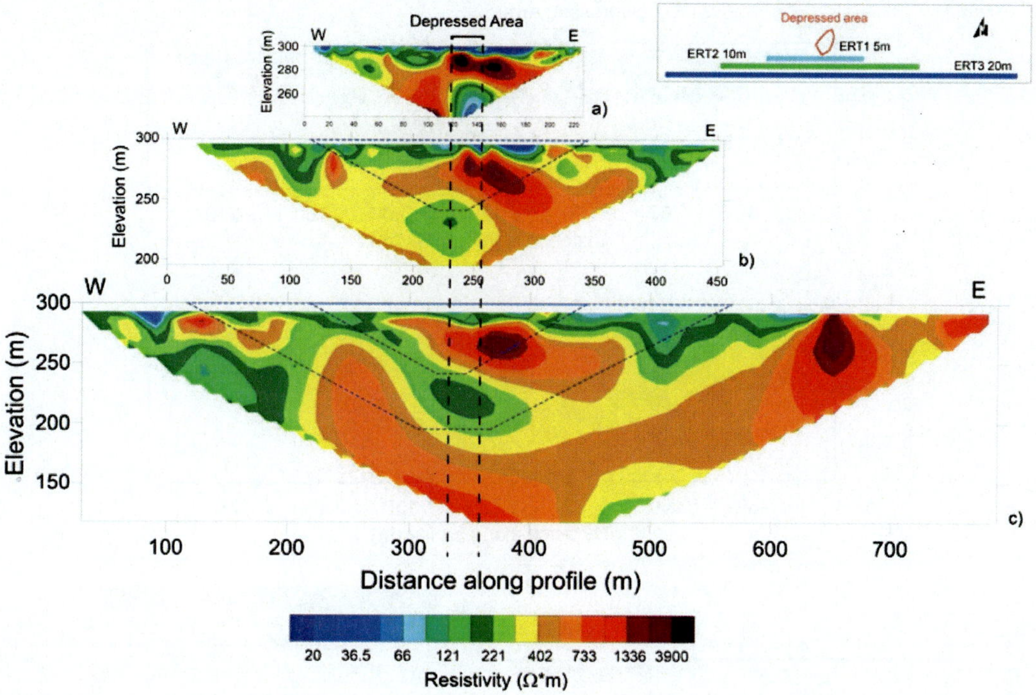


**Fig. 2** Subsurface resistivity models obtained by inverting the Dipole–Dipole geoelectrical data along profiles ERT1, ERT2, and ERT3. The inversion program used was RES2Dinv (Loke 2004). Panel **a** ERT 1, R.M.S = 6.53%; panel **b** ERT 2, R.M.S = 3.87%; panel **c** ERT 3, R.M.S. = 8.25%. Location of the geoelectrical profiles with respect to the depressed area of Gravaglione is shown in the upper-right corner inset (where ERT profiles do not overlap for graphical purposes). The vertical dashed black lines in the sections indicate the position at which the profiles cross the Gravaglione. On top of ERT3 and ERT2, dotted lines indicate the area investigated by ERT2 and ERT1, and by ERT1, respectively

fracturing condition and water presence. Out of this, scheme is the resistivity anomalies in the central part of each section and that visible in the large-scale tomography between 650 and 670 m along the profile.

The resistivity models are highly coherent when they relate to the same subsoil portion. A resistivity anomaly (>4 kΩ*m) is visible in the central part of all ERTs. However, the large scale, low resolution, ERT3 identifies a single resistivity anomaly well located between 250 and 280 m a.s.l, the higher resolution surveys (ERT1 and ERT2) reveal the presence of two different resistivity cores side by side with the eastern one slightly deeper and immerging eastward. These resistive anomalies could be associated to what seems to be a cave complex. As from the higher resolution resistivity image, there might be two distinct at least partially air-filled caves, but it is not possible to assess if they are connected or not. The presence of a further cave in depth is excluded by the large-scale ERT3 which does not show other relevant resistivity anomalies.

As regards the resistivity nucleus east of ERT3, this does not coincide with any morphological indication which might lead us to suppose the presence of another cave. Differently from the central resistivity anomalies which is clear recognizable also in the Wenner–Schlumberger and Wenner dataset, its presence is highlighted only in the Dipole–Dipole and in the Dipole–Dipole reverse sections. It cannot be hence excluded that it might be related to an artifact of the geoelectrical data inversion process. Further, investigations (high-resolution ERT) will be performed in the next future to assess its nature.

Once confirmed the cavity presence, the next step was to define its lateral extension. To this aim, other two geoelectrical profiles, with 5 m inter-electrode spacing, were acquired (see inset of Fig. 3):

- ERT4, a N120° oriented profile which crosses the depressed area north of the previously acquired ERTs, not intercepting this last;
- ERT5, a N30° oriented profile bordering the eastern side of the depressed area and crossing both the multi-scale ERTs and the N120° oriented one.

Finally, by plotting in three dimensions all tomographic results related to the high-resolution ERTs (5 m

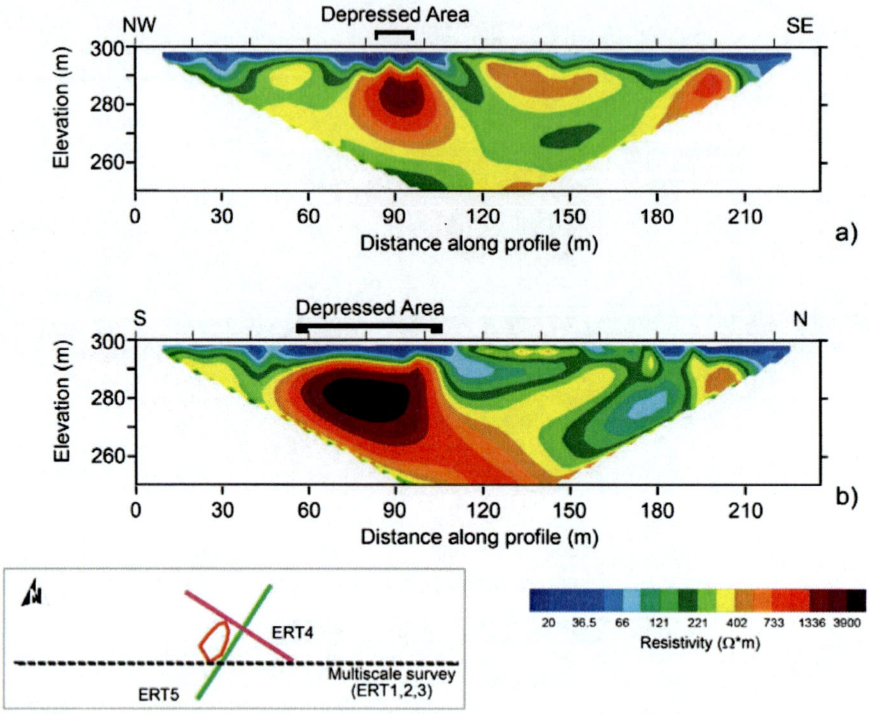


**Fig. 3** Subsurface resistivity models obtained by inverting the Dipole–Dipole geoelectrical data along profiles ERT4 and ERT5. Panel **a** ERT 4, R.M.S = 4.33%; panel **b** ERT 5, R.M.S = 5.29%. Location of the geoelectrical profiles with respect to the depressed area of Gravaglione (red line) is shown in the inset in the lower-left corner

inter-eletrodic spacing), it is possible to appreciate the coherence of the geophysical results and to have a first idea, even though not exhaustive, of the extension of the cavity (Fig. 4): in the figure, it was intentionally used a linear resistivity scale to highlight the cavity presence and to demonstrate how its presence is affecting the geophysical results.

Both the profiles show a general resistivity distribution like the one previously shown and described (Fig. 3). Where the ERT4 crosses the depressed area, an in depth resistivity anomaly can be observed. Its resistivity (<4 k Ω) let us suppose that the anomaly could be related to the presence of a smaller sector of the cavity (not well resolved by the ERT survey) or to a side-effect of the cavity presence. In this last case, it may be possible that ERT4 does not pass directly over the cavity, but is located in its near proximity.

ERT5 shows the presence of a large anomaly whose position, extension, and resistivity values are compatible with the ones previously interpreted as indicative of a cave presence.

A more classical approach, based on the acquisition of two orthogonal geoelectrical profiles, was adopted at the other site, corresponding to a sinkhole in the eastern sector

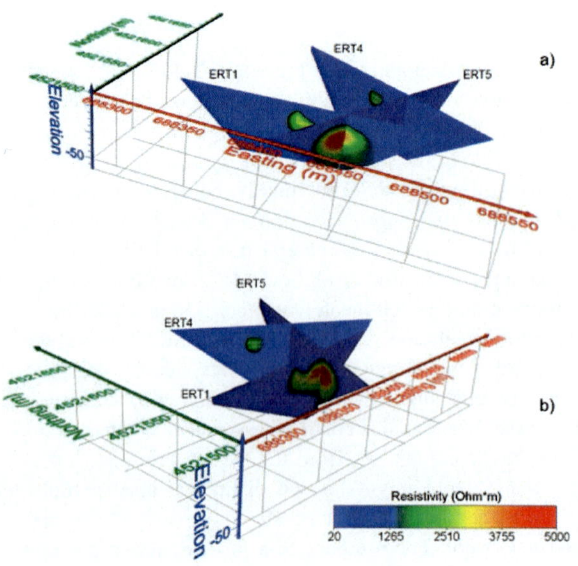


**Fig. 4** 3D plots of the geophysical results. The cavity is well highlighted by resistivity values well above the background. The Gravaglione depressed area, not indicated in the figure, is located between ERT1 and ERT4

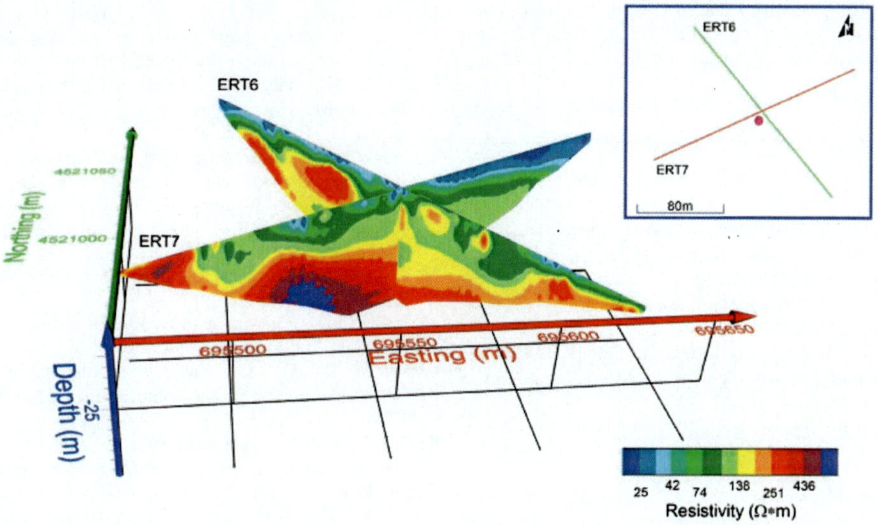


**Fig. 5** Sinkhole at the eastern sector of Canale di Pirro polje: subsurface resistivity models obtained by inverting the Dipole–Dipole geoelectrical data along profiles ERT6 and ERT7 (respectively showing R.M.S = 6.3%, and 4.8%). Location of the profiles with respect to the sinkhole (pink circle) is shown in the inset

of the Canale di Pirro polje. The geoelectrical profiles (ERT6 and ERT7), are centered on the area with evidence of ground sinking (see inset in Fig. 5). Considered the limited dimensions of the sinkhole (approximate diameter <10 m, see also Fig. 1), the inter-electrodic spacing was set to 5 m. Both profiles are 235 m long and reach a maximum investigation depth of ~50 m below the ground surface. Geoelectrical data were acquired by using Dipole–Dipole, Dipole–Dipole reverse, Wenner-Schlumberger, and Wenner configurations. Results of the joint inversion of the Dipole–Dipole and Dipole–Dipole data are here presented, considering that inversion of the other dataset produced similar outcomes. In analyzing Fig. 5, it has to be noted that the adopted color scale is different from that used in Figs. 2 and 3.

Generally, ERT images do not show the presence of resistivity anomalies possibly related to cavity presence. A possible geological interpretation of the resistivity distributions is based, as for the Gravaglione site, on the presence of a more conductive surficial topsoil layer overlaying a more resistive material, associated to the Altamura Limestone. This latter seems to be shallower in the SW sector of the investigated area and is characterized by (1) a not uniform resistivity distribution along the profile and (2) an increasing resistivity in depth. The presence of areas with varying resistivity could be associated to a different fracturing and saturation level of the Altamura Limestone formation, with the lower resistivity values associated to the higher degree of fracturing/saturation (Reynolds 1997; Schön 2015).

## 3 Conclusions

In this contribution, we have illustrated the first outcomes of geophysical analyses performed at the Canale di Pirro polje of central Apulia, in correspondence of two sites of particular interest. At Gravaglione, the main swallow hole in the polje, the survey was addressed to verify the existence of caves large enough to be explored by man, and therefore, the possibility to open a passage in the cover deposits. By adopting a multi-array and multi-scale strategy, based on the acquisition and joint inversion of geoelectrical data collected by using different arrays and with electrode spacing ranging from 5 to 20 m, a high-resolution model of the electrical subsurface structure was obtained. With an investigation depth up to ~150 m below the ground and a high surficial resolution, the model shows features which are compatible with the presence of large caves and provides useful information on their possible extension and on the thickness of the cover deposits.

The second examined site is a small sinkhole within a vineyard, with depth of about 3 m, circular shape and diameter of some 7 m. The survey was addressed to detect likely ways of water infiltration at the site, and the possible presence of underground voids and conduits in the limestone bedrock. In this case, no clear evidence of such features was identified, and the most likely explanation for the sinkhole at the ground surface is that the water running down from the nearby ridges at the northern boundary of the polje infiltrates within the complex network of fractures in the carbonate rock mass.

**Acknowledgements** This work was funded through Protocollo d'intesa con Regione Puglia per l'attuazione dell'art. 45 "Interventi per esplorazione dei fenomeni carsici," comma 1 della L.R. n. 45 del 30/12/2013.

## References

Cardarelli E, Cercato M, Cerreto A, Di Filippo G (2010) Electrical resistivity and seismic refraction tomography to detect buried cavities. Geophys Prospect 58:685–695

Cotecchia V, Scuro M (2010) Portrait of a coastal karst aquifer: the City of Bari. Aqua Mundi: 187–196

Das P, Mohanty PR (2016) Resistivity imaging technique to delineate shallow subsurface cavities associated with old coal working: A numerical study. Environ Earth Sci 75:661

Del Prete S, Iovine G, Parise M, Santo A (2010) Origin and distribution of different types of sinkholes in the plain areas of Southern Italy. Geodin Acta 23(1/3):113–127

Festa V, Fiore A, Parise M, Siniscalchi A (2012) Sinkhole evolution in the Apulian karst of southern Italy: a case study, with some considerations on sinkhole hazards. J Cave Karst Studies 74 (2):137–147

Gutierrez F, Parise M, De Waele J, Jourde H (2014) A review on natural and human-induced geohazards and impacts in karst. Earth Sci Rev 138:61–88

Kaufmann G (2014) Geophysical mapping of solution and collapse sinkholes. J Appl Geophys 111:271–278

Kaufmann G, Romanov D, Nielbock R (2011) Cave detection using multiple geophysical methods: unicorn cave, Harz Mountains, Germany. Geophysics 76(3):71–77

Leucci G, Parise M, Sammarco M, Scardozzi G (2016) The use of geophysical prospections to map ancient hydraulic works: the Triglio underground aqueduct (Apulia, Southern Italy). Archaeol Prospect 23:195–211

Liso IS, Chieco M, Fiore A, Pisano L, Parise M (2020) Underground geosites and caving speleotourism: some considerations from a case study in southern Italy. Geoheritage 12:13

Loke M (2004) Tutorial: 2-D and 3-D electrical imaging surveys. Geotomo Software: Penang, Malaysia

Margiotta S, Negri S, Parise M, Valloni R (2012) Mapping the susceptibility to sinkholes in coastal areas, based on stratigraphy, geomorphology and geophysics. Nat Hazards 62(2):657–676

Martınez-Lopez J, Rey J, Duenas J, Hidalgo C, Benavente J (2013) Electrical tomography applied to the detection of subsurface cavities. J Cave Karst Studies 75(1):28–37

Martinotti ME, Pisano L, Marchesini I, Rossi M, Peruccacci S, Brunetti MT, Melillo M, Amoruso G, Loiacono P, Vennari C, Vessia G, Trabace M, Parise M, Guzzetti F (2017) Landslides, floods and sinkholes in a karst environment: the 1–6 September 2014 Gargano event, southern Italy. Nat Hazard 17:467–480

Mochales T, Casas A, Pueyo E, Pueyo O, Román M, Pocoví A, Soriano M, Ansón D (2008) Detection of underground cavities by combining gravity, magnetic and ground penetrating radar surveys: A case study from the Zaragoza area, NE Spain. Environ Geol 53:1067–1077

Mossa M (2007) The floods in Bari: what history should have taught. J Hydraul Res 45(5):579–594

Parise M (2003) Flood history in the karst environment of Castellana-Grotte (Apulia, southern Italy). Nat Hazard 3(6):593–604

Parise M (2006) Geomorphology of the Canale di Pirro karst polje (Apulia, southern Italy). Zeitschrift Für Geomorphologie NF 147:143–158

Parise M (2011) Surface and subsurface karst geomorphology in the Murge (Apulia, southern Italy). Acta Carsologica 40(1):79–93

Parise M, Benedetto L (2018) Surface landforms and speleological investigations for a better understanding of karst hydrogeological processes: a history of research in southeastern Italy. In: Parise M, Gabrovsek F, Kaufmann G, Ravbar N (eds) Advances in Karst Research: theory, fieldwork and applications. Geological Society, London, sp publ 466, pp 137–153

Parise M, Gabrovsek F, Kaufmann G, Ravbar N (2018) Recent advances in karst research: from theory to fieldwork and applications. In: Parise M, Gabrovsek F, Kaufmann G, Ravbar N (eds) Advances in karst research: theory, fieldwork and applications. Geol Soc London, sp publ 466, pp 1–24

Pisano L, Zumpano V, Liso IS, Parise M (2020) Geomorphological and structural characterization of the "Canale di Pirro" polje, Apulia (southern Italy). J Maps 16(2):479–487

Reynolds JM (1997) An introduction to applied and environmental geophysics. Wiley, New York

Schön J (2015) Physical properties of rocks: fundamentals and principles of petrophysics, 2nd edn. Elsevier

Zumpano V, Pisano L, Parise M (2019) An integrated framework to identify and analyze karst sinkholes. Geomorphology 332:213–225

# Study of Rull Cave Dynamics to Understand the Complex Relationships Between Soil, Cave and External Atmosphere

C. Pla, S. Gil-Oncina, M. C. Ruiz, J. C. Cañaveras, S. Cuezva, A. Fernández-Cortés, S. Sánchez-Moral, and D. Benavente

**Abstract**

Rull Cave is a karstic cavity located in Vall d'Ebo (Alicante, Southeast of Spain) developed in massive Miocene conglomerates and Cretaceous limestones which consists of a single room with a rounded shape. It is a tourist cave with an average of 15,091 visitors per year. The environmental conditions in the cave's interior and exterior, as well as the physical properties above the cave, were measured from 2012 to 2021 to understand the gaseous dynamics and their relationships in this three-component system (soil-cave-atmosphere). Besides the presence of visitors, Rull Cave maintains stable values of mean temperature (16.1 °C) and relative humidity (97.6%). The cave presents two different gaseous stages in an annual cycle. In the study period, maximum average values of $CO_2$ and $^{222}Rn$ were reached in the hottest months (3966 ppm of $CO_2$ and 4185 Bq/m$^3$, respectively), while coldest months are related to the lowest gaseous concentrations (478 ppm of $CO_2$ and 368 Bq/m$^3$, respectively). Discrete sampling of $\delta^{13}C$ of $CO_2$ in cave, outdoor atmosphere and soil allows to characterize the spatial distribution and temporal variations of the gaseous concentration in background atmosphere, cave and soil air as well as the relationship between them. Results show that the dynamics of $^{222}Rn$ and $CO_2$ in the cave air follow different patterns defined by the complex relationships between external and internal factors. Findings permit lately to preserve the quality of the cave indoors and to understand the possible risk derived from the presence of indoors hazardous substances.

**Keywords**

Karst • Underground atmosphere • Gaseous exchange • Microclimate

C. Pla (✉)
Department of Civil Engineering, University of Alicante, 03690 Alicante, Spain
e-mail: c.pla@ua.es

S. Gil-Oncina · M. C. Ruiz · J. C. Cañaveras · D. Benavente
Department of Earth and Environmental Sciences, University of Alicante, 03690 Alicante, Spain
e-mail: sara.gil@ua.es

M. C. Ruiz
e-mail: mcandela.ruiz@ua.es

J. C. Cañaveras
e-mail: jc.canaveras@ua.es

D. Benavente
e-mail: david.benavente@ua.es

S. Cuezva
Department of Geology, Geography and Environment Science, University of Alcala, 28801 Madrid, Spain
e-mail: soledad.cuezva@uah.es

A. Fernández-Cortés
Department of Biology and Geology, University of Almeria, 04120 Almeria, Spain
e-mail: acortes@ual.es

S. Sánchez-Moral
Department of Geology, National Museum of Natural Sciences (MNCN-CSIC), 28006 Madrid, Spain
e-mail: ssmilk@mncn.csic.es

## 1 Introduction

Underground caves constitute one of the most characteristic components of a karst system. They develop within the existent porous network, in the voids and cracks. The study of caves can be approached from different perspectives due to their interest related to tourism, natural heritage, the exchange of matter and energy between the outside and the inside, the microclimatic and particular environmental conditions and the geochemical processes, among others (Bourges et al. 2014; Cuezva et al. 2011; Garcia-Anton et al. 2017).

The main factors responsible for cave formation are the dissolution processes that affect the karst system, which depend on the water infiltration and the geochemical


B. Andreo et al. (eds.), *EuroKarst 2022, Málaga*, Advances in Karst Science,
https://doi.org/10.1007/978-3-031-16879-6_23

reactions (Gombert 2002; Piccini and Mecchia 2009). Thus, the result of these processes will infer the formation of a shallow atmosphere that will have particular conditions concerning the geological, environmental and biotic features of the system.

In the vadose zone, subsurface caves commonly present high thermostability conditions as well as high gaseous concentrations ($CO_2$, $^{222}Rn$; among others) (Cuezva et al. 2011; Fernandez-Cortes et al. 2011) in comparison to the external atmosphere. Previous research demonstrated that $CO_2$ concentration comes from the overlying soil over the cave (Faimon et al. 2011), since soil emits the largest amount of $CO_2$ in the terrestrial ecosystem (Ryan and Law 2005). The transport of this greenhouse gas is performed by both gaseous diffusion and advection, related to the conditions on the soil-cave system and also by $CO_2$ dissolution in dripping waters (Frisia et al. 2011).

$^{222}Rn$ concentration in caves depends mainly on the exhalation of host rocks and soil and derives from radium decay. Depending on the relationships between different environmental and geological conditions, the exhalation rate and the transport mechanisms through the porous network, $^{222}Rn$ concentration will vary in the different shallow environments (Alvarez-Gallego et al. 2015; Lario et al. 2005).

Thus, the dynamics and the prevailing relationships of a system influenced by the outdoor atmosphere, soil and host rock will be directly responsible for the cave conditions, i.e., environmental conditions and gaseous concentrations (Garcia-Anton et al. 2017; Pla et al. 2016a).

Traditionally, karstic caves have been employed by human beings as they were shelter places due to their warmer interior conditions. In addition, they were also related to the water existence. Nowadays, these shallow environments provide a source of scientific information about, for instance, biological, geological, environmental and climatic features. They are vulnerable sites because their inherent environmental equilibrium is affected by the activities developed in the interior and the external-related atmosphere (Bourges et al. 2014).

Principal activities that can strongly affect the natural equilibrium are basically related to the human presence (Pla et al. 2016b). This presence can alter microclimatic conditions such as relative humidity, temperature, $CO_2$ concentration, microorganisms' development and particle dynamics that will be impacted in direct modification of the natural dynamics.

The presence of people inside these environments is quite common since caves are related to natural heritage. In this line, there are many examples of caves which conform the natural heritage of the places where they are located. Thus, they contain cultural treasures that attract the attention of tourists (Woo et al. 2019).

Because of that, the study of caves is significant and important, mainly when they are related to the presence of people. From one perspective, it is fundamental to know the conditions to protect these natural environments considering the effects that human presence can have on them. From the other, cave studies are considered a key factor to ensure the quality of the cave air, determining the absence of hazardous substances in the caves since $CO_2$ and $^{222}Rn$ affect human health and, in the presence of people, they must stay within certain maximum levels (Alvarez-Gallego et al. 2015; Pla et al. 2020).

In this study, we will focus on $^{222}Rn$ and $CO_2$ dynamics in Rull Cave, studying their evolution through a decade with the aim of understanding the relationships between external and internal factors to preserve the quality of the cave interior and evaluate possible risks derived from the presence of potentially hazardous substances within the cave.

## 2 Materials and Methods

### 2.1 Field Site

Rull Cave (38°48′40″N, 0°10′38″W) is located in Vall d'Ebo (Alicante), Southeast Spain. The cave is developed in both massive Miocene conglomerates and Cretaceous limestones (Pla et al. 2016a, 2017). It is a large room with a rounded shape and has an area of approximately 1535 $m^2$ and a volume of 9915 $m^3$. The maximum height inside the cave is around 20 m in the central chamber although it varies with location. The host rock within which Rull Cave is developed has a calculated variable thickness of 9.3–22.3 m. The entrance to the cave is located in its highest part. Thus, the cave is developed downwards from the outside. Multiple C3 plants, Mediterranean vegetation and microorganisms grow in the silty-silty loam soil over the host rock of the cave which constitutes the main source of the $CO_2$ concentration in the cave (Pla et al. 2016b, 2017).

The site has a Csa climate type (Köppen-Geiger Classification slightly modified, AEMET-IM 2011), which consists of warm annual temperatures with a dry and hot summer.

For the study period (November 2012–January 2022) average atmospheric temperature, relative humidity and pressure in Rull Cave site are 16.0 °C, 73% and 964.3 mbar. Average annual precipitation (2013–2021) is 431 mm.

Interior conditions in the cave (November 2012–January 2022) are defined by average temperature, relative humidity and pressure of 16.1 °C, 97.6% and 964.3 mbar, respectively. Average $CO_2$ and $^{222}Rn$ concentrations for this period are 2016 ppm and 1762 Bq/$m^3$. Rull Cave is a touristic cave which annually receives, on average, 15,091 people.

Soil average conditions for the period between February 2015 and January 2022 (soil temperature and volumetric water content) were 16.3 °C and 0.23 $m^3/m^3$, respectively.

## 2.2 Monitoring System

Environmental measurements in Rull Cave were performed with different weather stations and environmental probes which changed within the study period (Pla et al. 2016b, 2017). The present monitoring system inside the cave consists of one datalogger HOBO H22-001 (Onset Computer, USA) connected to the electrical supply but with security battery to ensure autonomy in case of power failure. Probes connected to the datalogger provide temperature and relative humidity (HMP45AC, Vaisala, Finland; accuracies of ± 0.2 °C and ± 2.0%, respectively), $CO_2$ concentration (GMP252, Vaisala, Finland; accuracy of ± 40 ppm) and barometric pressure (S-BPB-CM50 Sensor, Onset Computer, USA; accuracy of ± 3.0 mbar). In addition, a Radim 5 WP radon monitor (SSM&SISIE, Prague; accuracy of ± 12 $Bq/m^3$) measures $^{222}Rn$ concentration.

Outside conditions were measured with a H21 Hobo Weather Station (Onset Computer, USA) with S-THB-M002 temperature and relative humidity sensor (Onset Computer, USA; accuracies of ± 0.21 °C and ± 2.5%, respectively), S-BPB-CM50 barometric pressure sensor (Onset Computer, USA; accuracy of ± 3.0 mbar) and S-RGF-M002 Davis rain gauge sensor (Onset Computer, USA; accuracy of ± 4%). Soil conditions (temperature and volumetric water content) were measured with a HOBO U12 logger (Onset, USA; accuracy of ± 0.5 °C) and an ECHO EC-5 probe (Decagon Devices, USA; accuracy of ± 1–2%). Only from September 2019, additional long-term measurements of $^{222}Rn$ were performed in soil using Kodalpha films (LR115). From January 2014, additional measurements of air sampling in the outside atmosphere, soil and cave were also discretely performed to characterize the spatial distributions and temporal variations of $CO_2$ concentrations and its $\delta^{13}C$ signal by pumping the air and saving it in 1 l Ritter sampling bags. These samples were analyzed using a Picarro G2101-i analyser (California, USA; accuracy of ± 0.3‰ for $\delta^{13}CO_2$ after 5 min of analysis) that employs cavity ring-down spectroscopy (CRDS-WS) (Crosson 2008).

# 3 Results and Discussion

Rull Cave follows a stationary behavior with periodic annual cycles. In the cave, higher concentrations ($CO_2$ and $^{222}Rn$) are always existent in the hottest months of the year while in the coldest months cave gaseous concentration drops to minimum (Fig. 1). Temperature is a key factor in the cave dynamics since relationships between outside and interior temperature are responsible for changes in cave concentrations. For the study period, maximum outdoor temperatures reach 36.8 °C in summer falling down to −1.2 °C in winter. In contrast, interior temperature variations are within ± 0.5 °C from the average (16.1 °C) (Fig. 1a). Maximum temperatures in the cave are always reached in the coldest months due to the thermal conductivity of the host rock (Pla et al. 2016b).

Relationships between cave and outdoor temperatures are responsible for the increasing (recharge) or decreasing (discharge) of gaseous concentration in the cave due to density differences between cold and hot air masses. In an annual cycle, when outdoor temperatures are higher than cave temperatures, the colder and denser air in the cave indoors remains stagnant and causes the gaseous concentration to increase, reaching maximum values during summer (Fig. 1b). In contrast, the thermal inversion between atmospheric and cave temperatures, which occurs between October and November, produces advective ventilation through host rock fractures and brings the denser outside air into the cave, displacing the cave air containing maximum gaseous concentrations. This pattern is cyclic, and it is one of the mechanisms governing the cave dynamics (Pla et al. 2016b, 2017, 2020). In addition, soil conditions greatly influence cave dynamics. Figure 1c shows that soil conditions (soil temperature and volumetric water content) are related to outdoor conditions. Maximum values of soil temperature are reached nearly simultaneously to outdoor temperature. The amount of water present in soil is also directly related to the rainfall occurrence and the soil temperature. Thus, considering the relatively stationary rainfall behavior in the study area (more common in spring and autumn) higher values of volumetric water content are generally found in these periods. Soil conditions control plant respiration and the decomposition of the organic matter by soil microorganisms (Amundson et al. 1998) which are the producers of the biotic $CO_2$ on soil.

Lab experiments performed on Rull soil samples concluded that soil $CO_2$ production correlates to soil temperature and water content (Pla et al. 2016a). High $CO_2$ concentrations occur with high temperatures and moderate soil water content (lower than 0.5 $m^3/m^3$) which matches with the relation between $CO_2$ production (Fig. 1d) and soil conditions (Fig. 1c).

Generally, higher production rates of soil $CO_2$ are found in spring (April) and autumn (October). The high soil $CO_2$ concentrations determined by the discrete sampling in Rull Cave soil (Fig. 1d), match with the lighter values (decreasing values) of its $\delta^{13}C$ isotopic signal.

As established by Keeling (1958), the Keeling plot for the analysis of the $\delta^{13}CO_2$ isotopic signal confirms that the

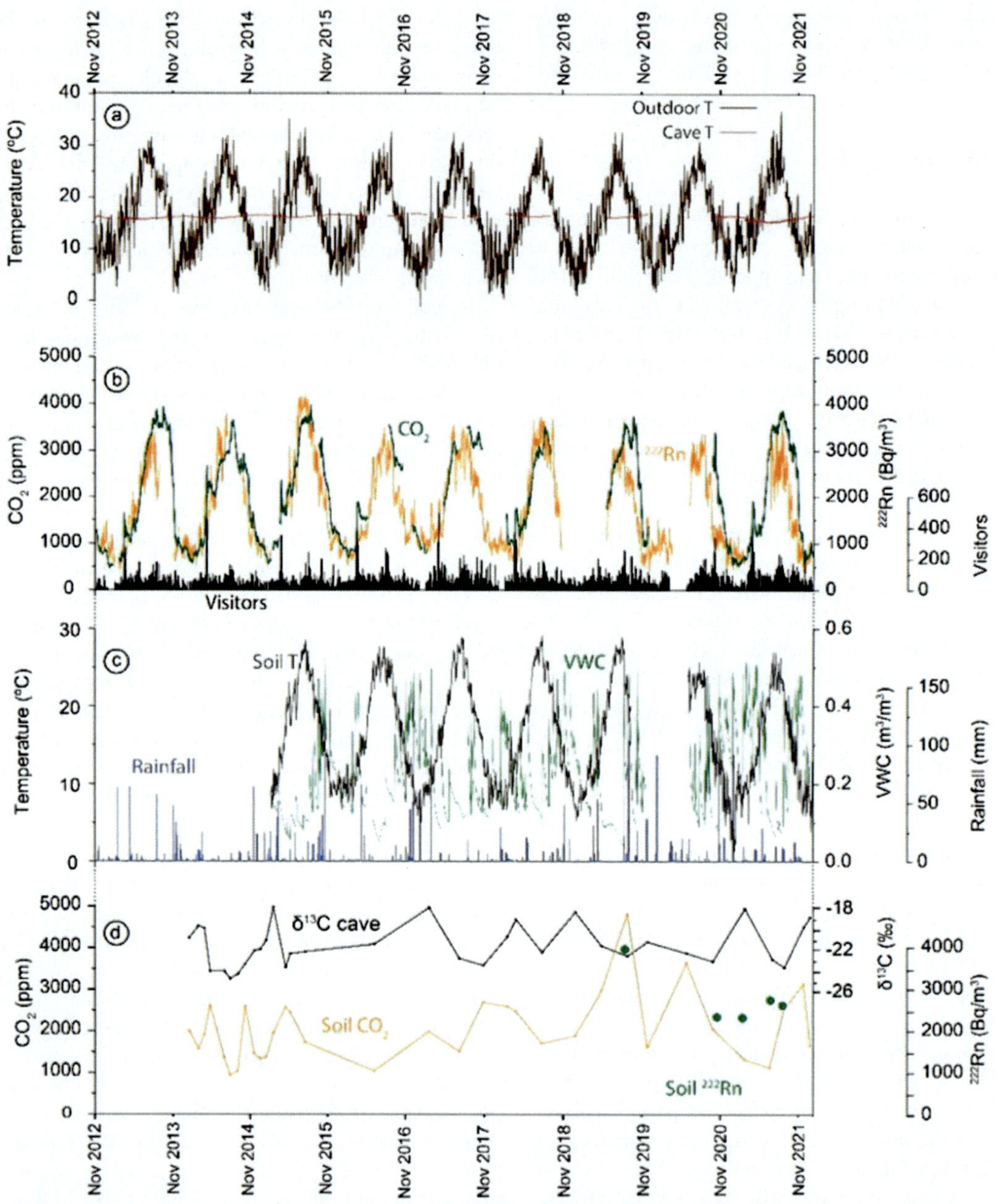


**Fig. 1** Environmental conditions in the cave, the outdoor atmosphere and soil for the study period, **a** Outdoor and indoor temperatures (°C), **b** $CO_2$ (ppm), $^{222}$Rn concentration (Bq/m$^3$) and visitors, **c** Soil conditions: soil temperature (°C) and volumetric water content (VWC, m$^3$/m$^3$), and rainfall (mm), **d** Soil $CO_2$ concentration, $\delta^{13}$C of $CO_2$ (‰) in cave and $^{222}$Rn concentration in the soil (Bq/m$^3$). Measurements of $CO_2$, $\delta^{13}$C (Ritter sampling bags) and $^{222}$Rn (Kodalpha films, LR115) in soil are discretely performed in the study period

carbon isotope chemistry of a soil is the result of plant root respiration and the decomposition of the organic matter by soil microorganisms. When C3 plants are predominant, $\delta^{13}CO_2$ is −27‰ (Amundson et al. 1998). The Keeling plot for Rull site, using outdoor, soil and cave air samples, demonstrates that the cave interior concentration proceeds from mixing the $CO_2$ of the outdoor atmosphere and the soil $CO_2$, and points to a dominant component of the soil $CO_2$ as highlighted by the y-intercept value (Fig. 2) (Garcia-Anton et al. 2017; Pla et al. 2016a).

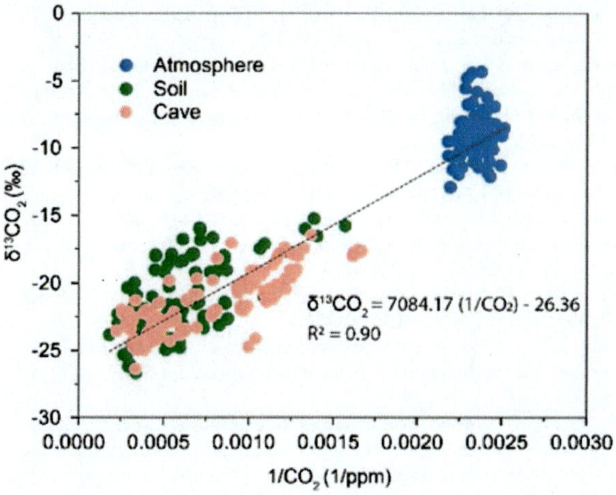


**Fig. 2** Keeling plot for the discrete sampling in rull soil, exterior air and cave air

Thus, decreases in the cave air $\delta^{13}CO_2$ (lighter values) indicate the predominance of soil $CO_2$ production while higher values of $\delta^{13}CO_2$ confirm the predominance of atmospheric air in the cave atmosphere (Fig. 1d). The produced soil $CO_2$ migrates by diffusion to the underground cave through the cracks, voids and pores, reaching the cave and contributing to increase $CO_2$ concentration (Pla et al. 2016a, 2017). Diffusion is regulated by soil concentrations and physical conditions. Both mechanisms, advection and diffusion, are responsible for gaseous concentration although one will be more important in absence of the other, with a dependence on the annual cycle.

$CO_2$ concentrations in Rull Cave also derive from degasification of dripping waters (Pla et al. 2020) although it is not predominant, and from the human presence since Rull Cave is a touristic place. Short term $CO_2$ disturbances inside the cave are noticeable during massive visits which mainly occur in the Easter period. However, baseline levels are rapidly recovered once the visits end. Thus, previous studies performed in the cave confirmed that visitors may not alter drastically the conditions of the interior cave atmosphere (Pla et al. 2016b).

The gaseous recharge and discharge stages of $CO_2$ are also followed by $^{222}Rn$ (Fig. 1b). Maximum concentrations of both gases are reached simultaneously, confirming the annual dependency of gas concentrations on the temperature gradient and its influence in the predominance of gaseous diffusion or advection. However, production and exhalation of $^{222}Rn$ depend, not only on the soil, but primarily on the host rock and sediments of the cave. Although $^{222}Rn$ measurements in Rull Cave soil have been performed for a short period compared to $CO_2$ soil measurements (Fig. 1d), they present a temporal variation related to soil conditions. If diffusion of $CO_2$ from soil to cave through the porous network of soil and rocks affects $CO_2$ cave concentration, diffusion may also affect the $^{222}Rn$ produced in soil and rocks through the vertical profile over the rock in Rull Cave. However, additional study of this process is necessary to establish the different roles of soil and the host rock in the production of $^{222}Rn$ which finally will concentrate in cave atmosphere.

Maximum average concentrations of $CO_2$ and $^{222}Rn$ in Rull Cave are 3966 ppm and 4185 Bq/m$^3$, respectively. The EH40/2005 establishes for $CO_2$ a long-term exposure limit (8 h) of 5000 ppm at workplace. For $^{222}Rn$, ICRP (2017) recommends, for the specific situations of indoor work involving exposures in tourist caves, a dose coefficient of 6 mSv per mJ h/m$^3$ which is equivalent, approximately, to 20 mSv per WLM (ICRP 2017). Attending to this, and considering monthly average values of $^{222}Rn$ concentration, one guide would not have to remain inside the Rull Cave more than 240 h in one year (20 h/month, on average). However, the exposure time could be adjusted more accurateley considering fewer hours in the periods with higher $^{222}Rn$ concentrations.

As demonstrated, microclimatic monitoring of Rull Cave compliments the official standards about air quality and assists in the determination of potentially hazardous concentrations.

## 4 Conclusions

Rull Cave dynamics depend on the complex relationships between the soil, cave and external atmosphere. This study demonstrates that environmental parameters in this multi-component system are responsible for the gaseous concentrations ($CO_2$ and $^{222}Rn$) in the cave air. Rull Cave follows an annual pattern with one recharge period and one discharge period. These periods occur as a consequence of the thermal relationships between interior and outdoor atmospheres that cause advective ventilation through the host rock fractures. Thus, in the warmest months colder and denser air in the cave indoors remains stagnant and causes the gaseous concentration ($CO_2$ and $^{222}Rn$) to increase, reaching maximum values. In the coldest season the advective ventilation brings the denser outside air to the cave interior, displacing the cave air containing maximum gaseous concentrations. Simultaneously, gaseous diffusion (dependent on soil and host rock pore network conditions) is also responsible of increasing gas concentration in Rull Cave. $CO_2$ produced in the soil migrates by diffusion to the underground cave through cracks, voids, and pores. Similarly, the gaseous diffusion also affects $^{222}Rn$ concentration in the cave, since $^{222}Rn$ exhaled from soil and rocks migrates, by differences in concentration, through the vertical profile over the cave host rock.

Results of this study confirm that the continuous monitoring of Rull Cave, soil, and the external atmosphere provides substantial information about the environmental situation of the cave atmosphere that ensures a proper conservation of the cavity. In addition, this information is significant and important considering that Rull Cave is a touristic cave in which official standards about air quality may be respected.

**Acknowledgements** This work was supported by the Spanish Ministry of Science, Innovation, and Universities [grant number RTI2018-099052-BI00] and Regional Governments of Comunidad Valenciana (Spain) [grant number AICO/2020/175].

## References

AEMET-IM (2011) Iberian Climate Atlas. Air temperature and precipitation (1971–2000) in Agencia Estatal de Metereologia (España), Instituto de Metereologia (Portugal) (eds)

Alvarez-Gallego M, Garcia-Anton E, Fernandez-Cortes A, Cuezva S, Sanchez-Moral S (2015) High radon levels in subterranean environments: monitoring and technical criteria to ensure human safety (case of Castanar cave, Spain). J Environ Radioactiv 145:19–29

Amundson R, Stern L, Baisden T, Wang Y (1998) The isotopic composition of soil and soil-respired $CO_2$. Geoderma 82:83–114

Bourges F, Genthon P, Genty D, Lorblanchet M, Mauduit E, D'Hulst D (2014) Conservation of prehistoric caves and stability of their inner climate: lessons from Chauvet and other French caves. Sci Total Environ 493:79–91

Crosson ER (2008) A cavity ring-down analyzer for measuring atmospheric levels of methane, carbon dioxide, and water vapor. Appl Phys B Lasers O 92:403–408

Cuezva S, Fernandez-Cortes A, Benavente D, Serrano-Ortiz R, Kowalski AS, Sanchez-Moral S (2011) Short-term $CO_2$ (g) exchange between a shallow karstic cavity and the external atmosphere during summer: role of the surface soil layer. Atmos Environ 45:1418–1427

EH40/2005, Workplace exposure limits. HSE Books (2018)

Faimon J, Ličbinská M, Zajíček P (2011) Relationship between carbon dioxide in Balcarka Cave and adjacent soils in the Moravian Karst region of the Czech Republic. Int J Speleol 41

Fernandez-Cortes A, Sanchez-Moral S, Cuezva S, Benavente D, Abella R (2011) Characterization of trace gases' fluctuations on a 'low energy' cave (Castañar de Íbor, Spain) using techniques of entropy of curves. Int J Climatol 31:17

Frisia S, Fairchild IJ, Fohlmeister J, Miorandi R, Spoetl C, Borsato A (2011) Carbon mass-balance modelling and carbon isotope exchange processes in dynamic caves. Geochim Cosmochim Ac 75:380–400

Garcia-Anton E et al (2017) Abiotic and seasonal control of soil-produced $CO_2$ efflux in karstic ecosystems located in Oceanic and Mediterranean climates. Atmos Environ 164:31–49

Gombert P (2002) Role of karstic dissolution in global carbon cycle. Glob Planet Change 33:177–184

ICRP (2017) Occupational Intakes of Radionuclides: Part 3. ICRP Publication 137. Ann. ICRP 46(3/4)

Keeling CD (1958) The concentration and isotopic abundances of atmospheric carbon dioxide in rural areas. Geochim Cosmochim Ac 13:322–334

Lario J, Sanchez-Moral S, Cañaveras J, Cuezva S, Soler V (2005) Radon continuous monitoring in Altamira Cave (northern Spain) to assess user's annual effective dose. J Environ Radioactiv 80: 161–174

Piccini L, Mecchia M (2009) Solution weathering rate and origin of karst landforms and caves in the quartzite of Auyan-tepui (Gran Sabana, Venezuela). Geomorphology 106:15–25

Pla C, Cuezva S, Garcia-Anton E, Fernandez-Cortes A, Cañaveras JC, Sanchez-Moral S, Benavente D (2016a) Changes in the $CO_2$ dynamics in near-surface cavities under a future warming scenario: Factors and evidence from the field and experimental findings. Sci Total Environ 565:1151–1164

Pla C, Galiana-Merino JJ, Cuezva S, Fernandez-Cortes A, Cañaveras JC, Benavente D (2016b) Assessment of $CO_2$ dynamics in subsurface atmospheres using the wavelet approach: from cavity—atmosphere exchange to anthropogenic impacts in Rull cave (Vall d'Ebo, Spain). Environ Earth Sci 75:446

Pla C et al (2017) Role of soil pore structure in water infiltration and $CO_2$ exchange between the atmosphere and underground air in the vadose zone: a combined laboratory and field approach. Catena 149 (Part 1):402–416

Pla C, Fernandez-Cortes A, Cuezva S, Galiana-Merino JJ, Cañaveras JC, Sanchez-Moral S, Benavente D (2020) Insights on climate-driven fluctuations of cave $^{222}$rn and $co_2$ concentrations using statistical and wavelet analyses. Geofluids 2020:8858295

Ryan MG, Law BE (2005) Interpreting, measuring, and modeling soil respiration. Biogeochemistry 73:3–27

Woo KS, Kim L, Ji H, Jeon Y, Ryu CG, Wood C (2019) Geological heritage values of the Yongcheon Cave (Lava Tube Cave), Jeju Island, Korea. Geoheritage 11:615–628

# Hydrological and Environmental Dynamics in Las Güixas Show Cave: Tourist Exploitation and Flood Risk Management

R. Giménez, M. Bartolomé, L. Ezquerro, G. Benito, M. Luetscher, and A. Moreno

**Abstract**

Show caves are great natural attractions and constitute important economic engines for particular rural areas. However, cave management requires knowledge of the cave dynamics to ensure adequate exploitation and conservation (e.g. number of visitors, amount of $CO_2$, other impacts). Show caves located close to the hydrological base level can be affected by a sudden rise of the water level in response to rainfall events, exposing touristic facilities, visitors and staff to flood risk. Here we present a monitoring campaign in Las Güixas show cave, a small tourist cave located in the Central Pyrenees, to guide in developing an adequate flood risk management plan (FRMP). Cave monitoring includes temperature, relative humidity, $CO_2$ concentration and water level measurements to characterize the different cave sections, as well as to assess the possible influence of the visitors on the cave dynamics. The tourist section is very well ventilated due to intense air flow associated with a chimney effect, and therefore $CO_2$ remains low and temperature shows high thermal oscillations throughout the year. During the maximum number of visits per day, the $CO_2$ generated by the tourists increases, however, is quickly reduced by the ventilation. Water level monitoring shows clear rises depending on the amount of rainfall and snowmelt in the catchment area. However, the water level does not always respond to given rainfall amounts in the same way, indicating that the water retained in the karst system plays an important role. Water can flood the tourist section forcing to cancel the following visits which generates economic losses. In addition, the higher water flow increases the natural inputs of $CO_2$ in the cave atmosphere due to degassing. Las Güixas cave monitoring shows that anthropogenic $CO_2$ emissions remain substantially lower than $CO_2$ concentrations during the flood events.

**Keywords**

Cave monitoring • Environmental conditions • Flood risk management • Karst hydrology • Natural heritage preservation • Show cave • Water level

## 1 Introduction

Caves are places with an important natural and cultural heritage assets. Caves enabled for tourism are accessible to large numbers of the public, and provide an opportunity spread knowledge and awareness of the underground environment. At the same time, touristic caves attract a large number of people, promoting the economic development in the places where they are located. Nonetheless, the exploitation of a cave for tourism involves a series of human impacts leading to alterations in its environmental balance. Thus, a detailed knowledge of the natural variability in the cave air (e.g. temperature, relative humidity (RH), and $CO_2$ concentrations) is necessary to evaluate the anthropogenic effects and establish the natural recovery time for the cave (Rivas et al., 2004). One of the most studied atmospheric cave parameter in show caves is $CO_2$. Under natural conditions, $CO_2$ concentrations in the cave atmosphere comes from several sources, such as soil respiration, exchange of air masses with the outside, dissolution of limestone or even $CO_2$ of deep origin (hypogenetic) (Lambert and Aharon 2011). $CO_2$ concentration is an essential parameter of

R. Giménez (✉) · M. Bartolomé · A. Moreno
Instituto Pirenaico de Ecología-CSIC, Madrid, Spain
e-mail: reiesgimenez@gmail.com; rgimenez.geo@gmail.com

M. Bartolomé · G. Benito
Museo Nacional de Ciencias Naturales- CSIC, Madrid, Spain

L. Ezquerro
Universidade Nova de Lisboa, Lisboa, Portugal

M. Luetscher
Swiss Institute of Speleology and Karstology (SISKA), La Chaux-de-Fonds, Switzerland


B. Andreo et al. (eds.), *EuroKarst 2022, Málaga*, Advances in Karst Science,
https://doi.org/10.1007/978-3-031-16879-6_24

underground atmospheres for cave heritage preservation, being widely studied together with temperature and relative humidity of air (e.g. Bourges et al. 2020). Besides, in show caves, an increase in $CO_2$ in relation to the number of visitors is typically observed (e.g. Lang et al. 2015). Thus, periodic $CO_2$ control is necessary in order to evaluate the effects of the touristic visits to the caves as well as to protect the visitors from high $CO_2$ levels. On the other hand, in caves with water flow, an accurate knowledge of the hydrological base level is desirable to understand the hydrological dynamics and determinate the response of the water levels to rainfalls. This fact becomes essential in the case of a show cave, where flood events caused by sudden rise of base level can produce economic losses and even expose human lives to risk. Here we present the preliminary results of cave monitoring (temperature, RH, $CO_2$, and water level) since July 2017 in Las Güixas, a small tourist cave located in the Central Pyrenees. The main goals of this study are (i) to establish the environmental parameters and understand the current hydrological dynamics of the cave, (ii) to characterize the different cave sections, and (iii) to assess the possible influence of the visitors on cave dynamics. The results of this research will inform improved cave management including issues related to its conservation and risks during the rainfall season.

## 2 Location and Site Description

Las Güixas cave is located in Villanua (Huesca Province), at the base of the Collarada massif (2886 m a.s.l) in the Central Pyrenees (Fig. 1A). This mountain area is characterized by a transitional Mediterranean—Oceanic climate, with a mean annual air temperature (MAAT) of 10 °C, while total precipitation is around 1100 mm. The cave is developed in Eocene-aged carbonate rocks (the Villanúa megabreccia) and it has created one of the main discharge springs of the endokarstic Collarada system. Las Güixas cave has been continuously occupied at least since the Bronze Age, and still maintains easy access from Villanúa through a historic passageway. The cave was first opened for tourism from 1929 to 1936, when it was closed to the public and used as a refuge and military prison. Tourism activity was resumed in 1996 and continues currently. Tourist visits have doubled during the last 7 years and about 28,000 people currently tour the cave each year, with touristic activity of the cave an important economic asset for the locality of Villanúa.

The cave extends for 1100 m of passages and galleries with a vertical extent of 67 m, being a relative shallow cave in which the bedrock thickness varies between 2 and 27 m. Three levels of galleries form the cave; (1) the lower level comprises the phreatic level, only accessible with diving techniques, and drains to the Aragon River; (2) the middle level is only hydrologically active during floods and corresponds to the area enabled for tourism. This level has two entrances, the main cave entrance (979 m a.s.l.) and a sinkhole; (3) the upper level comprises the highest galleries and has two more entrances (1016 and 1026 m a.s.l). These four entrances facilitate the bidirectional circulation of air in the cave through the middle and upper galleries. The middle level is connected with the permanently flooded lower level by a ramp and a siphon and works as a trop-plein a conduit that acts as an overflow tube during high water availability (Fig. 1C). When the water levels increase after rainfall or during snowmelt, the water ascends ~16 m from the siphon to the middle level and floods the touristic area forcing cancelation of tours, ending in the western drain located in the lowest elevations of the middle level. The upper level is currently disconnected to the hydrological activity derived from floods. The cave is well decorated and speleothems tyoes include flags, stalagmites, flowstones and stalactites, as well as clastic deposits, as the result of breakdown processes and floods.

## 3 Methods

Cave monitoring includes temperature, RH, $CO_2$ and water level measurements. Location of sensors and control points are shown in Fig. 1B. Temperature and RH were recorded with four temperature sensors (Hobo U23-001) at a 1 h interval. $CO_2$ was measured continuosly using a $CO_2$ sensor (HD37VBTV.1 - GHM Group) located in the Cathedral room. In addition, monthly $CO_2$ measures were carried out with a $CO_2$ hand-meter AZ-001 in 9 sites along the cave from November 2018 to March 2020 and outside of the cave. Two water level sensors (Hobo U20L-02) record changes in the water height at the siphon and atmospheric pressure. The water level monitoring was completed by three buoys located on the access ramp to the siphon that alert via SMS about the flow stage. Punctually, flow rate measurements were taken during flood events using salt gauging methods. Furthermore, local atmospheric variables (temperature, precipitation, wind and RH) were obtained from nearby meteorological station.

## 4 Results and Discussion

### 4.1 Temperature and RH Variations

The temperature shows similar trends in most parts of the cave (sensors *T*-1, *T*-2, and *T*-3), defining a thermal behavior (thermal zone A) for the middle gallery and part of the upper galleries (Fig. 1B). These areas had a mean temperature during the monitoring period of ~9.1 °C and high thermal

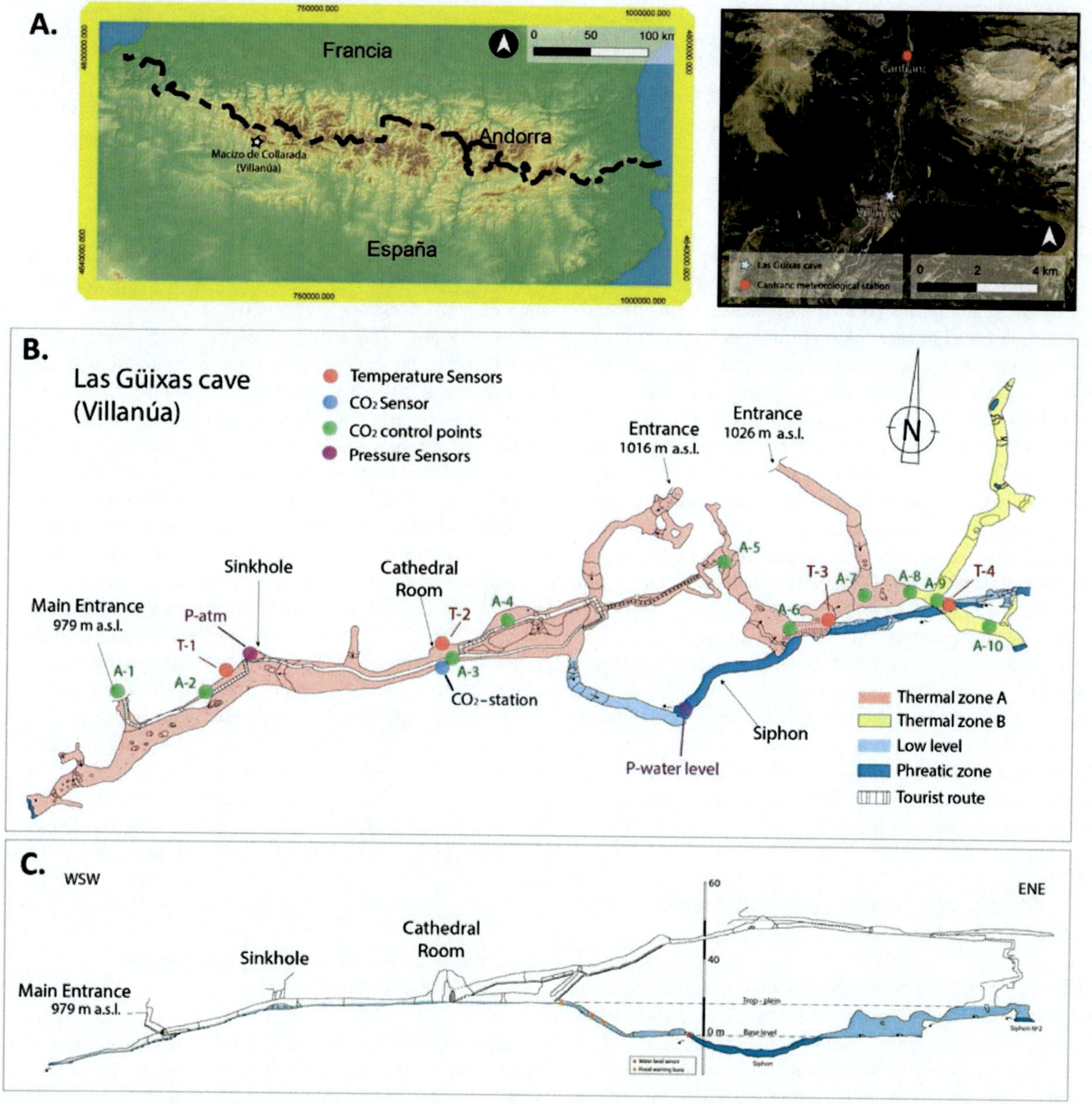


**Fig. 1** A: Situation of the Collarada Massif in the Pyrenees (left) and the Las Güixas cave in Villanúa village (right). B: Sketch of the Las Güixas cave with the location of the measurement points and the thermal zones. C: Topographic cross-section of the Las Güixas cave, the water and the flooding levels are indicated with dashed lines. Topography modified from Bengoetxea et al. (1994).

oscillations up to 13.8 °C (Table 1) due to the air currents between entrances (chimney effect) that induce a strong ventilation. The innermost part (sensor *T*-4) exhibits a different behavior and shows relative thermal stability (thermal zone B), with a mean temperature of 10.7 °C and small temperature oscillations (0.6 °C), being slightly warmer than the thermal zone A.

Along the zone A, there is a trend toward higher temperatures as height increases (Table 1). During winter, seasonal cryospeleothems formed from ice are found close to the main entrance, in the sinkhole and in the cathedral area, where the strong ventilation occurs, indicating occasional drops of temperature below 0 °C. The temperature oscillations in thermal zone A track closely to the outside

**Table 1** Summary of the mean, maximum and minimum values of temperatures and RH recorded in the Las Güixas cave during the monitoring survey

| | Temperature (°C) | | | Relative humidity (%) | | |
|---|---|---|---|---|---|---|
| Sensor position | Mean | Maximum | Minimum | Mean | Maximum | Minimum |
| *T*-1 | 8.2 | 12.8 | 3.7 | – | 100 | – |
| *T*-2 | 8.9 | 15.3 | 1.5 | 99.8 | 100 | 83 |
| *T*-3 | 10. 2 | 16.9 | 4.3 | 99.2 | 100 | 75.7 |
| *T*-4 | 10.7 | 11.3 | 9.4 | 100 | 100 | 100 |

variations, with small delay of 2–3 h. Conversely, in thermal zone B temperature variations are smaller exhibiting a delay of 7 days. Similarly, this behavior (delay and attenuation) has been observed in other caves (Smerdon et al., 2004).

Regarding RH, significant differences between the defined thermal zones are observed. In the lower part of the thermal zone A (the touristic zone; sensors *T*-1 and *T*-2) there are important RH drops during autumn and winter. These oscillations are probably related to the entry of cold and dry air. On the contrary, in the upper gallery (sensor *T*-3), RH drops are intensified in summer likely in relation to the great temperature increases in this area. Finally, in the innermost part (sensor *T*-4), the RH remains constant close to 100%.

Thus, our temperature and RH observations suggest that temperature changes in the thermal zone A are mainly controlled by advection as consequence of ventilation processes. On the contrary, the thermal stability in the thermal zone B indicates a limited advection effect due to a low air circulation associated to a lack of direct connection with the outside. Thus, temperature in this zone may be controlled by heat conduction through the rock.

## 4.2 $CO_2$ Natural Dynamics and Influence of Touristic Visits

Monthly $CO_2$ records from different stations along the cave (Fig. 1B) define two sectors similar to those described in the thermal zones. In zone A (stations 2–7), $CO_2$ remains low during the whole year. In contrast, $CO_2$ concentration is higher in the upper galleries (station 8 to 10), increasing toward the innermost part of zone B (up to 7600 ppm). $CO_2$ data recorded continuously in the Cathedral Room (zone A) has shown a low concentration of ~420 ppm, but visits results in significant increases in $CO_2$ concentrations adding more than 100 ppm. However, the observed $CO_2$ peaks are generally reduced during the next hour after the visit, recovering their background level. This suggests a quick elimination of the anthropic $CO_2$ in relation to the effective ventilation in this area (Fig. 2). On the other hand, natural $CO_2$ peaks (~900 ppm) represent sudden increases in the $CO_2$ concentration produced by the degassing of water during flood events, being up to 5 times higher than those anthropically produced. Similarly to the anthropogenic increases, those $CO_2$ values are easily eliminated by ventilation after the flood event (Fig. 2B).

The visits are organized in groups of 20–35 people with a guide that tour the cave for one hour, reaching 30,000 visitors per year in 2019. The usual regimen of visits to the cave is two visits per day, normally spaced throughout the day (12 am and 6 pm) with an extension to four visits on the weekends, depending on demand. However, during high touristic seasons (e.g. Christmas, Easter and Summer holidays), the number increases up to 6–8 visits per day during several consecutive days, resulting in high air $CO_2$ concentrations that need more time to recover back to normal. In March 2020, the cave was closed due to lockdown derived from the COVID-19 pandemic during three months and the tourist activity was resumed with reduced groups due to government restrictions in June 2020. The low number of visitors since the lockdown coincides with a progressive decrease of $CO_2$ concentration in background values (Fig. 2a). This break in the number of visits as well as the reduction of $CO_2$ values at the outside atmosphere during 2020 due to national lockdowns at global scale (Le Quéré et al. 2020), may explain the observed $CO_2$ reduction in the cave.

## 4.3 Water Level Monitoring: Response to Precipitation

The data recorded by the water level sensor provide clear evidence of a good response of water level to the rainfall events. The time of response to rainfall is highly variable and it depends on the amount of rainfall and snowmelt, which are controlled by temperature and rainfall amount. The maximum flow rate measured in the cave was 900 L $s^{-1}$ and occurred 8 h after a ~60 mm rainfall event. However, after similar rainfall amounts, the water level does not always respond in the same way, suggesting the water retained in the lower level plays an important role. The highest water level annually, which involves the flooding of the touristic gallery, takes place mainly during the spring, when the melting of the snow is combined with the usual rainfall increase. However, some spring-flood events can also occur without any relation to rain events, indicating that daily

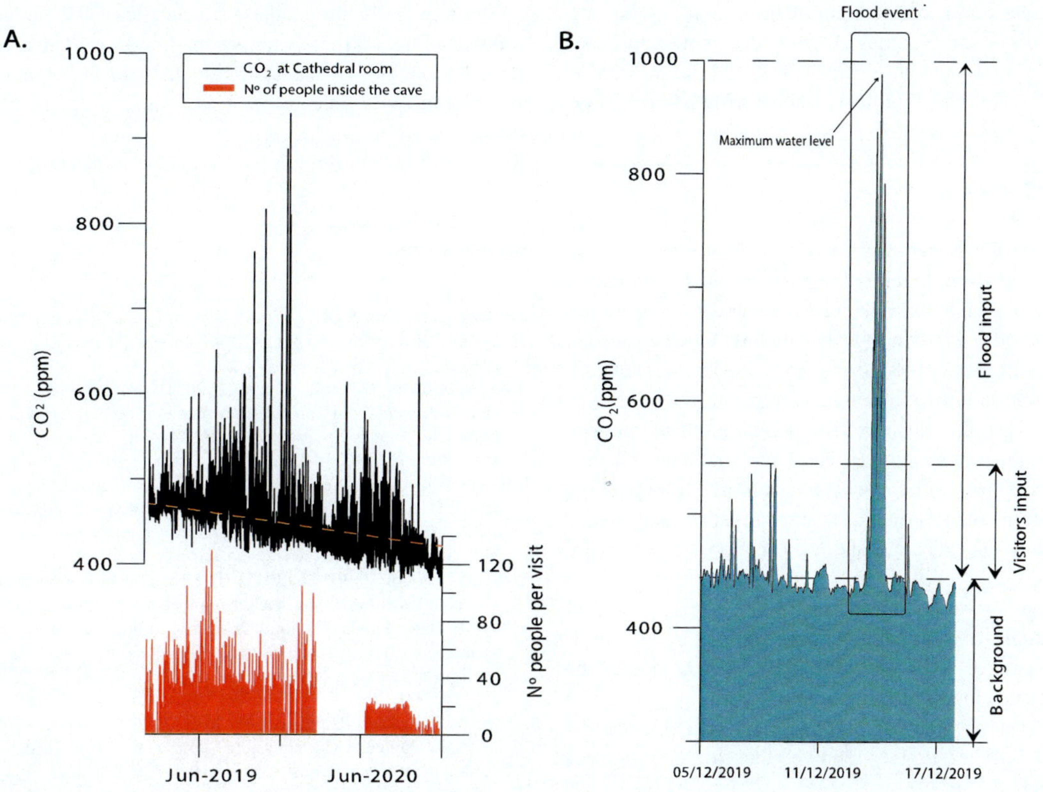


**Fig. 2** A: Hourly variation of the $CO_2$ at the Cathedral room and its correlation with number of visitors inside the cave, B: Detail of $CO_2$ variations during several days on December 2019, coinciding the lower increments in concentration with visits and the highest $CO_2$ peak with a significant flood event

snowmelt is important enough to raise the groundwater level that peaks in the cave at night. During the summer and autumn, a low level is maintained with little water retained in the system and only scattered events with a high amount of precipitation may generate a rise in the water level to flood the cave (Fig. 3).

The occurrence of extreme precipitation events can produce damage to the cave facilities leading to large economic losses and even human risks, if a sudden level increase takes place during a visit. As an example, a huge and drastic flood destroyed the cave facilities in the autumn of 2012. During this event, the main entrance to the cave worked as a trop-plein within 3-days of intense rainfalls (>230 mm). After repairing the damage caused in the cave, a flood warning system based on buoys at different heights were installed in the ramp that connect to the siphon and directly connected to the guides' center. Currently, once the lower warning turns on, visits are immediately canceled before the cave floods, to avoid human risks. The cave remains closed to the public while the alarm system remains active, even when the water level descends back down to the siphon area, as long as rainfall is forecast. This situation may last for

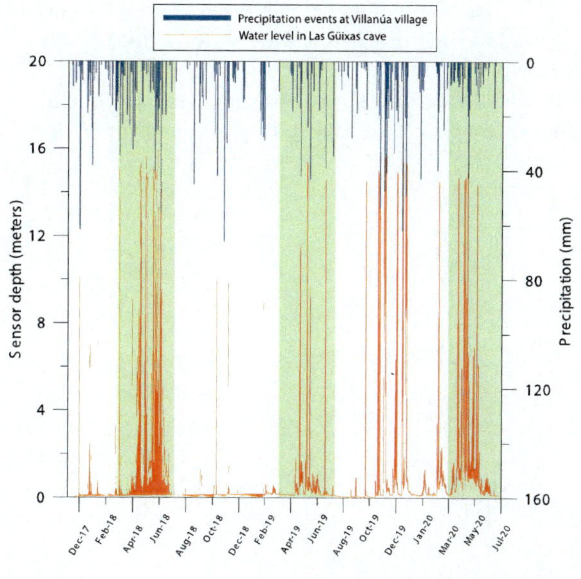


**Fig. 3** Water level changes in the cave compared to rainfall events in Villanúa village. Shaded areas indicate the spring season

several weeks and generates important economic losses due to the absence of incomes. Conversely it exemplifies a responsible management of the cave with a priority on safety, increasing the confidence and security of the visitors.

## 5 Conclusions

Show caves have become important geotouristic targets worldwide but its management requires an adequate monitoring of physical and environmental parameters to ensure sustainable and safe conditions. Our key findings based on analysis from Las Güixas cave monitoring are summarized below:

Cave morphology and entrances regulate the ventilation regime through the air advection, which controls the variability of temperature and RH. These air parameters allow us to define two areas: zone A, characterized by high variability of temperature and RH due to strong ventilation; and zone B, with the low air exchange and more constant temperatures and humidity.

The spatial variability has also correlated with $CO_2$ concentrations, distinguishing similar zones. Changes in the $CO_2$ are control by water degassing during floods and anthropogenic inputs from visitors. Anthropogenic $CO_2$ emissions are substantially lower than those induced by floods and both are rapidly removed due to the strong ventilation regime.

The response of water levels in the karst system is complex and it is influenced by the water storage in the Collarada karst system. During the spring, high water levels are related to snowmelt and the increase in rainfall amounts throughout the season. On the other hand, during summer, the water level remains low and the cave is flooded only after heavy rainfall events.

Considering the touristic use of some caves located in the epiphreatic zone, as at Las Güixas cave, precise hydrological studies (e.g. hydraulic models) are necessary to establish and understanding of the dynamics of flooding. Thus, hydrological modeling of the cave will help to anticipate flood hazards and risk management actions during the rainfall season improving the security of the visitors and, additionally, avoiding economic losses as much as possible.

## References


Bengoetxea A, García D, García JM, Yzaguirre I (1994) Exploraciones del G.E.R.S en el Macizo de la Collarada (Huesca). Subterránea 2:8–14

Bourges F, Genty D, Perrier F et al (2020) Hydrogeological control on carbon dioxide input into the atmosphere of the Chauvet-Pont d'Arc cave. Sci Total Environ 716:136844. https://doi.org/10.1016/j.scitotenv.2020.136844

Lambert WJ, Aharon P (2011) Controls on dissolved inorganic carbon and $\delta^{13}C$ in cave waters from DeSoto Caverns: Implications for speleothem $\delta^{13}C$ assessments. Geochim Cosmochim Acta 75:753–768. https://doi.org/10.1016/j.gca.2010.11.006

Lang M, Faimon J, Ek C (2015) The relationship between carbon dioxide concentration and visitor numbers in the homothermic zone of the Balcarka Cave (Moravian Karst) during a period of limited ventilation. International Journal of Speleology 44: https://doi.org/<p>http://dx.doi.org/https://doi.org/10.5038/1827-806X.44.2.6%3C/p%3E

Le Quéré C, Jackson RB, Jones MW et al (2020) Temporary reduction in daily global $CO_2$ emissions during the COVID-19 forced confinement. Nat Clim Chang 10:647–653. https://doi.org/10.1038/s41558-020-0797-x

Rivas A, Cabezas J, Carrasco F, et al. (2004) Las cuevas turísticas españolas: Un recurso natural de gran interés ecológico, económico y social. In: Andreo B, Durán JJ (eds) Investigaciones en sistemas kársticos españoles. Publicaciones del Instituto Geológico y Minero de España. Serie Hidrogeología y Aguas Subterráneas 12:367–384

Smerdon JE, Pollack HN, Cermak V, et al (2004) Air-ground temperature coupling and subsurface propagation of annual temperature signals. J Geophys Res Atmospheres 109. https://doi.org/10.1029/2004JD005056

# Understanding Morphosedimentary Changes and Extreme Past Floods: The Case of Ojo De Valjunquera Cave (Iberian Range, Spain)

G. Pérez-Villar, M. Bartolomé, G. Benito, A. Medialdea, M. Luetscher, R. L. Edwards, and A. Moreno

**Abstract**

During a flood inside a cave, sands and silts are transported through the cave system and deposited as (i) a coating of this detritus on speleothem surfaces and (ii) a flood bed accumulated on the cave passages which are protected from the main water stream. After the floodwater level drops, speleothem deposition is restarted and detrital coatings get trapped inside the carbonate laminae, while detrital flood bed sequences may contain clay laminae and water drops features at their contacts. When this process is unique and continuous (in the case of speleothems) record of paleofloods in a particular region is obtained. Typically, speleothems and detrital sequences close to usual groundwater table, record ordinary floods. On the contrary, sequences deposited in areas above the usual epiphreatic zone are expected to register extraordinary floods. Here, we present the geomorphosedimentary and chronological (U/Th, OSL) information derived from the Ojo de Val Junquera Cave (Iberian Range, Spain). The cave comprises two sub-horizontal levels connected by shafts and ramps. The lower level corresponds with an ephemeral spring and includes six siphons. Stalagmites from the upper level show single sand layers and brownish bands in relation to past low/high-frequency flood episodes, respectively. U/Th ages show that speleothems grew during Marine Isotope Stages (MIS) 9, 8, 7, 6, 3, 2 and 1. In the lower level, at the epiphreatic zone, a poorly consolidated detrital infill is formed by sands and gravels, indicating that the cave outlet was once blocked by these sediments, with an age (OSL) of 377 ± 39 ka (MIS 10–11). The study of detrital sediment facies and their chronology allow discerning between extreme and regular floods to provide high resolution records of extreme floods under climate variability up to a millennial scale in the Western Mediterranean.

**Keywords**

Speleothems • Paleofloods • Sediments • Morphosedimentary records

G. Pérez-Villar (✉) · M. Bartolomé · A. Moreno
Instituto Pirenaico de Ecología-CSIC, Madrid, Spain
e-mail: perezvillarguillermo@gmail.com

M. Bartolomé · G. Benito
Museo Nacional de Ciencias Naturales-CSIC, Madrid, Spain

A. Medialdea
Centro Nacional de Investigación Sobre La Evolución Humana-CENIEH, Burgos, Spain

M. Luetscher
Swiss Institute of Speleology and Karstology (SISKA), La Chaux-de-Fonds, Switzerland

R. L. Edwards
University of Minnesota, Minnesota, USA

## 1 Introduction

Fluvial and lake sediments are the most common geologic records to study past extreme flood events (e.g., Benito et al. 2015; Corella et al. 2016). However, the study of detrital layers trapped in speleothems is growing as a proxy to reconstruct palaeofloods in a particular area (Gázquez et al. 2014; González-Lemos et al. 2015a, b; Denniston et al., 2015; Denniston and Luetscher 2017; Bartolomé et al. 2021a). Indeed, speleothems may represent the most adequate geological archive to improve flood hazard assessment due to the high-resolution dating of the interlayering carbonates and their continuity, despite which their use as proxy for flooding is still largely unexploited. Accurate flood chronologies are obtained from well-dated U disintegration series ($^{238}U$-$^{234}$U-$^{230}$Th) covering a wide temporal range (500 ka). A limitation to this approach is the complex response of karst systems to rainfall, which mostly depends upon rainfall thresholds and therefore, do neither show a linear nor gradual reaction (Jeannin 2001).


B. Andreo et al. (eds.), *EuroKarst 2022, Málaga*, Advances in Karst Science,
https://doi.org/10.1007/978-3-031-16879-6_25

During a cave flood, sands and silts are transported through the cave system and deposited on (i) the cave walls and speleothems (stalagmites, stalactites and flowstones) and (ii) as detrital flood beds on sedimentary stacks typically on blind galleries and conducts protected from the main water stream. Once the flood is over, speleothem deposition is restarted, and detrital coatings are trapped inside the speleothems. Detrital flood beds are preserved on gallery bottoms or forming inset attached to the cave wall, typically forming sedimentary stacks with contacts marked by clay layering, slablike deposits and water drop features. Hence, speleothems located close to the water table usually record minor or ordinary floods while those located higher, preserve a record of extreme flood events. However, discerning the flood magnitude is not straightforward since karst hydrology depends upon multiple factors such as land use, changes in the sediment supply and other preconditioning circumstances (Denniston and Luetscher 2017). For example, temporary changes in the cave geometry associated to the deposition of flood deposits or, on the contrary, related to the scouring of the cave passages during the flood event; both cases may complicate the classification of past floods in ordinary or extraordinary (e.g. Bättig and Wildberger 2007; González-Lemos et al. 2015a, b; Denniston and Luetscher 2017; Bartolomé et al. 2021a). Therefore, speleothems located in elevated areas with respect to the river may in some cases record ordinary floods after a sediment mass flow.

Here, we present a preliminary analysis of morphosedimentary, and chronological data collected in the Ojo de Valjunquera cave, located in the Iberian range (NE Spain). The general objectives are (i) to study the flood beds infilling the cave passages and (ii) associated these infill sequences with detrital layers coating the cave speleothems.

## 2 Study Site

The Ojo de Valjunquera (OdV) cave (41° 45′ 20.64″ N; 1° 37′ 57.84″ *W*, ~700 m a.s.l) is located between the Ebro Basin and the highlands of the Iberian Range (Fig. 1a), closed to the Moncayo Natural Park. Moncayo peak (2316 m a.s.l) acts as an orographic barrier that favors storm formation. The annual rainfall is ~600 mm mostly concentrated in spring and autumn. Folded Triassic materials (Buntsanstein) dominate the Moncayo highland, while well-karstified carbonate rocks are in the lowland separated by the Talamantes Fault. Tertiary polymictic sands and conglomerates are located downstream to the OdV cave. The cave develops within Jurassic limestone and dolostone. The Talamantes fault (Fig. 1a) area plays an important hydrogeological role, where many streams infiltrate in the contact between the Buntsanstein and the lower Jurassic (LIAS aquifer) (San Román et al. 1989). Only the elevated areas from Moncayo peak glaciated during the Quaternary. On the contrary, the lowlands never have been affected directly by the glaciers.

The OdV cave (total length of ~1033 m) comprises two sub-horizontal levels with phreatic and vadose sections (Gisbert and Pastor, 2006). Odv probably had two entrances (Fig. 1a) located at similar elevation. Currently, the entrance located downstream of OdV main entrance is completely blocked by sediments. At 0.5 km from OdV, the water from seasonal springs infiltrates few meters away, and a groundwater borehole provides drinking water supply to Ambel village (Fig. 1b). On the other hand, a borehole (227 m deep, 2006 year) located ~ 3 km downstream from the cave indicates a water level 143 m below the surface (629 m a.s.l, Fig. 1b) (Bartolomé et al., 2021a). The lower cave level corresponds to an ephemeral spring (Fig. 1b), only active during heavy rainfalls. This lower level includes six siphons, shafts and ramps connected to the upper level, which is + 11 m above the cave entrance. This level is reached by water during some extreme events, such it has been evidenced in the paleoflood record from modern stalagmites (Bartolomé et al. (2021a)). Both levels after 650 m converge to form a single cave passage. The cave contains speleothem formations, as well as detrital sequences resulting from fluvial activity in the cave. In general, the galleries size varies from 4 to 6 m in diameter, although narrow conduits are 0.5 min diameter. Hydrologically, the cave entrance acts as a *trop-plein*, and the spring is active following rainfalls above 60 mm typically remaining active for more than a month. Bartolomé et al. (2021a) analyzed the cave dynamics during heavy rainfalls that occurred from March to May of 2018 and 2020. In one of the rainfall events (>60 mm), the water level rose up by ~ 4 m between the Plaza sector and the siphon 3 (Fig. 1). Moreover, several old signatures with date (etchings), made over the wall sediments located in the lower positions in the de cave, present signs of reworking. On the contrary, those located in higher positions indicate these areas have not been affected by floods since many years (Bartolomé et al. 2021a).

## 3 Material and Methods

A general exploration of the cave was carried out focused on characterizing the cave geomorphology. Some deposits, including carbonated and detrital sediments, were studied from stratigraphical, sedimentological and chronological point of view. Several broken stalagmites from the upper gallery were recovered for *U*-Th dating (Fig. 2). Around thirty milligrams of carbonate powder were sampled using a

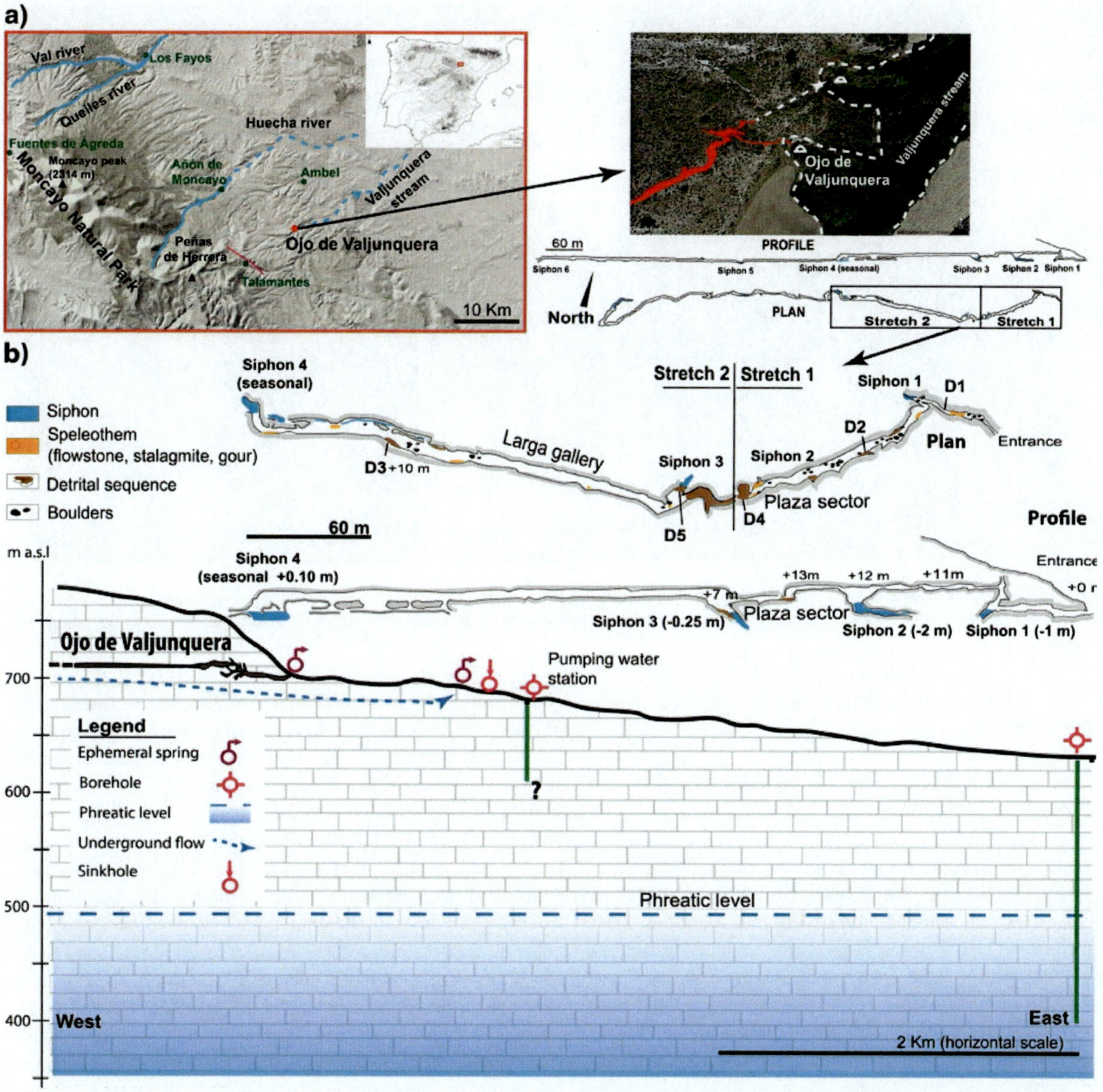


**Fig. 1** **a** Location of Ojo de Valjunquera cave the main rivers, aerial view of Valjunquera stream (©Google Earth) and cave topography modified from Gisbert and Pastor (2006). Talamantes fault is the pink line, **b** Zoom of studied stretches in OdV cave, approximate location of the main cave deposits and selected samples, and schematic cross-section. Stretch 1 and 2 represent the first ~ 650 m of the cave.

microdrill in a continuous air flow chamber at the Pyrenean Institute of Ecology (IPE-CSIC, Jaca). The carbonate powder was processed in the Isotope Geochemistry Laboratory at University of Minnesota (USA), following the protocol described in Edwards et al., (1987). A key detrital deposit (Fig. 2a) located close to the main cave entrance was dated by OSL. For OSL, an opaque PVC tube was introduced using a hammer in a sandy section in *D*1 sequence (Fig. 2a). Up to 200 mg of detrital material around the sample were collected to measure the humidity and the environmental dose. The OSL was measured on small multi-grain aliquots of quartz grains (sizes 180–250 μm) extracted from each sample. Dose recovery tests were carried out to assess the optimal heating conditions and the suitability of the SAR protocol (Murray and Wintle, 2003). Environmental dose rate has been calculated using DRAC (Durcan et al., 2015)

**Fig. 2** **a** *D*1 detrital sequence at the entrance gallery. **b** Detail of unit 2 in *D*1 formed by pebbles and angular clast. **c** Remnants of *D*1 deposit closed to siphon 1 in the lower gallery. **d** Sand-carbonate nodules stuck to the bedrock. **e** Sand-carbonate nodules accumulation at the beginning of larga gallery

from the concentration of radionuclides present in the matrix around the sample.

## 4 Results

Cave deposits comprise carbonate and detrital sequences (Fig. 1b). Occasionally small bat guano accumulations, as well as charcoal remains introduced thorough the karst system from recent forest fires, are found constituting organic deposits. Clastic deposits are dominant in the cave. Boulder accumulations (Fig. 1b) generated by gravitational processes are present in the first meters in the upper gallery. Sandy deposits coat the boulder chaos and appear also in the blind sectors in the upper gallery as a result of water level rise during flooding. A centimetric to millimetric sandy layer covers speleothems, walls, and ceiling all along the cave, denoting that at some point water filled up the cave completely. In some cases, modern looking stalagmites cover detrital sediments at different locations along the cave. A total of five main detrital sequences were recognized (*D*1, *D*2, *D*3, *D*4 and *D*5, Fig. 1b). In the lower gallery, close to the current cave entrance, the *D*1 ($\sim$150 cm of thickness, Fig. 2a) deposit covers the walls and reaches the cave ceiling. Two clearly different units were identified in *D*1. Unit 1 is formed by $\sim$100 cm of sands and silts. Regarding grain size, this unit is formed by 48.2% of sand, 47.2% of silt (coarse + fine silt) and 4.7% of clay. The Unit 2 comprises $\sim$50 cm of polimictic, centimetric well-rounded gravels, and angular clast derived from cave walls (Fig. 2b). *D*1 deposit appears forming small sediment patches until siphon 1, indicating the lateral extension of *D*1 was $\sim$ 30 m in the past (Fig. 2c). The OSL age derived from Unit 1 was 377 $\pm$ 39 ka (MIS 11–10). The upper gallery deposits *D*2 ($\sim$40), *D*3 ($\sim$120), *D*4 ($\sim$170) and *D*5 ($\sim$40 cm) are distributed from the access point of the upper gallery to the siphon 4 at different heights from the cave entrance. In general, these sequences are formed by rhythmic sand and silt layers (cm to mm in thickness) (*D*3) and slack water deposits formed by coarse silts and sands (*D*2, *D*4 and *D*5). Sequences *D*3 and *D*4 rich carbonate-sand nodules levels were found. Remains of similar nodules are located at the beginning of the Larga gallery (Fig. 2d, e), around 2–3 m higher than the top of D5 deposit. Finally, at the cave entrance, a small detrital sequence (30–40 cm) shows that the cave entrance was smaller in the past. People accessed the cave crawling on the floor sediments (Alberto Gomollón personal communication), while nowadays it is not necessary.

Carbonate speleothems occur mainly in the upper gallery and are well-developed in some areas. U/Th ages indicate speleothem growth during MIS 9, 8, 7, 6, 3, 2 and 1. A major speleothem growth took place during MIS 8–7. In general, the stalagmites in the upper gallery consist of clean carbonate and individual detrital layers, and brownish areas related to flood events at different time periods (Fig. 3).

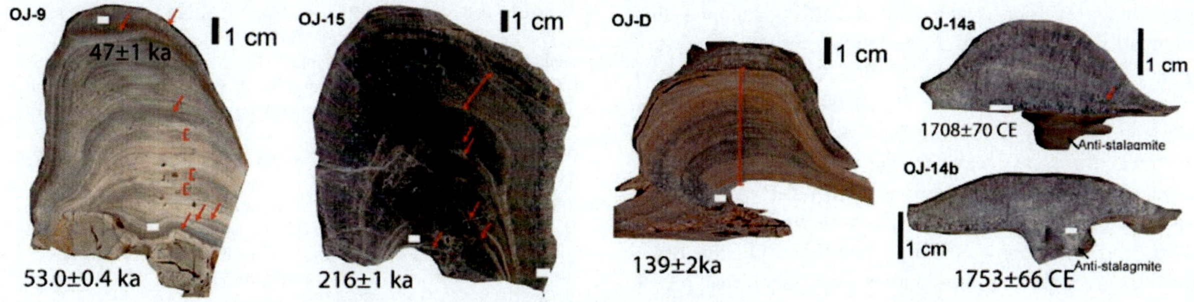


**Fig. 3** Different stalagmites from MIS 7, MIS 6 and MIS 3 and modern with single flood events and brownish areas. Red arrows indicate a single flood event, while red square bracket corresponds to brownish areas formed by multiple flood events

For example, the stalagmite Oj-*D* (139 ± 2 ka, MIS 6) contains multiple detrital layers from floods, while other stalagmites which grew during the MIS 7 (216 ± 1 ka), MIS 3 (53.0 ± 0.4–47 ± 1 ka) and the Holocene combine individual detrital layers and brownish areas formed by multiple-floods. The modern speleothems (Oj-14*a* and Oj-14*b*, Fig. 3) grew over a detrital layer that covers the Larga gallery. Both of them show anti-stalagmites[1] formed by the impact of drips over anon-consolidated sandy material. The ages indicate that those speleothems started growing at 1708 ± 70 CE (Oj-14*A*) and 1753 ± 66 CE (Oj-14*b*) (Bartolomé et al., 2021a).

## 5 Discussion

The input of sediment or erosion associated to flood events can introduce temporary changes in the cave geometry (Denniston and Luetscher, 2017). These changes (in the volume and location of many deposits) may affect the hydraulic head rise, making elevated and/or further areas from the cave stream water to be easily affected by an ordinary flood event. Below, we discuss how the morphosedimentary observations could affect the hydraulic head rise in the past, and therefore, to the flood record in the stalagmites.

The first morphosedimentary evidence that probably affected the rise of hydraulic head, this area have not been glaciated, is related to D1 sequence. This deposit suggests that cave entrance could be blocked during MIS-11-MIS-10. In such a situation, it is likely that the upper gallery would be flooded much more easily than. Although the duration of blockage is difficult to establish, these unconsolidated sediments are easily detached if they were exposed several times to water flow. Possible evidence of speleothem growth during that period, must be taken into account. Regarding the sediment supply, this could be related to soil degradation at the transition from warm-wet to cold-dry conditions (MIS-11-MIS-10). A similar mechanism has been proposed for sediment supply in caves in non-glaciated areas (Arriolabengoa et al., 2015).

The second evidence of temporal changes in cave geometry is related to the presence of sand-carbonate nodules in a position 2–3 m higher than the D5 deposit at the beginning of Larga gallery. These nodules indicate that sediments reached, at least, that height in the past. The sand was scoured probably during a large flood event, leaving the nodules in the area. Unfortunately, we do not have yet a chronological framework for D5 sequence. Perhaps the nodules could be related to a major lateral extension of D5 in the past. Another small change in cave geometry is related to a small deposit located at the cave entrance that evidence the small enlargement of cave entrance in recent years related to the cavers.

There is no evidence of successive blockages of the entrance, such as D1, although they cannot be totally discarded. Thus, although the presence of flood layers in pre-Holocene stalagmites opens an opportunity to analyze the frequency of floods in different time windows, the flood record from these early periods should be taken with caution in terms of characterizing extreme floods. It follows from this study that a morphosedimentary and chronological framework of the different clastic deposits whenever possible is mandatory to interpret the flood magnitude recorded in the stalagmites. Preliminary ages suggest significant stalagmite development during MIS 8 and MIS 7. This is in agreement with the development of tufa in the nearby Val and Añamaza rivers (Arenas et al., 2014; Sancho et al., 2019), as well as, in other rivers of the Iberian Range (Sancho et al., 2015) and the formation of stalagmites the Recuenco cave (Pérez-Mejías et al., 2017). Flood layers indicate that recurrent floods affected the cave during that period. Furthermore, the stalagmite displaying many flood

[1] Speleothem that forms due to the impact of the drips over a poorly consolidated sediment. The drip generates a hole, which is filled with carbonate. This type of speleothem usually appears in the base of some stalagmites pointing a growth towards the substrate.

layers that grew during the end of MIS 6 supports this statement. During this period, speleothem growths and tufa deposits formed in N-NE of Iberia (Stoll et al., 2013; Sancho et al., 2015; Bartolomé et al., 2021b) suggesting a climate amelioration during MIS 6b. This evidence, together with the results from OdV cave, makes this cold period quite different in the climate conditions from other cold periods associated to glacial stages (ej. MIS 4) in which speleothem growth was limited by cold/dry conditions. On the other hand, and given the morphosedimentary observations, recent speleothems are the most interesting samples because (i) the flood layers could correlate with instrumental/ historical flood records, and (ii) the evidence of sedimentary changes is better preserved than the older ones, and the cave geometry. Besides has remained relatively constant.

## 6 Conclusions

We summarize our preliminary findings related to morphosedimentary changes and floods in the OdV cave below:

- The cave outlet could be blocked by sediments during MIS 11–10. Thus, water could rise more easily than in the present situation. Therefore, speleothems with a record of flooding during that period may reflect ordinary floods even though they are located above today´s high-water level.
- Preliminary ages indicate that speleothem growths in the upper gallery occurred during MIS 9, 8, 7, 6, 3, 2 and 1. All speleothems have individual detrital layers or brownish bands related to past floods. The presence of these records opens an opportunity to study past floods in pre-Holocene times. However, changes in cave geometry due to sedimentary changes that affected the height of hydraulic head in the past are not discarded.

The morphosedimentary and chronological study (distribution, volume, age) of detrital sequences in caves improves our understanding on how and when changes in cave geometry occurred, to better comprehend the type of flooding recorded in speleothems. This morphosedimentary and chronological exercise helps to define changes which occurred in terms of cave section or blockage, which is a necessary step before performing a hydraulic modeling of the cave for present and past conditions.

## References


Arenas C, Vázquez-Urbez M, Pardo G, Sancho C (2014) Sedimentology and depositional architecture of tufas deposited in stepped fluvial systems of changing slope: Lessons from the Quaternary Añamaza valley (Iberian Range, Spain). Sedimentology 61:133–171. https://doi.org/10.1111/sed.12053

Arriolabengoa M, Iriarte E, Aranburu A, Yusta I, Arrizabalaga A (2015) Provenance study of endokarst fine sediments through mineralogical and geochemical data (Lezetxiki II cave, northern Iberia). Quaternary International, Quaternary of the Western Pyrenean Region 364:231–243. https://doi.org/10.1016/j.quaint.2014.09.072

Bartolomé M, Benito G, Luetscher M, Badules-Iglesias J, Pérez-Villar G, Edwards RL, Moreno A (2021a) The potential of Ojo de Valjunquera cave (NE of Iberia) sediments for paleoflood reconstructions. Cuaternario y Geomorfología 35 (3–4), 11–28. https://doi.org/10.17735/cyg.v35i3-4.89413

Bartolomé M, Sancho C, Benito G, Medialdea A, Calle M, Moreno A, Leunda M, Luetscher M, Muñoz A, Bastida J, Cheng H, Edwards RL (2021b) Effects of glaciation on karst hydrology and sedimentology during the Last Glacial Cycle: The case of Granito cave, Central Pyrenees (Spain). CATENA 206:105252. https://doi.org/10.1016/j.catena.2021.105252

Bättig G, Wildberger A (2007) Ein Vergleich des Hölloch-Hochwassers vom August 2005 mit seinen Vorgängern. Stalactite 57:26–34

Benito G, Macklin MG, Panin A, Rossato S, Fontana A, Jones AF, Machado MJ, Matlakhova E, Mozzi P, Zielhofer C (2015) Recurring flood distribution pat- terns related to short-term Holocene climatic variability. Sci Rep 5:16398. https://doi.org/10.1038/srep16398

Corella JP, Valero-Garcés BL, Vicente-Serrano SM, Brauer A, Benito G (2016) Three millennia of heavy rainfalls in Western Mediterranean: frequency, seasonality and atmospheric drivers. Sci Rep 6. https://doi.org/10.1038/srep38206

Denniston RF, Luetscher M (2017) Speleothems as high-resolution paleoflood archives. Quatern Sci Rev 170:1–13. https://doi.org/10.1016/j.quasci-rev.2017.05.006

Denniston RF, Villarini G, Gonzales AN, Wyrwoll KH, Polyak VJ, Ummenhofer CC, Lachniet MS, Wanamaker AD, Humphreys WF, Woods D, Cugley J (2015) Extreme rainfall activity in the Australian tropics reflects changes in the El Niño/Southern Oscillation over the last two millennia. Proc Natl Acad Sci USA112, 4576–4581. https://doi.org/10.1073/pnas.1422270112

Durcan JA, King GE, Duller GAT (2015) DRAC: Dose Rate and Age Calculator for trapped charge dating. Quat Geochronol 28:54–61. https://doi.org/10.1016/j.quageo.2015.03.012

Edwards RL, Chen JH, Wasserburg GJ (1987) 238U-234U-230Th-232Th systematics and the precise measurements of time over the past 500.000 years. Earth Planet Sci Lett 81:175–192

Gázquez F, Calaforra JM, Forti P, Stoll H, Ghaleb B, Delgado-Huertas A (2014) Paleoflood events recorded by speleothems in caves. Earth Surf Proc Land 39:1345–1353. https://doi.org/10.1002/esp.3543

Gisbert M, Pastor M (2006) Cuevas y simas de la provincia de Zaragoza. Centro de Espe- leología de Aragón

González-Lemos S, Jiménez-Sánchez M, Stoll HM (2015a) Sediment transport during recent cave flooding events and characterization of speleothem archives of past flooding. Geomorphology 228:87–100. https://doi.org/10.1016/j.geo-morph.2014.08.029

González-Lemos S, Müller W, Pisonero J, Cheng H, Edwards RL, Stoll HM (2015b) Holocene flood frequency reconstruction from speleothems in northern Spain. Quatern Sci Rev, Novel Approaches to and New Insights from Speleothem-Based Climate Reconstr 127:129–140. https://doi.org/10.1016/j.quas-cirev.2015.06.002

Jeannin PY (2001) Modeling flow in phreatic and epiphreatic Karst conduits in the Hölloch Cave (Muotatal, Switzerland). Water Resour Res 37:191–200. https://doi.org/10.1029/2000WR900257

Murray AS, Wintle AG (2003) The single aliquot regenerative dose protocol: potential for improvements in reliability. Radiation Measurements, Proceedings of the 10th international Conference on Luminescence and Electron-Spin Resonance Dating (LED 2002) 37, 377–381. https://doi.org/10.1016/S1350-4487(03)00053-2

Pérez-Mejías C, Moreno A, Sancho C, Bartolomé M, Stoll H, Cacho I, Cheng H, Edwards RL (2017) Abrupt climate changes during Termination III in Southern Europe. PNAS 114:10047–10052. https://doi.org/10.1073/pnas.1619615114

San Román J, Sánchez J, Martínez FJ (1989) El drenaje subterráneo del macizo del Mon- cayo: aspectos hidrológicos e hidroquímicos. Tvriaso IX 203–224.

Sancho C, Arenas C, Vázquez-Urbez M, Pardo G, Lozano MV, Peña-Monné JL, Hellstrom J, Ortiz JE, Osácar MC, Auqué L, Torres T (2015) Climatic implications of the quaternary fluvial tufa record in the NE Iberian Peninsula over the last 500 ka. Quatern Res 84:398–414. https://doi.org/10.1016/j.yqres.2015.08.003

Sancho C, Bartolome M, Arenas C, Aranbarri J, Moreno A, Chen H, Edwards RL (2019) Cronología de los depósitos tobáceos del río Val (Cordillara Ibérica, provincia de Soria). En: Libro de resúmenes. XV Reunión Nacional de Cuaternario (1–5 Julio, 2019) 285–288.

Stoll HM, Moreno A, Mendez-Vicente A, Gonzalez-Lemos S, Jimenez-Sanchez M, Dominguez-Cuesta MJ, Edwards RL, Cheng H, Wang X (2013) Paleoclimate and growth rates of speleothems in the northwestern Iberian Peninsula over the last two glacial cycles. Quatern Res 80:284–290. https://doi.org/10.1016/j.yqres.2013.05.002

# Identification of Near-Surface Karst Cavities Using the Posterior Population Expansion Inverse Method Applied to Electrical Resistivity Data

Manon Trottet, Przemyslaw Juda, Arnulf Schiller, and Philippe Renard

**Abstract**

Traditional inverse methods used to interpret electrical resistivity tomography (ERT) measurements are not able to properly characterize karst environments due to the strong contrast between air or water filled conduits and the surrounding matrix. In fact, these inversion methods were originally conceived to interpret rather uniform lithologies. The posterior population expansion (PoPEx) method is an inverse method designed to deal with data presenting abrupt variations such as those encountered in a karst system. The advantage of this method is that it directly returns maps indicating the probabilities of encountering conduits instead of the usual deterministic set of resistivity values that one has to interpret manually. This method is tested here to invert synthetic 2D ERT data of karst systems. At this stage, PoPEx is computationally demanding and needs to be standardized. However, the results presented in this paper show that it is capable of properly identifying simple and synthetic caves. Further, research is needed to test its applicability in the field.

**Keywords**

Electrical resistivity tomography • Probabilistic inversion • Multiple-point statistics

## 1 Introduction

Limestone environments are subject to significant karstification leading to the formation of large cavities. In many situations, it is necessary to be able to detect them as accurately as possible. For example, near-surface cavities can seriously impact any kind of construction through subsidence or collapse. Therefore, identifying them at an early stage may avoid important material and economical losses (Parise et al 2015). It may also be important to characterize the karst subsurface to locate, for example, where to drill new boreholes for the drinking water supply.

Geophysical techniques, and among them geoelectrical techniques, are broadly used methods to characterize efficiently the subsurface (Bechtel et al 2007; Chalikakis et al 2011). Although widely used and very effective in unconsolidated media, geoelectric methods are not fully effective in karst environments. In fact, locating the exact location of a conduit remains one of the most challenging tasks in karst research (Zhu et al. 2011).

Most traditional geophysical inversion methods are based on a deterministic approach and do not consider prior geological knowledge. Moreover, their most frequent numerical schemes assume that the spatial distribution of the petrophysical parameters (e.g., the electrical resistivity) is as smooth as possible. In the case of karstic systems, this can be an issue as conduits are discrete objects displaying strong contrasts with the underlying limestone.

Among the inverse methods that can deal with discrete problems is the posterior population expansion method (PoPEx) (Jäggli et al. 2017, 2018). This method was specifically developed to account for categorical distributions of the unknown parameters, i.e., the electrical resistivity when applied to ERT data. Moreover, it allows the integration in the inversion scheme of prior geological knowledge. Models of the possible distribution of the electrical resistivity values are defined before running the inversion.

M. Trottet (✉) · P. Juda · P. Renard
Stochastic Hydrogeology Group, University of Neuchâtel, Neuchâtel, Switzerland
e-mail: manon.trottet@gmail.com

P. Renard
e-mail: philippe.renard@unine.ch

A. Schiller
Geological Survey of Austria, Vienna, Austria


B. Andreo et al. (eds.), *EuroKarst 2022, Málaga*, Advances in Karst Science,
https://doi.org/10.1007/978-3-031-16879-6_26

The method can be used for a broad range of applications in complex, geologically realistic, and discrete model space settings. Until now it has been tested on hydrogeological problems. This paper presents its first application on a geophysical data set. The study aims to investigate if the method can provide reasonable results when it is applied to interpret electrical resistivity tomography (ERT) data. The study is based on synthetic data sets for which the exact locations of the conduits are known and can be used to verify the quality of the results. Application to real field data is out of the scope of this paper and will be studied in future works.

## 2 Methods

### 2.1 The Posterior Population Expansion Method

All the mathematical theory underlying the PoPEx method is described in detail in Jäggli et al (2017) and Jäggli et al (2018). Here, for sake of brevity, only the main ideas are outlined. PoPEx is a generic probabilistic Bayesian inversion method. Its inversion scheme is based on a set of geological models that are generated using a geostatistical method and extended iteratively. At each iteration, the existing set of samples is used to learn, in a statistical sense, the relationship between the model parameters and the variables.

As for any Bayesian inversion technique, the first step is to define the statistical model of the prior distribution of the unknown parameter values (i.e., the electrical resistivity). This prior distribution expresses what we know about the underground before collecting the geophysical data. In karst systems, we will express the fact that we know that we expect to find cavities in limestone formations with a certain spatial geometry derived a priori from analogous sites and geological studies.

To express this knowledge in statistical terms, a multiple-point statistics (MPS) approach is used. Then, each geological model that is generated is transformed into petrophysical parameters. In this paper, to keep the problem very simple, we assumed that this relation is deterministic, but this is not an obligation.

The models are then used to compute the forward geophysical response which is compared to the observation data. This allows to define a misfit function and evaluate the likelihood by assuming a certain statistical distribution of the errors. Results are stored in an ensemble, and statistics from the ensemble are computed to check which features in the model space are likely to produce a good fit between the observations and the calculated data. This knowledge is then used to preferentially sample certain locations and parameter values and generate new models by conditioning the geostatistical simulations with this information. In this process, the algorithm progressively learns how to generate new models that are more likely to fit the data.

The PoPEx method generates a large ensemble of stochastic simulations and returns either the simulations having the highest probabilities of fitting the data or the probability maps of finding a conduit at a certain location.

### 2.2 Prior Parameter Distribution

To apply the PoPEX inversion method to ERT data, a statistical model of all the possible geometries of the encountered cavities is needed in order to define a prior parameters distribution. To this aim, the direct sampling (DS) MPS method (Mariethoz et al. 2010) is used.

This method is based on a training image (TI). The training image is a conceptual model that represents the general structures and the variability of the geometries that are expected to be found in the studied area. From the training image, the DS method learns the statistics of the pattern and produces stochastic simulations (realizations) that resemble the training image. The TI must be larger than the realizations so that the algorithm can deduce statistics. Figure 1a shows the training image used in this study. Size of the TI is 80 by 1000 m. Realizations of 15 m depth by 100 m long are produced from it. This TI is a concept derived from field observations and adapted to represent the expected variability of the geometry of karst conduits.

Figure 1b, c shows few examples of the many possible realizations obtained from the training image 1a with the DS algorithm. When no conditioning data is given, realizations can be very different and allow exploring a wide uncertainty space. If conditioning data, such as observations of the rock type at the surface or along a borehole are available, the simulations can account for that information (compare Fig. 1a, b).

### 2.3 Forward Model and Traditional Inversion

There are many codes and software available for inverting geophysical ERT data. Here, the pyGIMLi open-source python package (Rücker et al. 2017) is used for two different purposes.

Firstly, data acquisition is simulated through the forward ERT responses of the reference models to obtain sets of apparent resistivity values used as input for the inversion schemes. Responses are calculated using 72 electrodes

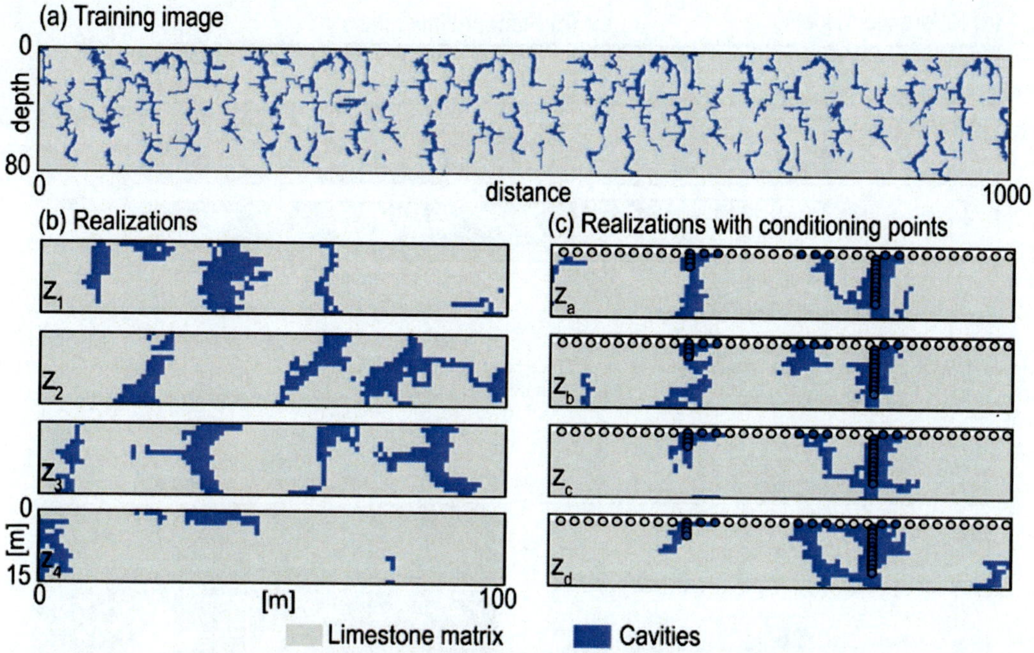


**Fig. 1** Prior synthetic cross-sections that represent karst cavities. **a** is the training image. **b** is a set of four simulated realizations showing the variability that can be expected at the scale of this cross-section. **c** are a set of four realizations conditioned with observation points (the circles) located at the surface and along a borehole. In cases **c** the variability of the simulations is reduced as simulations are constrained by more data

regularly placed at the surface with a dipole–dipole setting. One example of the results of this forward computation is shown in Fig. 2. Secondly, a forward simulator is needed to run the PoPEx inversion. Finally, the standard inversion is also performed with pyGIMLi in order to compare its results with the results obtained with PoPEx.

## 3 Synthetic Case Study

In order to assess the PoPEx method, synthetic reference models are defined as the unknown truth of the numerical experiment. To this aim, three DS realizations (references 1, 2, and 3) were chosen (Fig. 3a). These references were selected because they show situations with different densities and complexities of conduits allowing to assess the performance of the methods in different situations.

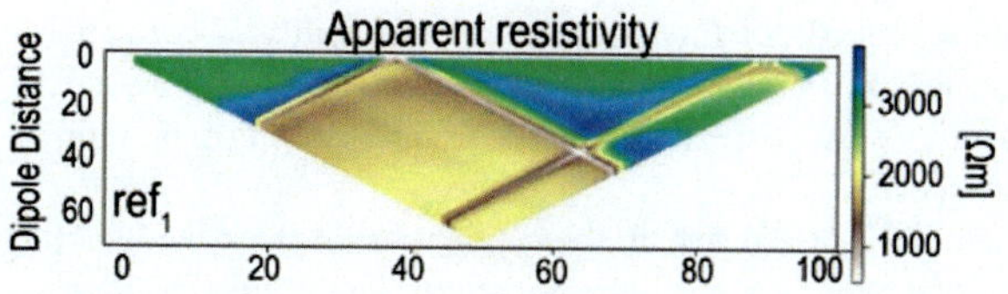


**Fig. 2** Result of the forward ERT computation of Reference 1 using pyGIMLI

## 4 Results

### 4.1 Results Obtained with Standard Inversion Methods

Figure 3b shows the spatial distribution of the resistivity values obtained when a standard inversion is applied to the three reference models. For *reference 1*, we see a strong resistivity anomaly centered around the location of the main cavity and a smaller anomaly corresponding to the smaller cave close to the surface on the right side. In the second case, the anomalies are also located properly, but some artefacts are visible. Finally, for the denser and more complex geometry, it becomes much more difficult to relate the inverted resistivity fields to the exact geometry of the caves. In all cases, the exact geometry of the boundary of the caves is difficult to infer only based on the resistivity fields.

### 4.2 Results Obtained with PoPEx

Results of the PoPEx inversion are illustrated in Figs. 4 and 5. They are expressed either in terms of cross-sections showing the probability (between 0 and 1) of finding a cave at each location (Fig. 4) or as individual simulations

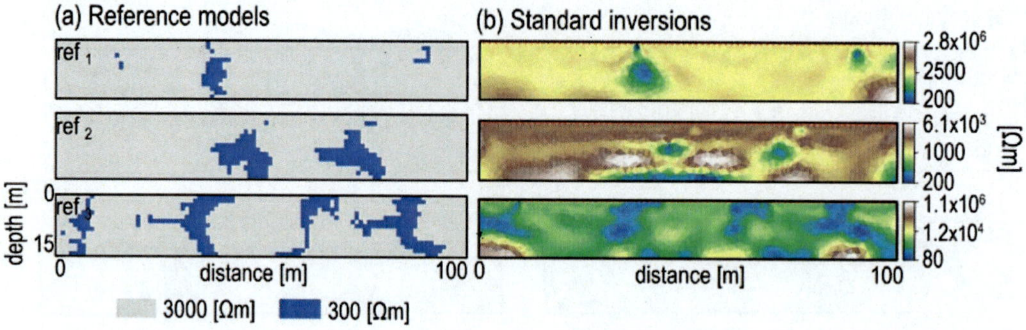


**Fig. 3** Reference models and results of the standard inversion. **a** Three reference cross-sections, the cavities are represented in blue and the limestone matrix in grey. **b** Results of the standard inversion for these three reference cases

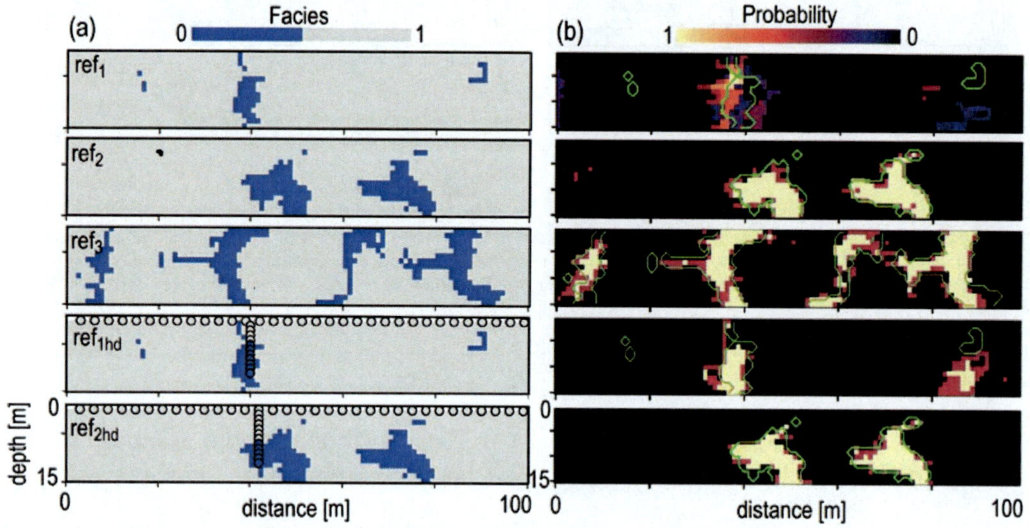


**Fig. 4** Results of the PoPEx inversion. **a** Unknown references and **b** The resulting probabilities of the occurrence of a cave. The green lines represent the position of the true caves. The first three examples represent the same references as in Fig. 3. In addition, the last two examples illustrate how geological observations can be added to constrain the inversion and improve the results

(Fig. 5.). A probability of 1 means that there is 100% of chance to find limestone at this location.

Figure 4a shows the reference models. For references 1, 2, and 3, the inversion is performed without conditioning data, whereas references 1hd and 2hd include conditioning data at the surface and along a vertical line mimicking a possible borehole log. These conditioning points are shown with grey and blue circles in the reference models.

Figure 4b shows the probability maps obtained after the inversion for these models. A probability of 1 indicates that there is a 100% probability of finding a conduit at this location. The shapes of the true cavities are represented with a green line in the probability maps to facilitate the analysis of the results. Figure 4b shows that all the large and main cavities are detected and rather well represented in the probability maps. However, small cavities such as the one located in the upper right corner in *Reference 1* are not always detected properly.

In all the references, the uncertainty zones (zones with intermediate probabilities) are essentially occurring at the edges of the cavities. In most cases, the central part of the cavity is detected and delineated with a probability of 1, whereas outside of the cavities, the probability is correctly identified as 0. In between, intermediate probabilities are indicating some uncertainty on the exact position of the limit of the cave.

The only case in which the central part of the cave is not identified with a very high probability is the first example (*Reference 1*). In that case, the shape of the cavity is not well determined and seems to be slightly shifted from its actual position. It can be observed that adding conditioning data (*Reference 1hd*) improves significantly the quality of the predictions for the largest cave. The shape of the detected cavity is more accurate, and there are much more locations where the probability of finding a conduit is 1. However, in that case, artefacts appear, PoPEx identifies wrongly the position and the size of the small cavity on the right side. It finds that the cavity should be deeper than it is, and it is overconfident about the results giving a wrong probability of 1 instead of 0. For *Reference 2hd*, adding the conditioning data improves the results (without creating artefacts) only slightly because they were already good without conditioning data.

For *Reference 3,* results obtained with PoPEx (Fig. 4b) are much better than results obtained with the traditional inversion (Fig. 2b), even if the number of cavities is rather large.

Finally, Fig. 5b shows five of the individual simulations obtained for the five test cases. The PoPEx method generates a large ensemble of stochastic simulations, from which it produces the probability maps discussed above. Each simulation is associated with a posterior probability value. In Fig. 5b, the simulations having the highest probabilities are displayed. This figure shows that the geometries are close but slightly different from the true unknown reality (Fig. 5a).

An important point to notice with these individual realizations is that they can be used in a Monte Carlo procedure for probabilistic risk assessment studies. For example, it would be straightforward to use them as input in a geo-mechanical model to study how a construction project could be damaged if built on such type of ground. By repeating the mechanical computation with the different karst simulations, one could estimate for example the probability of collapse of the building from these results.

## 5 Discussion and Conclusion

The aim of the study was to assess if the PoPEx method, coupled with the direct sampling MPS algorithm, could be used to invert synthetic ERT geophysical data to detect near-surface karst cavities. Results showed that applied to simple and synthetic models, the method is able to detect the cavities, which shows the benefits of pursuing this strategy. Shape of the conduits is determined with a higher accuracy than with a classical inversion method.

An interesting feature of the approach is the possibility to integrate surface or borehole observations to condition the prior distribution of the electrical resistivity values and thus combine geological observations, geological concept and the geophysical data. The prior distribution of the electrical resistivity is performed using the MPS approach. One advantage of the MPS approach is its flexibility allowing to include a conceptual geological knowledge in the inversion process and generate rapidly many very different models. Here, only simple models where tested, but different levels of complexity could be easily studied. Indeed, it is rather easy to modify and increase step by step the complexity of the geological models: starting with very simple ones, as the examples presented above, and adding complexity by simply changing the training image.

Although results are very encouraging and PoPEx provides many advantages, results presented here must be taken

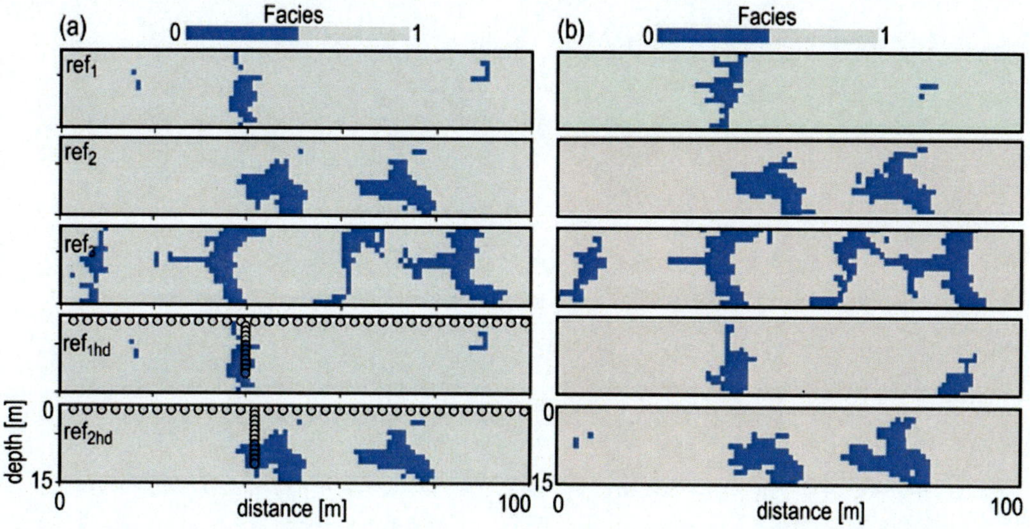


**Fig. 5** Results of the PoPEx inversion. **a** The five unknown references and **b** the best individual simulations belonging to the posterior ensemble for the five different cases

with care. Two important points must be highlighted. First, results show low uncertainty in the probability maps. This behaviour could be due to the way the likelihood is computed. Different ways to compute it were tested and, a tempered likelihood was selected, but that aspect requires further sensitivity analysis. The second point is that the synthetic models used as references models were generated using the same conceptual geological model (the same training image) as the one used to define the prior distribution for the inversion. Reference models and the one used to define the prior distribution of the parameters are hence rather similar. Quality and robustness of the results still need to be evaluated if these images are different and if the training image does not contain enough variability to allow the MPS simulation algorithm to model the actual cave geometry. These cases must be further studied, especially when considering real field data. Another possible impact on the quality of the results is the variability of petrophysical parameters and the number of cavities in the rock matrix. This variability can be incorporated into the simulations, but its impact on the accuracy of the results must be tested.

To conclude, results showed that the PoPEx approach is able to improve the detection of cavities in simple and synthetic cases. The major drawbacks of the method are that it requires longer computing times than a standard inversion, and a conceptual model of the cave geometries needs to be defined. However, the results obtained with this method are very encouraging.

**Acknowledgements** The authors would like to thank A. Neven for his help in setting up the forward problem with pyGIMLi, J. Straubhaar for his support with the MPS model, as well as two anonymous reviewers who helped improve the paper. The project was funded by the Austrian Academy of Science via the FlowCast project.

## References

Bechtel TD, Bosch FP, Gurk M (2007) Methods in karst hydrogeology, vol 26. CRC Press, Taylor & Francis Group; chap 9. Geophysical methods, pp 171–199

Chalikakis K, Plagnes V, Guerin R, Valois R, Bosch FP (2011) Contribution of geophysical methods to karst-system exploration: an overview. Hydrogeol J 19(6):1169–1180

Jäggli C, Straubhaar J, Renard P (2017) Posterior population expansion for solving inverse problems. Water Resour Res 53(4):2902–2916

Jäggli C, Straubhaar J, Renard P (2018) Parallelized adaptive importance sampling for solving inverse problems. Front Earth Sci 2018:203

Mariethoz G, Renard P, Straubhaar J (2010) The direct sampling method to perform multiple-point geostatistical simulations. Water Resour Res 46(11)

Parise M, Closson D, Gutierrez F, Stevanovic Z (2015) Anticipating and managing engineering problems in the complex karst environment. Environ Earth Sci 74(12):7823–7835

Rücker C, Günther T, Wagner FM (2017) pygimli: an open-source library for modelling and inversion in geophysics. Comput Geosci 109:106–123

Zhu J, Currens J, Dinger J (2011) Challenges of using electrical resistivity method to locate karst conduits a field case in the Inner Bluegrass Region, Kentucky. J Appl Geophys 75(3):523–530

# Use of Terrestrial LiDAR Scanner for Monitoring of Ice Thickness in Ice Caves; Examples from Slovenia

M. Blatnik, J. Obu, J. Košutnik, and F. Gabrovšek

**Abstract**

Melting of cave ice is one of the most common indicators showing consequences of increasing air temperatures and changing precipitation patterns. The most notable reports of ice shrinking are known from polar regions and mountain glaciers, however significant loss has also been observed in karst (ice) caves. Amongst 14,169 registered caves in Slovenia there are 230 with perennial ice. We present observations of cave microclimate and ice volume changes in 10 ice caves. Initially visual estimation of ice volume changes were made, followed by continuous air temperature measurements and topographic measurements using a tape measure and fixed reference points. Recently, measurements have been improved with the use of terrestrial laser scanning. This method enables detailed measurements of the cave geometry, ice surface and detection of changes of the position of individual objects (in this case rocks). Two ice caves with relative easy access (Velika Ledena Jama v Paradani and Snežna Jama na Planini Arto) were chosen for more detailed observations. In both caves scanning has been done for two year twice a year, in spring and in autumn. First analyses showed significant loss (up to 50 cm) during summer melting period and weak ice accumulation during winters. Preliminary results encourage further measurements to improve understanding of seasonal and long-term ice dynamics and its relation to climate changes. Periodic sampling of ice and analysis of its chemical and isotopic composition will provide additional information about characteristics and past development of ice.

**Keywords**

Karst • Cave • Ice • Terrestrial Lidar scanner • Climate changes

M. Blatnik (✉) · F. Gabrovšek
Karst Research Institute ZRC SAZU, Titov trg 2, 6230 Postojna, Slovenia
e-mail: matej.blatnik@zrc-sazu.si

F. Gabrovšek
e-mail: gabrovsek@zrc-sazu.si

J. Obu
University Oslo, Problemveien 7, 0315 Oslo, Norway
e-mail: jaroslav.obu@geo.uio.no

J. Košutnik
University of Nova Gorica, Vipavska cesta 13, 5000 Nova Gorica, Slovenia
e-mail: jurekosutnik@gmail.com

## 1 Introduction

Ice caves are special types of caves with microclimate characteristics favourable for ice accumulation. Usually, these are caves located in higher altitudes, characterized with general low air temperatures and more abundant snowfall. Another important reason for ice accumulation and its preservation is an airflow pattern, which is established in spacious vertical caves with larger entrances or caves with multiple entrances at different altitudes (Persoiu and Lauritzen 2017). During winter periods in multiple entrance caves, colder and denser outside air enters the cave through lower entrances, gradually warms up in the cave, and exits the cave through higher entrances. As a result, negative temperatures are present at lower entrance, causing freezing of percolated water and therefore accumulation of ice. During warm periods, outside air temperature is relatively higher and airflow circulation switches: Colder, so denser, air in the cave sinks and exits the cave through lower entrance and passes ice at lower temperature, causing only minor ice melt. At the higher entrance, warm outside air is sucked into the cave (Badino 1995; Luetscher and Jeannin 2004; Persoiu and Onac 2012; Covington and Perne 2015). These conditions cause that the mean air temperatures close to lower cave entrances are lower than the mean outside air temperatures of the surrounding, which are often reflected by inversion of vegetation belts (Mihevc 2021). In spacious


B. Andreo et al. (eds.), *EuroKarst 2022, Málaga*, Advances in Karst Science,
https://doi.org/10.1007/978-3-031-16879-6_27

vertical caves with a single entrance, winter airflow dynamics is similar, whereas during summers they act as a cold air trap, because warmer and lighter outside air is not able to enter the cave (Persoiu and Onac 2012; Covington and Perne 2015).

There are several studies describing ice caves characteristics. They mostly describe climatic characteristics, like temperature and airflow dynamics (Badino 1995; Luetscher and Jeannin 2004; Persoiu and Onac 2012; Covington and Perne 2015). However, in the last decades, other topics are also a point of interest. Such as descriptions of ice cave fauna and microorganisms (Itcus et al. 2018; Mulec et al. 2021), ice composition (Persoiu et al. 2011; Carey et al. 2018), age of ice (Kern et al. 2018), geochemical processes (Skoglund et al. 2010; Colucci et al. 2017), and within global warming studies: the shrinking of ice. The latter has been noticed in most known ice caves in Europe (Persoiu and Pazdur 2011; Buzjak et al. 2014; Colucci and Guglielmin 2019; Persoiu et al. 2021), as well as in Slovenia (Mihevc 1993, 2021; Košutnik 2016; Blatnik and Košutnik 2021; Blatnik et al. 2021), for example in caves Velika ledena jama v Paradani and Snežna jama na planini Arto, which are subjects of the present study (Fig. 1).

The aim of this study was to make detailed temporal and spatial ice thickness observations using a terrestrial laser scanner. Such technology has been already used in cave environment, usually to get better view on cave geometry (Konsolaki et al. 2021), mapping geological structures (Novak 2022), and also precise measurements of large speleological objects (Walters 2017; Walters and Zupan Hajna 2020), which are not easy to survey with other techniques. Performance of laser scanners for ice content mapping has not yet been reported in scientific literature.

## 2 Study Area

Snežna jama na planini Arto is a 1327 m long mainly horizontal cave located in Kamnik-Savinja Alps with an entrance at 1504 m a.s.l. The cave originally developed in a phreatic and epiphreatic environment, but the terrain has been uplifted for about 1 km in about 2 million years, so that the current base level is 1000 m below the cave (Häuselmann et al. 2015). In winter, air enters the main entrance and follows the first 300 m of passages until a large chamber, where it ascends a vertical passage (chimney) and unexplored airflow pathways to the surface. The presence of ice is limited to this section. The rest of the cave is isolated from active air circulation and therefore warmer (Zupan Hajna et al. 2008). Permanent ice is present on the floor of the first spacious chamber and small sections of the side passage (Fig. 2). In some other parts, ice occurs only during winters at sporadic places below active drip water parts of the ceiling. Due to ice formations and richness of speleothems, especially moonmilk, the cave is open for tourism. Mean annual temperature in the remote (part of the cave, where air circulation is minimal, is about 4.5 °C (Zupan Hajna et al. 2008). At meteorological station Krvavec, located at 1740 m a.s.l., 20 km to the SW, the mean annual air temperature between 1980 and 2010 was 3.4 °C, with a mean January temperature of −3.7 °C and a mean July temperature of 12.0 °C (ARSO 2022a).

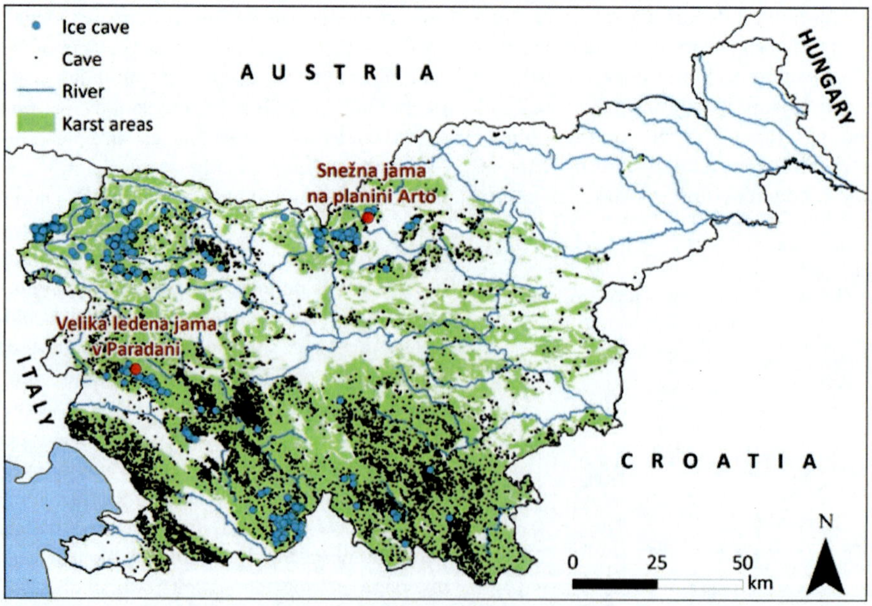


**Fig. 1** Karst areas in Slovenia and distribution of caves, ice caves and studied caves. *Source of data* Atlas Okolja (2022)

**Fig. 2** Laser scanning in an ice chamber in Snežna jama na planini Arto (left) and Velika ledena jama v Paradani (right)

Velika ledena jama v Paradani is 7311 m long and 858 m deep cave on the Trnovski Gozd plateau. It is characterized by a large entrance in the bottom of a karst depression and numerous deep shafts parallel to each other. The lowest part reaches groundwater level. During exploration, between 1970 and 1990s, ice was noticed up to 150 m below the entrance (Mihevc 2021), but nowadays exists only in the entrance chamber and side passages (Fig. 2) down to 50 m. The accumulation of ice was explained by the presence of inaccessible higher positioned entrances (fractures in the rock), so during winter period cold outside air is entering the cave through known entrance and flowing out through fractures (Mihevc 2021). Due to the large amount of ice, between the 1850s and 1950s the cave was used for ice harvesting, but with no significant negative effect on the current amount of ice. The closest comparable meteorological station is Vojsko (1067 m a.s.l., 6 km to the SW), which has registered a mean annual temperature of 6.4 °C, with a mean January temperature of −2.1 °C and a mean July temperature of 15.7 °C in the period 1971–2000 (ARSO 2022b).

## 3 Methods

Ice levels were measured with the terrestrial laser scanner Riegl® VZ-2000i (Fig. 2) which enables a measurement range 2500 m (at 50 kHz) and highest frequency of 500.000 points per second (at 1200 kHz). At each scan, the scanner rotates and records a 360° point cloud (Fig. 3). During the procedure, the scanner must be in a stable position. Multiple measurements are integrated into a single point cloud (Riegl 2022).

During observations of selected ice caves the highest frequency 1.2 MHz have been used, which enables up to 500.000 measurements per second and 600 m range of measurements. To identify the rate of melting during warm periods of the year and accumulation of ice during cold periods, the ice level was scanned twice a year: before the start and at the end of winter. In the first two years three observations were performed in Snežna jama na planini Arto, on 16th July 2020, on 19th November 2020 and on 11th May 2021 (Figs. 4 and 5). The focus was the entrance part of the cave, with a large chamber and small side passage, both covered with the ice on the floor (Figs. 2 and 6). In the cave Velika ledena jama v Paradani, two observations were performed; on 22nd July 2020 and on 9th May 2021 (Figs. 7 and 8).

During individual observations between 20 and 30 scans were made, which were joined together into a single point cloud. Point clouds were co-registered using stable cave walls. Because ice and wet surfaces are prone to false reflections, a reflection filter tool was been used. The final result was filtered point clouds showing cave walls and ice surface. For better comparison of ice surfaces at different time periods, points that are presenting ceilings and walls were removed in order to keep only the floor covered with ice (Fig. 3). All described operations were made using the RiScan software (Riegl 2022), which is produced for analysing scans obtained with Riegl® laser scanners. In the last step ArcGIS 10.3 software was used, in which filtered point clouds were transformed into digital models of elevation

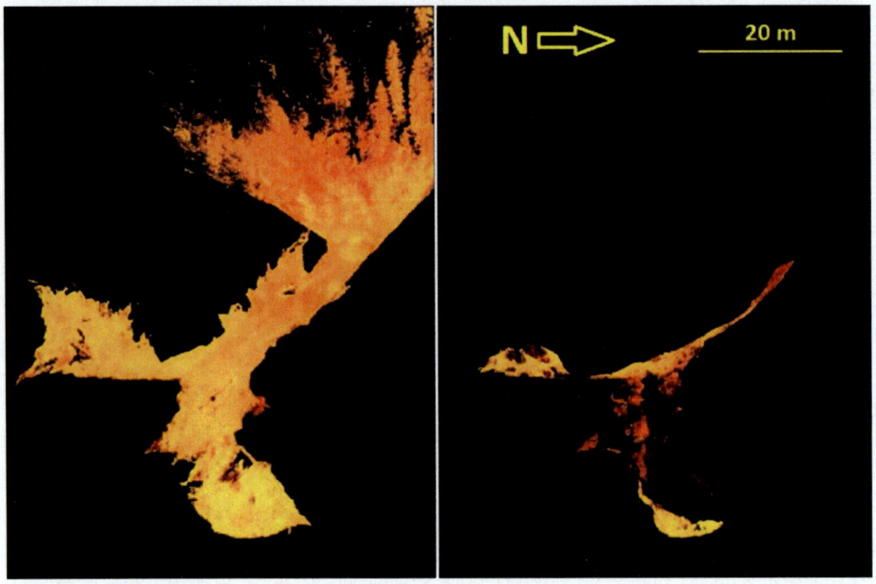


**Fig. 3** Left—complete point cloud of Velika ledena jama v Paradani. Right—selection of points showing ice surface in Velika ledena jama v Paradani

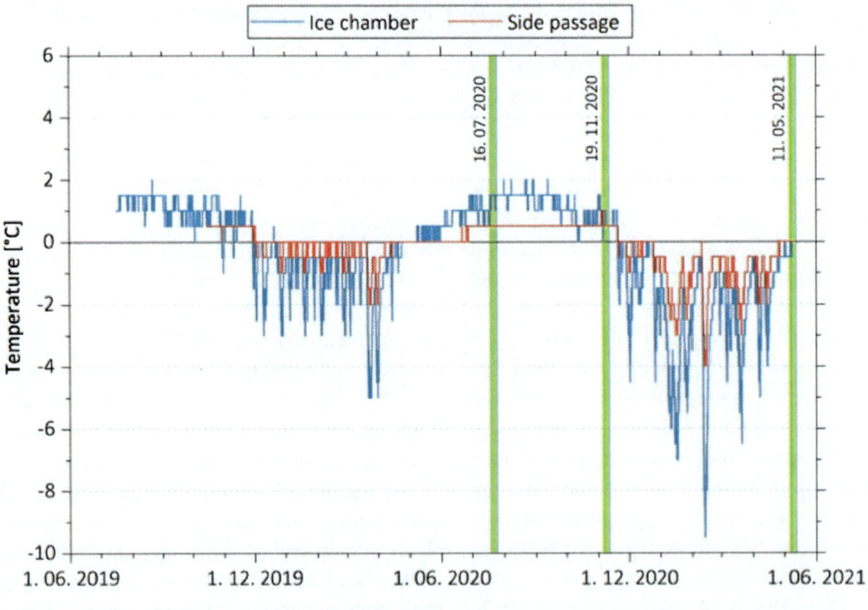


**Fig. 4** Air temperature dynamics in Snežna jama na planini Arto between June 2019 and June 2021 and periods of analysed scans

with 10 cm horizontal resolution. These datasets were used to make detailed comparisons of ice level change (Figs. 6 and 8) and also calculation of ice volume change between different periods.

For a better analysis of scans in terms of meteorological conditions air temperature measurements at different positions between cave entrances and chambers with ice have been gathered (Fig. 4). For this purpose temperature loggers iButton DS1922L with a measuring range from −40 to +85 °C, temperature, a resolution of 0.0625 °C and accuracy of 0.5 °C (Maxim Integrated 2021) were used. Temperature measurements have been made at 1 h intervals since the 15th January 2011 in cave Velika ledena jama v Paradani and since the 26.11.2011 in Snežna jama na planini Arto.

## 4 Results and Discussion

According to the latest measurements, the area with ice surface in Snežna jama na Planini Arto covers about 600 $m^2$. Between 16th July and 19th November 2020 the average ice level dropped by about 20 cm, which is equivalent to 110 $m^3$ of volume reduction. The largest decrease, of up to 40 cm, was in the western part of the chamber, caused by direct inflow of water from the chimney. Notable changes of ice level were also detected in the passage connecting the ice chamber with the side passage, where the ice surface is inclined (Figs. 5 and 6). In the following time period between 19th November 2020 and 11th May 2021, which was mostly represented by the cold period, the level of ice

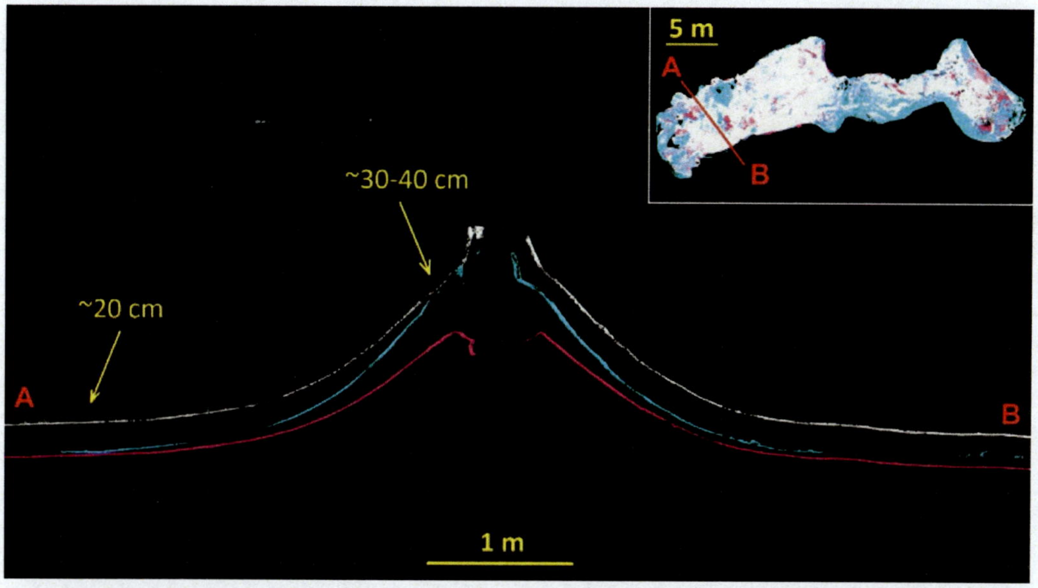


**Fig. 5** Change of ice level in the chamber in Snežna jama na planini Arto on the basis of three scan observations (white = 16th July 2020, purple = 19th November 2020, blue = 11th May 2021)

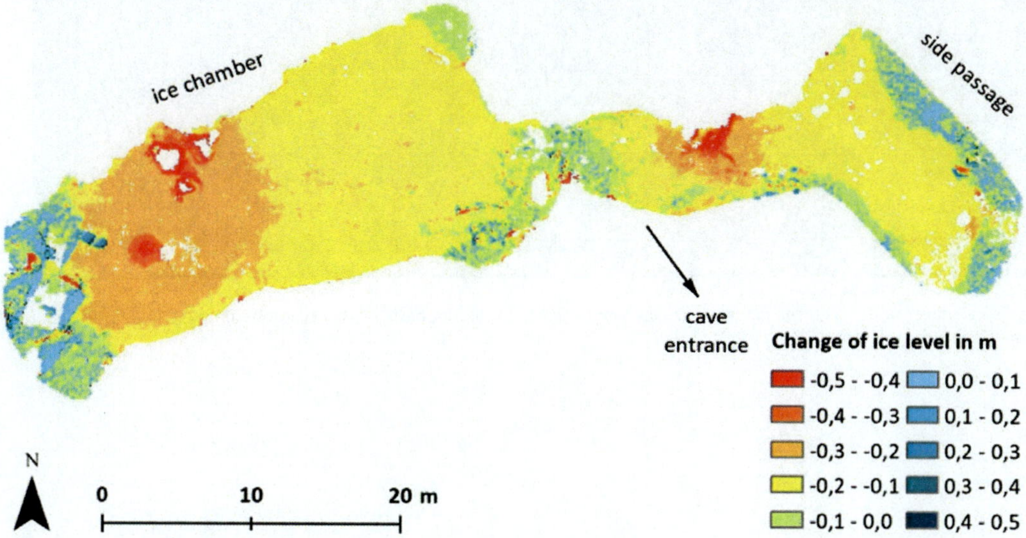


**Fig. 6** Change of ice level in metres in the chamber in Snežna jama na planini Arto between 16th July and 19th November 2020

on average increased by about 6 cm, meaning that about 35 $m^3$ of ice accumulated. The ice accumulation is spatially uneven, with up to about 30 cm of accumulated ice just below the chimneys and low decrease of ice level in some other areas (Figs. 5 and 6).

In the cave Velika ledena jama v Paradani, the ice surface is also about 600 $m^2$ large. Two scans were performed: on the 22nd July 2020 and the 9th May 2021, which comprised almost the whole warm period in 2020 and the whole following winter. Comparison of the point clouds showed that in this time period, both increase and decrease of ice levels was present (Figs. 7 and 8). The level of ice increased in the western part of the ice chamber below the group of chimneys, where a group of ice stalagmites and ice waterfalls developed. Up to 20 cm increase of ice level was recorded also in the lowest part of side passage on east (Fig. 8). In the rest, shrinking was present: on the flat surface of the ice chamber up to 20 cm, and on the vertical profile up to 50 cm respectively (Figs. 7 and 8). Decrease of level can be observed also on the rock slope between the cave entrance and ice chamber, indicating that ice is present below. Altogether, ice melting prevailed, resulting in about 105 $m^3$ loss of ice.

Described scanning campaigns in caves Snežna jama na planini Arto and Velika ledena jama v Paradani showed large seasonal changes of the ice level, which are characterized by intensive ice melting in warm periods and weak accumulation of ice during cold periods. The general trend is therefore shrinking of ice, which was noted also during historic visits of the caves.

Point cloud comparison showed that the highest change is in the vicinity of water inlets from chimneys. During winters, drip water from chimneys accumulates and freezes on the floor below. During these periods, quite large ice stalagmites, sometimes also ice columns connecting floor and ceiling can develop (Fig. 2). With increase in lateral distance from chimneys (or the landing point of water) the accumulation of ice decreases. During warm periods, ice melting is present in all parts of ice chambers due to positive air temperatures (Fig. 4). However the most intensive melt of ice is also below chimneys due to drip water from the surface.

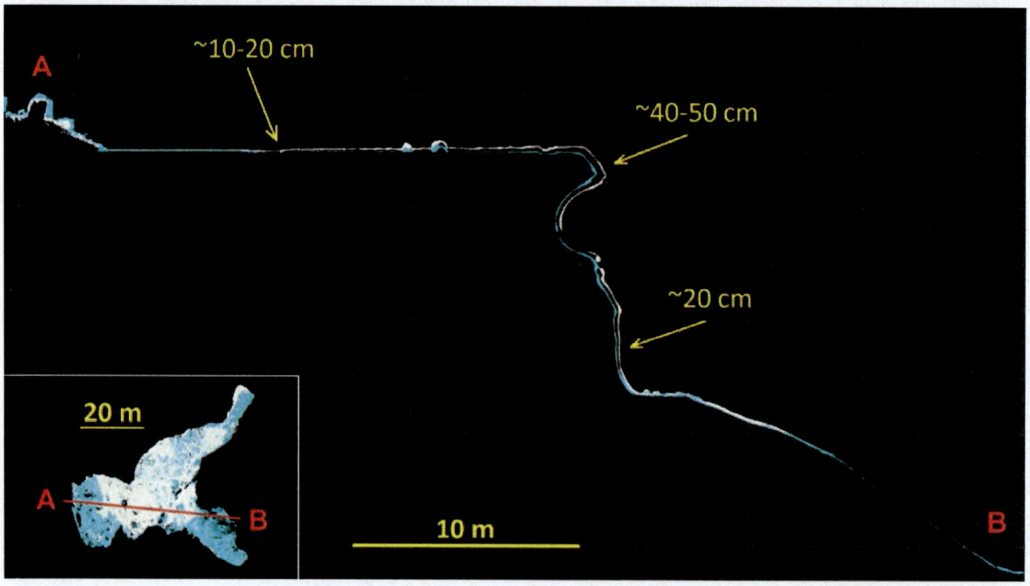


**Fig. 7** Change of ice level in the chamber in Velika ledena jama v Paradani on the basis of two scan observations (white = 22nd July 2020, blue = 9th May 2021)

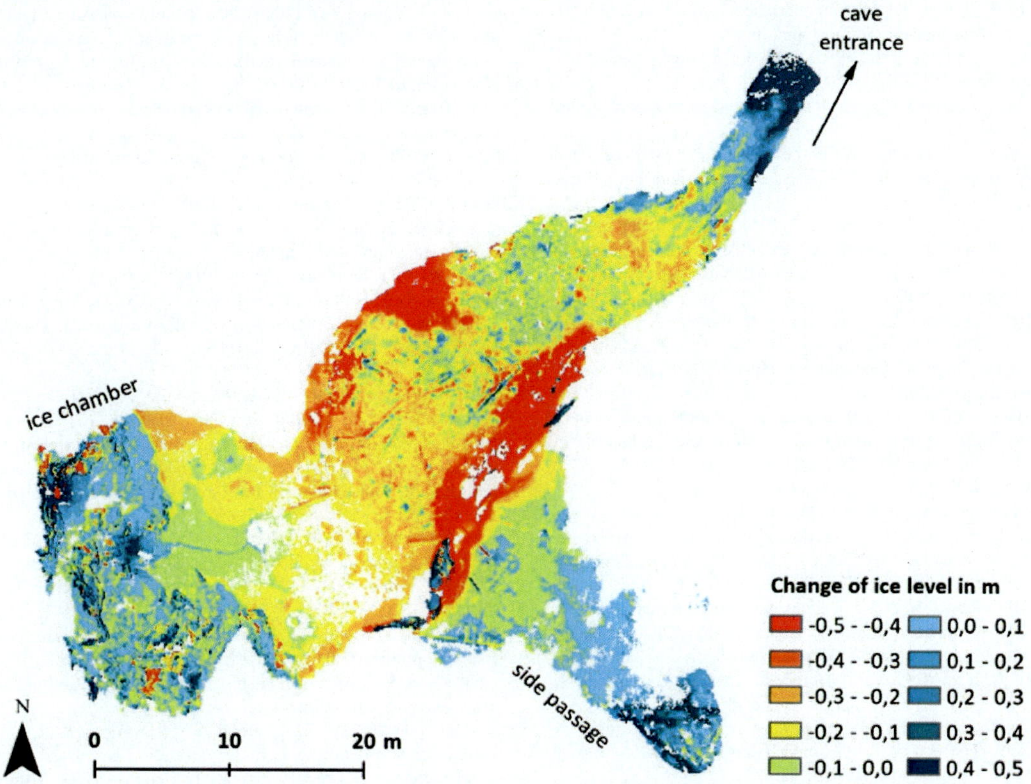


**Fig. 8** Change of ice level in metres in the chamber in Velika ledena jama v Paradani between 22nd July 2020 and 9th May 2021

## 5 Conclusions

Preliminary measurements in caves Snežna jama na planini Arto and Velika ledena jama v Paradani with a terrestrial laser scanner showed large spatial and seasonal variations of ice thickness with predominant decrease. This means that generally ice melt during warm parts of the year is more intensive than ice accumulation during cold periods. Variations are of bigger extent in the vicinity of chimneys which are supplying chambers with percolated water. At some places also rock movements have been detected, indicating that ice is present below them, causing active slope processes. Laser scanning is therefore a useful tool for detailed observation of ice level changes with some shortcomings, which sometimes occur because of problematic reflection of measurements on ice and especially on wet surfaces. The best way to avoid this is to do scanning during cold periods, when water is completely frozen, but unfortunately this is not always possible due to difficult access of caves during winter conditions.

**Acknowledgements** This study also takes place within the Karst Research programme (P6-0119), financially supported by the Slovenian Research Agency. Equipment has been supported by projects “Development of research infrastructure for the international competitiveness of the Slovenian RRI space—RI-SI-EPOS and RI-SI-LifeWatch.” The operation is co-financed by the Republic of Slovenia, the Ministry of Education, Science and Sport and the European Union from the European Regional Development Fund. Many thanks also to Jonathan Gabris for reviewing the English text.

## References

ARSO (2022a) Klimatološka povprečja 1981–2010 Krvavec. [Online] Available from: https://meteo.arso.gov.si/uploads/probase/www/climate/table/sl/by_location/krvavec/climate-bnormals_81-10_Krvavec.pdf. Accessed 14 Jan 2022a

ARSO (2022b) Klimatološka povprečja 1971–2000 Vojsko. [Online] Available from: http://meteo.arso.gov.si/uploads/probase/www/climate/table/sl/by_location/vojsko/climate-normals_71-00_vojsko.pdf. Accessed 14 Jan 2022b

Atlas Okolja (2022) Okoljski prostorski podatki. [Online] Available from: http://gis.arso.gov.si/atlasokolja/profile.aspx?id=Atlas_Okolja_AXL@Arso. Accessed 14 Jan 2022

Badino G (1995) Fisica del clima sotterraneo-memorie IIS, vol 7. Istituto Italiano di Speleologia, Bologna, p 136

Blatnik M, Košutnik J (2021) Ledene jame kot okoljski laboratorij. Planinski Vestnik 121(10):11–15

Blatnik M, Obu J, Košutnik J (2021) Ice thickness measurements in ice caves using terrestrial LiDAR scanner; examples from Slovenia. In: 28th international karstological school "classical karst"—regional karstology: general and local aspects, abstracts and guidebook. Postojna, p 78

Buzjak N, Dubovečak V, Paar D, Bočić N (2014) The influence of karst topography to ice cave occurrence—example of Ledena jama in Lomska duliba (Croatia). In: 6th international workshop on ice caves, Idaho Falls, pp 17–23

Carey AE, Zorn M, Tičar J, Lipar M, Komac B, Welch SA, Smith DF, Lyons WB (2018) Glaciochemistry of cave ice: Paradana and Snežna Caves, Slovenia. Geosciences 94(9):12. https://doi.org/10.3390/geosciences9020094

Colucci RR, Guglielmin M (2019) Climate change and rapid ice melt: suggestions from abrupt permafrost degradation and ice melting in an alpine ice cave. Prog Phys Geogr Earth Environ 43(4):561–573. https://doi.org/10.1177/0309133319846056

Colucci RR, Luetscher M, Forte E, Guglielmin M, Lenaz D, Princivalle F, Vita F (2017) First alpine evidence of in situ coarse cryogenic cave carbonates (CCCCOARSE). Geogr Fis e Din Quat 40(1):53–59. https://doi.org/10.4461/GFDQ2017.40.5

Covington MD, Perne M (2015) Consider a cylindrical cave: a physicist's view of cave and karst science. Acta Carsologica 44 (3):363–380. https://doi.org/10.3986/ac.v44i3.1925

Häuselmann P, Mihevc A, Pruner P, Horaček I, Čermak S, Hercman H, Sahy D, Fiebig M, Zupan Hajna N, Bosak P (2015) Snežna jama (Slovenia): interdisciplinary dating of cave sediments and implication for landscape evolution. Geomorphology 247:10–24. https://doi.org/10.1016/j.geomorph.2014.12.034

Itcus C, Pascu M-D, Lavin P, Persoiu A, Iancu L, Purcarea C (2018) Bacterial and archaeal community structures in perennial cave ice. Sci Rep 8(15671):1–14. https://doi.org/10.1038/s41598-018-34106-2

Kern Z, Palcsu L, Pavuza R, Molnar M (2018) Age estimates on the deposition of the cave ice block in the Saarhalle Dachstein-Mammoth Cave (Mammuthöhle, Austria) based on $^{3}$H and $^{14}$C. Radiocarbon 60(5):1379–1389. https://doi.org/10.1017/RDC.2018.96

Konsolaki A, Vassilakis E, Gouliotis L, Kontostavlos G, Giannopoulos V (2021) High resolution 3D modelling of subsurface morphological structures of Koutouki Cave, Greece. Acta Carsologica 49(2/3):163–178. https://doi.org/10.3986/ac.v49i2-3.7708

Košutnik J (2016) Ice mass balance in selected caves in Slovenia. In: International workshop on ice caves. Program guide and abstracts, pp 51–52

Luetscher M, Jeannin P-Y (2004) The role of winter air circulation for the presence of subsurface ice accumulations: an example from Monlési ice cave (Switzerland). Theor Appl Karstol 17:19–25

Maxim Integrated (2021) DS1922L—product details. [Online] Available from: https://www.maximintegrated.com/en/products/ibutton-one-wire/data-loggers/DS1922L.html. Accessed 14 Jan 2022

Mihevc A (1993) Ledene in snežne jame. Geografski Obzornik 40 (4):32–35

Mihevc A (2021) Ice in caves and its effect on thermal inversion and permafrost in the case of the Velika ledena jama v Paradani, Smrekova draga and neighbouring dolines. Acta Carsologica 50 (2/3):241–251. https://doi.org/10.3986/ac.v50i2-3.10495

Mulec J, Oarga A, Holko L, Pašić L, Kopitar AN, Eleršek T, Mihevc A (2021) Microbiota entrapped in recently-formed ice: Paradana Ice Cave, Slovenia. Sci Rep 11(1993):1–12. https://doi.org/10.1038/s41598-021-81528-6

Novak U (2022) Uporaba nove tehnologije za geološko kartiranje površja krasa in kraških jam. In: Zbornik del. 26. srečanja Slovenskega združenja za geodezijo in geofiziko, in print

Perşoiu A, Pazdur A (2011) Ice genesis and its long-term mass balance and dynamics in Scărişoara Ice Cave, Romania. Cryosphere 5:45–53. https://doi.org/10.5194/tc-5-45-2011

Persoiu A, Onac BP (2012) Ice in caves. In: White W, Culver DC (eds) Encyclopedia of caves. Elsevier, Amsterdam, pp 399–404

Persoiu A, Lauritzen SE (eds) (2017) Ice caves. Elsevier, p 729

Persoiu A, Buzjak N, Onaca A, Pennos C, Sotiriadis Y, Ionita M, Zachariadis S, Styllas M, Košutnik J, Hegyl A, Butorac V (2021) Record summer rains in 2019 led to massive loss of surface and cave ice in SE Europe. Cryosphere 15:2383–2399. https://doi.org/10.5194/tc-15-2383-2021

Persoiu A, Onac BP, Wynn JG, Bojar A-V, Holmgren K (2011) Stable isotope behavior during cave ice formation by water freezing in Scărişoara Ice Cave, Romania. J Geophys Res Atmos 116 (D2):1449–1465. https://doi.org/10.1029/2010JD014477

Riegl (2022) RIEGL VZ®-2000i—long range, very high speed 3D terrestrial laser scanning system. [Online] Available from: http://www.riegl.com/uploads/tx_pxpriegldownloads/RIEGL_VZ-2000i_Infosheet_2019-09-02.pdf. Accessed 14 Jan 2022

Skoglund R, Lauritzen SE, Gabrovšek F (2010) The impact of glacier ice-contact and subglacial hydrochemistry on evolution of maze caves: a modelling approach. J Hydrol 388(1):157–172. https://doi.org/10.1016/j.jhydrol.2010.04.037

Walters R, Zupan Hajna N (2020) 3D laser scanning of the natural caves: example of Škocjanske jame. Geodetski Vestnik 64(1):89–103. https://doi.org/10.15292/geodetski-vestnik.2020.01.89-103

Walters R (2017) Largest cave chambers in the world—3D scanning project. UIS Bull 59(2):52–53

Zupan Hajna N, Mihevc A, Pruner P, Bosak P (2008) Palaeomagnetism and magnetostratigraphy of karst sediments in Slovenia. Carsologica 266

# Evidences of Past and Present Hypogenesis in the Serrezuela De Carratraca Massif (Málaga, Southern Spain)

S. R. Durán-Laforet, J. J. Durán-Valsero, R. Morales-García, and P. A. Robledo-Ardila

**Abstract**

The Serrezuela de Carratraca Massif is a small carbonate massif located in the South of Spain. It corresponds to an outcrop of a geological unit formed by Triassic dolomitic marbles and limestones, belonging to the Alpujárride Complex of the Betic Cordillera. The surface of this karstic massif lacks well-developed exokarstic features, however, there are several known caves, with significant dimensions, as well as a spring with relatively high flow rates. Ardales Cave (1522 m of horizontal development) is a maze-type cave, with a great amount of cupolas inside, in the walls and ceilings. It is situated above the water table, without current hydrological activity. Sima Gorda (121 m of vertical development) is a rising shaft type cave, that reaches the actual water table and that contains different gases in its underground atmosphere ($H_2S$, $CO_2$, and methane). The main spring, located near Sima Gorda, is the Baños de Carratraca spring (Carratraca Baths), a well-known sulphurous spring used for centuries as a spa. These characteristics, as well as the presence of large amounts of pyrite transformed into iron oxy-hydroxides (limonite) in the Triassic marbles, suggest the possible common origin of all of these phenomena, linked to a deep regional groundwater flow. The Serrezuela Massif is a region of deep vertical water discharge that conditioned the hypogenic speleogenesis of the caves in the massif, induced by the mix of deep water and local recharge water. The presence of recent travertines near the spring suggest that the processes of dissolution–precipitation are still active.

**Keywords**

Ardales cave • Cave hypogenesis • Speleogenesis • Sulphurous spring

## 1 Introduction

The purpose of this paper is to give new evidence and discuss the possible speleogenesis of the karstic caves in the Serrezuela (Málaga, Southern Spain), understanding the geological context of the study area and stablishing the regional and local hydrogeological system.

New knowledge advances about the existence of many hypogenic caves and their morphological characteristics (Klimchouck 2009, 2012) have taken place in the last decade and must be considered. Recently, in the North of the province of Málaga (Southern Spain), a series of hypogenic caves have been described in the Jurassic limestones of the Camorra Range, in Mollina. The genesis of these caves is linked to the processes of hyper-karstification caused by sulphuric acid generated by the reduction of Triassic sulphates (González Ramón et al. 2018, 2021).

## 2 Geological and Tectonic Features

Geologically, Serrezuela de Carratraca despite being small is a complex location, as it is located between the limit of the External and Internal zones of the Betic Range, where outcrops of diverse lithologies are located. These lithologies have different paleogeographic precedence, belonging to different units and geologic complexes (Martín Algarra et al. 2009a, b).

S. R. Durán-Laforet (✉)
Department of Geology, Center of Hydrogeology of the University of Malaga (CEHIUMA), 29071 Malaga, Spain
e-mail: sduran@uma.es

J. J. Durán-Valsero · R. Morales-García
Instituto Geológico y Minero de España (IGME-CSIC), Calle de Ríos Rosas, 23. 28003, Madrid, Spain

P. A. Robledo-Ardila
Instituto Geológico y Minero de España (IGME-CSIC). Administración Periférica del Estado. Carrer de Felicià Fuster, 7. 07006, Palma de Mallorca, Spain


B. Andreo et al. (eds.), *EuroKarst 2022, Málaga*, Advances in Karst Science,
https://doi.org/10.1007/978-3-031-16879-6_28

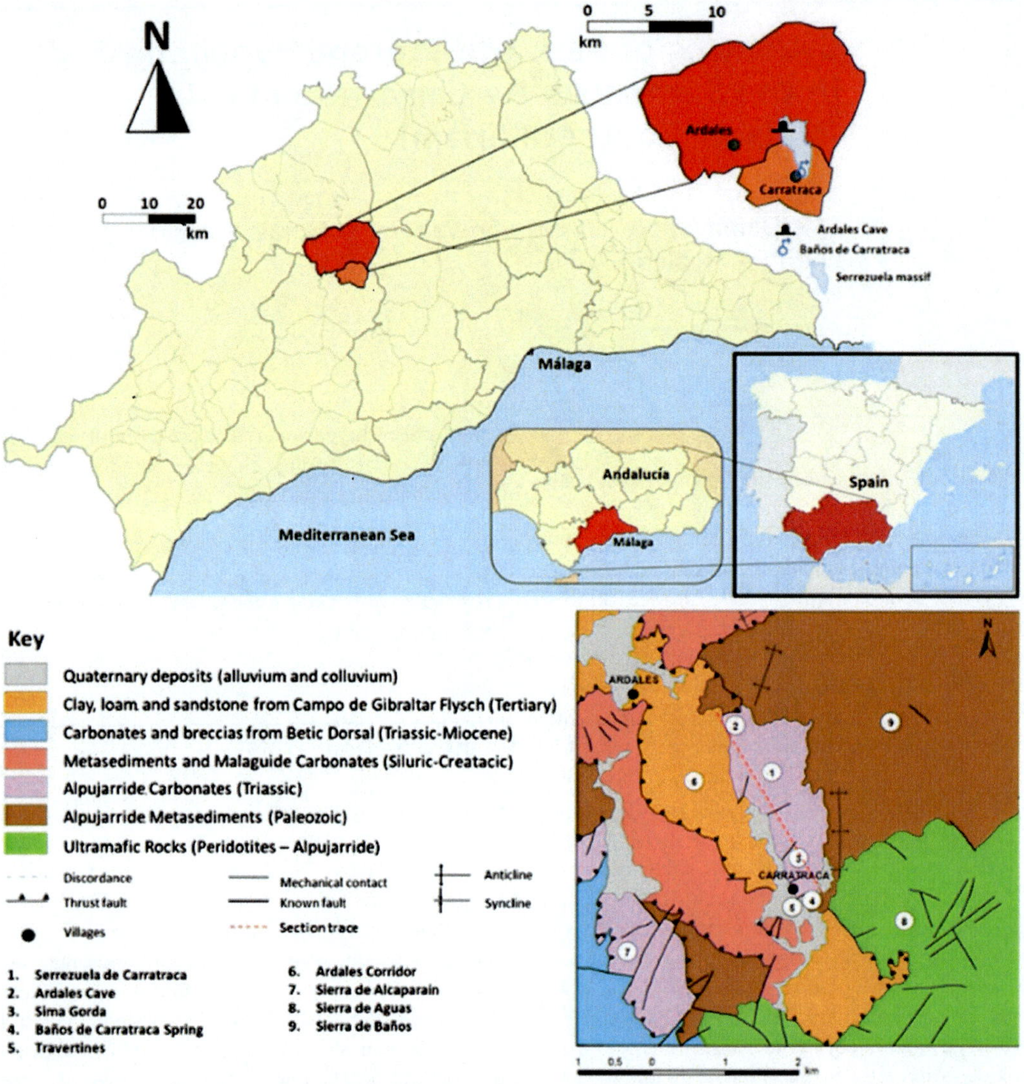


**Fig. 1** Location map of the study area (top), with the area corresponding to Ardales (red) and Carratraca (orange), the Serrezuela de Carratraca massif, the Ardales cave and the Baños de Carratraca Spring. Geologic map of the surroundings of the Serrezuela massif (bottom) (ITGE 1990, modified)

The Serrezuela de Carratraca Massif itself is an isolated, 2 km$^2$ outcrop of mainly carbonate lithologies. Tectonically, it is the northeast corner of a great syncline structure that strikes NW–SE composed by the following ranges: Sierra de las Nieves, Sierra Prieta and Sierra de Alcaparaín, where a series of materials are involved, mostly the ones of the Nieves Unit (Betic Dorsal) and the Yunquera Unit (Alpujarride Complex). This last one is superposed over the first one, and it appears in the sunken lips of the faults that strike roughly NW–SE. The eastern side of Alcaparaín and the Serrezuela itself are the last sunken blocks of the structure, where materials from the Alpujárride Complex, the Maláguide Complex and the Campo de Gibraltar Flysch are piled, as can be seen in Fig. 1.

The outcropping materials of the Serrezuela and its surroundings correspond to the three formations stablished by Martín Algarra (1987) in the Yunquera Unit of the Alpujárride Complex: Paleozoic metamorphic sediments in the base, Triassic white massive dolomitic marbles, and Triassic greyish-blue levels of limestones, with some intercalation of marbles.

The limits of the permeable carbonate materials are constituted by the Alpujarride metamorphic sediments in the East, and the clay, loams, and sandstones of the Flysch that fill the Ardales corridor towards the West. In the southern point of the Serrezuela, a small travertine outcrop is located, not represented by the geological map (1:50,000) from ITGE (1990). The geomorphologic position of the travertines dates the deposit from the Quaternary, with an origin linked to the groundwater drained from the Baños de Carratraca Spring, main spring of the massif.

The more notable penetrative fracture of the Serrezuela Massif consists of a reverse fault that divides the massif in two different subsectors, characterized by a lithological change as well as a different hydrogeological behaviour. This fault strikes roughly SW–NE, similar to the foliation of the surrounding peridotites in the southern side of the Massif (Gervilla et al. 2019).

The main structure of the massif is a syncline fold, with a verticalized eastern limb. It is in this limb where the white Triassic dolomitic marbles appear, whereas in the western side, the blue-greyish limestones appear, composing the normal limb of the fold. This whole structure is strongly fractured but the fractures are not as penetrative as those that affect the whole massif, it is also affected by secondary folds, the most notable ones being the anticlinal ones, as the secondary synclines have a wider wavelength.

## 3 The Endokarstic Morphologies in the Serrezuela

In the Serrezuela de Carratraca Massif well-developed exokarstic features are practically non-existant, however, there are some minor examples. Some *lapiace* fields are located in the highest points of the Cerro de la Calinoria (near the current entrance of the Ardales Cave) and the surroundings of the Serrezuela Peak, and there is evidence of a relatively large sinkhole, entrapped by the river network called Llanos de los Arenalejos. These features are the only significant exokarstic ones in the massif. However, there are good examples of well-developed endokarstic features. The most important ones are Ardales Cave, (also known as Doña Trinidad Cave, or Calinoria Cave) and Sima Gorda (also known as Sima Cadete, Sima de la Ermita or Sima de Carratraca) located in the northern corner of the massif and the southern point of the Massif, respectively (Ramírez-Trillo 1995).

Ardales Cave is the biggest one in the area, reaching 1577 m long and a depth of 34 m from the current entry; the cave has a practically horizontal development, located in two levels. It is characterized by a maze-like configuration of its rooms and galleries, with a structural control, with two main patterns: the stratification dip, to the NE, and the fractures, with a strike of 10°–20° NE.

In this cave, there are three main types of rooms and galleries: big halls with collapse structures and chimneys in the ceilings; straight galleries following the fracture patterns; and inclined galleries developed following the dip of the strata.

This cave was the first show cave open for the public in Spain, with the tours linked to the touristic thermal baths located in Carratraca, the Baños de Carratraca Spring (Vallejo and Durán 1999). Geologically, it has been studied by Durán (1992), Durán et al. (1992), Durán and López (1995), López et al. (1995), and Durán (1996), who have stablished the vadose position of the cave in the aquifer, close to the current phreatic level, the relationship with the local geological structure (dipping to the NE) and nearly vertical fractures that strike to the north (N10°–20° E), and the existence of detrital sediments with Middle Pleistocene fauna and three generations of speleothems, ranging from the start of the Upper Pleistocene up to now.

These authors have also suggested the relationship between the endokarstification of the Serrezuela and the presence of the sulphurous waters of the Baños de Carratraca spring, located south of the massif, draining the aquifer, with a temperature of the emerging waters higher than the average temperature of the local atmosphere and with a hydrochemistry linked to the presence of disseminated sulphides in the Triassic dolomitic marbles that make most of the aquifer.

It was not until carrying out a series of representative geomorphologic sections of the cave, in 2014 and 2015, when the most abundant feature of the cave was described. With exception to the important volume of speleothems, mostly in the biggest rooms, and detrital materials in gravity-induced cones, linked to collapses and filling of paleo-entries; the most important feature is the erosive morphologies that carve certain parts of the ceilings and walls of the cave, along with the different rooms and galleries.

Some examples can be found in the sections located in Fig. 2. These morphologies are mostly *cupolas* (usually less than 1 m in size), commonly linked to the existence of fractures that influence the vertical development of the cave, and mostly ending very abruptly, without continuity. Sometimes, these morphologies can also be found in the walls of linear galleries, or in the ceilings of galleries that follow the strata dip.

## 4 Hypogenic Criteria and Features

Klimchouck (2009) established that the main criteria for identifying hypogenic caves are the morphological and the hydrogeological features. The most important features of the geomorphological criteria is the spatial distribution pattern of the caves (maze caves, spongework mazes, irregular chambers, isolated passages, crude clusters of passages,

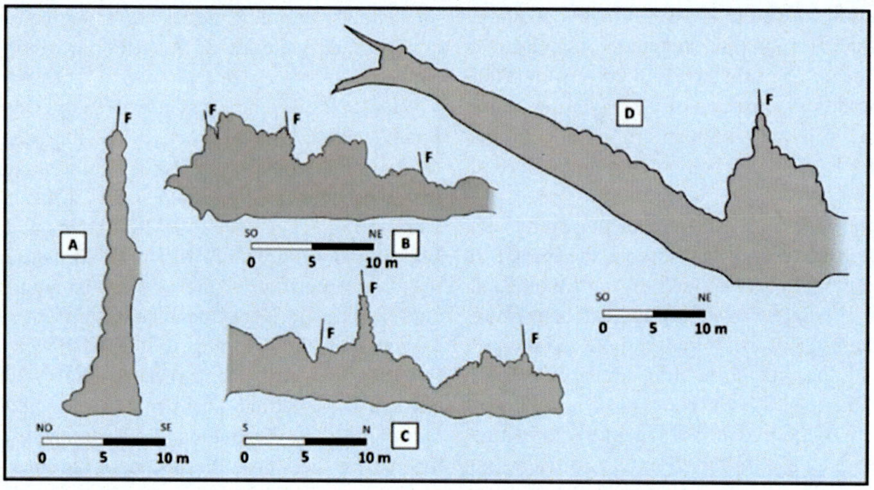


**Fig. 2** Representative geomorphologic sections of some of the halls and corridors of the Ardales cave, where some of the characteristic hypogenic speleogenesis morphologies can be seen. **a** Well to the Upper Galleries. **b** Sala del Lago (Lake Hall). **c**: Sala de la Olla (Upper Galleries) **d** Galería del Calvario (Calvario Gallery). In all the sections, the fracture control can be seen (F). The tendency of flat floors of the halls and galleries corresponds to the phreatic level (**a**, **b** and **d**), and, in some cases (**d**), the corrosion cupolas in strata planes can be seen

inclined passages, collapsed shafts, and rising shafts), and the meso-morphologic ones, specifically the so-called Morphologic Suite of Rising Flow (MSRF), where feeders, master passages, rising wall channels, parietal cupolas, arches, dome pits, outlets, dead ends and cupolas stand out as the most significative ones.

As for the hydrogeological diagnostic criteria, the most important ones suggest the presence of dissolution processes in confined conditions, specifically related to the relation with the discharge areas of intermediate to regional flow systems.

This same author indicates that in caves with an extensive geological history the coexistence of hypogenic features and later epigenic ones may occur, and it is indeed a frequently observed phenomenon. This circumstance may mask the hypogenic features of the caves, making the already difficult identification of these features even more difficult, as the epigenic development of speleothems and detrital sediments cover the walls, floors, and ceilings of the caves.

## 5 Hypogenesis in the Serrezuela

In the specific case of the caves located in the Serrezuela massif, both types of hypogenetic identification criteria can be found. First, the spatial pattern of two of the main caves of the massif corresponds to a maze cave (Ardales Cave) and a rising shaft (Sima Gorda). In the Ardales Cave, additionally, the most abundant erosive meso-shapes are the cupolas, not only in the ceilings, but also in the walls. Osborne (2004) points that these types of morphologies are the most characteristic of the caves with hydrothermal origin, as well as the caves generated by the interaction with sulphuric acid. Secondly, the Serrezuela massif has a structural disposition linked to a small-scale horst system, with a risen northeast area of the regional structure (Sierra de las Nieves-Sierra Prieta-Sierra de Alcaparaín-Serrezuela de Carratraca). This structural construction conditions the hydrogeological system, indicating that the Serrezuela works as a discharge area of a regional deep flow system.

Studies carried out in the Carratraca Baths, the main spring that discharges the Massif, located in the southern part (Durán et al. 1998), indicate the presence of a mix of water, by isotopic and geochronological analysis ($^{2}H$, $^{18}O$ and $^{14}C$); one from deep flows with a geochemical signature that indicate that the recharge area took place in areas with a higher altitude than the Serrezuela, and with a long transit time (thousands of years) and a more modern water, coming from the local recharge, with a recharge area limited by the surface of the massif itself, identified by the exokarstic features found.

In other places of the Serrezuela, a series of morphologies compatible with a hypogenic origin has been found, such as the ones located in the surroundings of the fracture zone of the Serrezuela Peak, visible in the slopes of the road to the windmill field located in the area. This enclave is characterized by the presence of vertical shafts, with a diameter of ten to twenty centimetres, and a visible length of around three to five metres, and in some cases these channels reach the surface and have evidences of deep flows (such as small cupolas, dead ends, associated mineralizations, sparitic calcite crystals). In Fig. 3, the location of the main caves of the

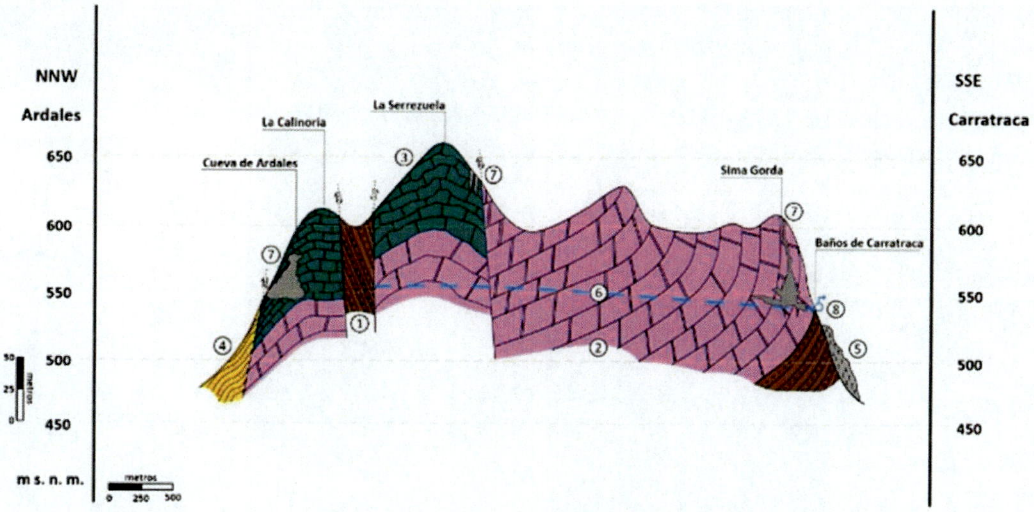


**Fig. 3** Schematic hydrogeological section of the Serrezuela de Carratraca massif, with the positions of the Ardales cave, Sima Gorda and the Baños de Carratraca Spring. Key: 1. Palaeozoic Micaschists; 2. White Triassic Marbles; 3. Grey Triassic Limestones; 4. Clay, loam, and sandstone (Flysch); 5. Quaternary Travertines; 6. Approximated position of the current phreatic level; 7. Karstic cavities with hypogenic evidence; 8. Baños de Carratraca Spring

Serrezuela Massif as well as the location of the lesser morphological evidences linked to the fracture areas can be found.

The Carratraca Baths spring has been known for centuries for the special hydrochemical characteristics of the water (Salgado y Guillermo 1860; De Prado y Valle 1861; Prolongo 1873). There is a high concentration of dissolved sulphates (210 mg/L on average) and a high content of $H_2S$. There are also high concentrations of sulphide, fluoride, arsenic, cyanide, boron, and iron. The most frequent concentration of sulphide is around 10 mg/L (Vallejo and Durán 2001).

The presence of these elements is compatible with the existence, within the Triassic dolomitic marbles of frequent mineralization of pyrite, as disseminated crystals, cubic crystals of diverse size, ranging from millimetre to more than 10 cm, as well as irregular masses and fracture fillings. The most part of the near-surface pyrite appears to be altered, transformed into iron oxides and hydroxides, but on some occasions, it is possible to find remains of the primary mineralization (Fig. 4).

Exceptionally, fractures filled with gypsum have been observed, with a possible origin in the transformation of the pyrite to the iron oxi-hydroxides (Fig. 4).

All these evidence clearly point to an artesian speleogenesis (Klimchouk 2012), linked to the existence of deep ascending flows, enriched in $CO_2$ and $H_2S$, coming from the alteration of the primary iron sulphides of the mineralizations found it the surroundings of the massif (mostly pyrite). These gases have been detected currently in the underground atmosphere of Sima Gorda, specifically in the deepest area, close to the current phreatic level (water table), drained by the Carratraca Baths.

The interpretation of these types of evidences as a produce of hypogenic phenomena is frequent in the context of the Mediterranean alpine mountain ranges, related to tectonic structures such as the one described in this paper and to the presence of thermal waters or the action of sulphuric acid from mineralizations in the speleogenesis (Audra 2008; Frumkin and Fischendler 2005; Goldscheider et al. 2010; De Waele et al. 2016), even to the point that Klimchouk (2015) mentioned a new paradigm in the origin of many karstic caves.

In the case of the Serrezuela, a first hypogenic stage exists, probably very extended in time, linked to the late moments of the alpine orogen (possibly Pliocene). In a later moment (possibly from the Middle Pleistocene onward, indicated by the age of the detrital sediments of the Lake Hall (Sala del Lago) in the Ardales Cave), the evolution of the cave changed to an epigenic phase, ending the connection with the phreatic level linked to the ascending hydrogeologic flows. It is from this point that the surface processes started to take place, such as collapses of near-surface shafts, development of new cave entries, presence of allochthonous detritic sediment, development of several generations of speleothems in their respective Upper Pleistocene climatic stages (the only ones dated so far).

Meanwhile, currently the main spring that acts as a draining area (discharge) of the Serrezuela aquifer is the Baños de Carratraca, that produced its own sedimentary

**Fig. 4** Photographs of some of the most important geological, geomorphological, and hydrogeological aspects of the Serrezuela. **a** White Triassic Marbles; **b** Grey Triassic Limestones; **c** Altered pyrite mass (oxy-hydroxides), with its centre non-altered; **d** Limonite-Pyrite crystal in the Triassic marble, with a fracture filled with gypsum (alteration); **e** Hypogenic chimneys in one slope; **f** Detail of the corrosion cupolas inside Ardales Cave; **g** Sima Gorda; **h** Detail of the outcrop of travertines in Carratraca

record represented by a relatively small travertine deposit located within the low areas of the Carratraca village itself, that had been unidentified and did not appear in the geological maps so far, so no geological study or dating has been carried out.

## 6 Conclusions

The Ardales Cave is the most important cave of the Serrezuela de Carratraca Massif, with a total length of more than 1500 m, distributed within a multilevel maze-like network. The Serrezuela is a carbonatic massif that makes the eastern side of the geologic structure that strikes Southwest-Northeast and dips to the North, constituted by dolomitic marbles and limestone from the Triassic Period, belonging to the Alpujarride Complex of the Betic Moutain Range. This massif is drained by a sulphurous water spring, the Carratraca Baths, with waters that have some thermal character and a singular chemical composition, rich in sulphates, and dissolved gases ($CO_2$, methane, and $H_2S$).

The geologic structure, the chemistry of the water from the Carratraca Baths, the presence of high quantities of pyrite transformed into iron oxi-hydroxides in the Triassic rocks, the existence of vertical caves that reach the phreatic level (Sima Gorda) with high levels of gases of deep origin in the underground atmosphere and the presence of certain endokarstic shapes (particularly the high number of cupolas) in the galleries and halls of the Ardales Cave and other enclaves of the massif, lead to the hypothesis of the hypogenic origin (speleogenesis) of the complex of caves and endokarstic shapes existing in the Serrezuela (specially Ardales Cave), generated by ascending fluids with high dissolving potential. The water drained in the Carratraca Baths represent a mix of these regional fluids and the local recharge.

## References

Audra P (2008) The sulfuric hypogene speleogenesis: processes, cave pattern, and cave features. Berliner Höhlenkundliche Berichte 26:5–30

De Prado y Valle C (1861) Aguas de Carratraca. Revista Minera XII:449–466. Madrid.

Durán JJ (1992) La cueva de ardales. Geología. En: Cueva de Ardales: su recuperación y estudio. Ayuntamiento de Ardales, pp 46–56

Durán JJ (1996) Los sistemas kársticos de la provincia de Málaga y su evolución: contribución al conocimiento paleoclimático del Cuaternario en el Mediterráneo occidental. Tesis doctoral. Universidad Complutense de Madrid, p 409

Durán JJ, López J (1995) El karst de la Serrezuela y la Cueva de Ardales: aspectos geológicos, geomorfológicos e hidrogeológicos. En: Ayuntamiento de Ardales—Grupo Andaluz del Cuaternario (AEQUA). Geología y Arqueología prehistórica de Ardales, pp 47–53

Durán JJ, Grün R, Ford DC (1992) Cueva de Ardales. Geocronología evolutiva desde el Pleistoceno Superior hasta la actualidad y su relación con los cambios paleoclimáticos. En: Cueva de Ardales: su recuperación y estudio. Ayuntamiento de Ardales, pp 57–66

Durán JJ, Baeza J, Vallejo M, Anglada R (1998) Estudio hidrogeológico de los Baños de Carratraca. ITGE-Banesto. Informe inédito. Madrid

Frumkin A, Fischhendler I (2005) Morphometry and distribution of isolated caves as a guide for phreatic and confined paleohydrological conditions. Geomorphology 67:457–471

Gervilla F, González-Jiménez JM, Hidas K, Marchesi C, Piña R (2019) Geology and metallogeny of the upper mantle rocks from the Serranía de Ronda. Sociedad Española de Mineralogía, p 124

Goldscheider N, Mádl-Szőnyi J, Erőss A, Schill E (2010) Review: thermal water resources in carbonate rock aquifers. Hydrogeol J 18:1303–1318

González Ramón A, Pedreira Parias A, Martos Rosillo S, Jiménez de Cisneros-Bencelá C, Ruiz Constán A (2018) Espeleogénesis de las cavidades de la Sierra de Mollina (Málaga, sur de España): implicaciones hidrogeológicas y geomorfológicas. En: Robledo PA, Durán JJ (eds) Cuevas: la flecha del tiempo, de la prehistoria a la actualidad. ACTE. Madrid, pp 15–28

González-Ramón A, Pedrera A, Martos-Rosillo S, Jiménez de Cisneros C, Ruiz-Constán A, Gázquez F (2021) Constraints on the evolution of sulfuric acid speleogenesis within carbonate rocks partially covered by evaporites (Sierra de Mollina, southern Spain). Geomorphology 390

Instituto Tecnológico Geominero de España (1990) Mapa Geológico de España a escala 1:50.000. 2ª Serie (MAGNA), Hoja 1038—ARDALES. Mapa y memoria. Madrid, p 56

Klimchouk A (2009) Morphogenesis of hypogenic caves. Geomorphology 106:100–117

Klimchouk A (2012) Speleogenesis, hypogenic. In: Encyclopedia of caves, pp 748–765

Klimchouk A (2015) The karst paradigm: changes, trends and perspectives. Acta Carsologica 44(3):289–313

López J, Durán JJ, Arribas A (1995) Génesis, evolución y geocronología de la Cueva de Ardales. En: Ayuntamiento de Ardales—Grupo Andaluz del Cuaternario (AEQUA). Geología y Arqueología prehistórica de Ardales, pp 55–70

Martín Algarra A (1987) Evolución geológica alpina del contacto entre las zonas internas y las zonas externas de las cordilleras béticas. Tesis Doctoral Universidad de Granada. Granada, p 1171

Martín Algarra A, Mazzoli S, Perrone V, Rodríguez-Cañero R, Navas-Parejo P (2009a) Variscan tectonics in the malaguide complex (Betic Cordillera, Southerm Spain): stratigraphic and structural alpine versus pre-alpine constraints from the ardales area (Province of Málaga) I. Stratigraphy. J Geol 117(3):241–262

Martín Algarra A, Mazzoli S, Perrone V, Rodríguez-Cañero R (2009b) Variscan tectonics in the malaguide complex (Betic Cordillera, Southerm Spain): stratigraphic and structural alpine versus pre-alpine constraints from the ardales area (Province of Málaga) II. Structure. J Geol 117(3):263–284

Prolongo P (1873) La historia de los copos de azufre que salen mezclados con las aguas del manantial de los baños de Carratraca: Sulfuraria carracatrensis. Memoria de la Sociedad Malagueña de Ciencias Físicas y Naturales. 27 páginas manuscritas.

Ramírez-Trillo F (1995) Grandes cavidades en la provincia de Málaga. Espeleotemas 5:71–94

Salgado y Guillermo J (1860) Monografía de las aguas sulfo, selénido hídricas, arseniadas, bicarbonatadas alcalino-térreas, metálicas, de Carratraca. Imprenta de Manuel Minuesa, pp 267

Vallejo M, Durán JJ (1999) Serrezuela de Carratraca (Málaga, Southern Spain): an small spot with a diverse geological heritage. In: Barettino D, Vallejo M, Gallego E (eds) Towars the balanced managemend and conservation of the geological heritage in the new millenium. Madrid, pp 374–377

Vallejo M, Durán JJ (2001) Mixture of deep and meteoric waters in a karst spring: an analytical study using hydrochemical, isotopic and radiocative techniques. Present state and future trends of karst studies. International hydrological programme, tecnical documents in hydrology, vol 1(49). UNESCO, París, pp 197–204

De Waele J, Audra P, Madonia G, Vattano M, Plan L, D'Angeli IM, Bigot JY, Nobécourt JC (2016) Sulfuric acid speleogenesis (SAS) close to the water table: examples from southern France, Austria, and Sicily. Geomorphology 253:452–467

# Microstratigraphic Analysis of a Speleothem from the Nerja Cave (Málaga, Southern Spain)

C. Jiménez de Cisneros, G. Loncomilla, A. González-Ramón, and C. Liñán-Baena

**Abstract**

The microstratigraphic analysis in speleothems allows to identify and characterize the stratigraphic elements and limiting surfaces. The textural variations are related to the availability of dripping water, the nature of the flow, and the transport of water and/or evaporation. The different phases of growth therefore respond to the different physicochemical conditions that existed during their formation and even to possible post-sedimentary processes, constituting true paleoclimatic indicators. Fluid inclusions formed during crystal growth may harbor existing karst water in the cave during precipitation. Verification by petrographic studies of the genetic relationship between the fluid inclusion and the surrounding calcite allows defining the type of fluid inclusions based on the temporal relationship with the calcite (primary or secondary), the spatial relationship with the calcite crystals (inter- or intra-crystalline), and their morphology. In this work, a microstratigraphic–petrographic characterization of the carbonate fabrics and textures is presented, as well as the type of fluid inclusions of a calcitic–aragonitic stalagmite from the Cueva de Nerja (Málaga, Southern Spain). The most recognized fluid inclusions appear in the columnar textures which are formed in conditions close to equilibrium.

C. Jiménez de Cisneros (✉) · G. Loncomilla
Instituto Andaluz de Ciencias de La Tierra (CSIC-UGR), Avd. de las Palmeras 4, 18100 Armilla, Granada, Spain
e-mail: concepcion.cisneros@csic.es

A. González-Ramón
Instituto Geológico y Minero de España (CSIC), Urb. Alcázar del Genil, 4. Edf. Zulema bajo, 18006 Granada, Spain

C. Liñán-Baena
Fundación Cueva de Nerja, Instituto de Investigación, Carretera de Maro, s/n, 29787 Nerja, Malaga, Spain

C. Liñán-Baena
Departamento de Ecología y Geología, Facultad de Ciencias, Universidad de Málaga, Campus de Teatinos s/n, 29071 Malaga, Spain

**Keywords**

Fabrics • Fluid inclusions • Paleoclimate • Speleothem • Nerja cave

## 1 Introduction

Speleothems are good indicators of the climatic conditions during their deposition. Variations in their chemical and mineralogical characteristics, their growth rate, among others, depend on the environmental variables inside the cave (temperature, drip rate, and water chemistry) and reflect the environmental conditions outside (mean annual temperature, precipitation, atmospheric circulation, and even vegetation). The dating of these materials by $^{230}Th/^{234}U$ makes it possible to place the climatic information in a precise time frame.

Stalagmites are generally the most commonly used speleothems because of their simple geometry and continuity of crystalline growth (relatively fast). Their tendency to precipitate under conditions close to isotopic equilibrium facilitates paleoclimatic reconstructions, due to the slow evaporation and degassing of $CO_2$ (McDermott 2004).

Small amounts of air and water (from meteoric water infiltrating through epikarst/vadose zone discontinuities) may be trapped as intra-crystalline inclusions in the calcite during the developmental phases of the speleothem and indicate the external temperature of the cave (Scheidegger et al. 2010). Fluid inclusions in speleothems show a remarkable variability in shape and size, reflecting a wide range of mechanisms capable of fluid entrapment during crystal growth. Indeed, speleothem grown mechanisms provide valuable insight on the interpretation about the formation of fluid inclusions (López-Elorza et al. 2021).


B. Andreo et al. (eds.), *EuroKarst 2022, Málaga*, Advances in Karst Science,
https://doi.org/10.1007/978-3-031-16879-6_29

## 2 Geographic and Geological Context

The study was carried out on a fragment of a speleothem (Stalagmite NE) from the Nerja Cave (Málaga), located on the floor of the Hall of Columnas de Hercules, in the High Galleries (Fig. 1). The Nerja Cave is developed on dolomitic marbles belonging to the Middle Triassic that can reach a thickness of 400 m, very diaclastic and with a saccharoidal aspect in some areas. On top of these marbles, other limestone marbles appear, with calcschist intercalations, attributed to the Upper Triassic (Carrasco et al. 1998). The karstification process that led to the formation of the cavity occurred during the Pliocene and Pleistocene (Carrasco et al. 1998). During the Quaternary, with alternating temperate and warm periods, large volumes of chemical precipitation deposits were formed (Jiménez de Cisneros et al. 2003; Jiménez de Cisneros and Caballero 2013).

The cave has a surface area of 35,000 $m^2$ and an approximate volume of 350,000 $m^3$, with a maximum drop of 68 m, so its topographic development is practically horizontal. It is located in the unsaturated zone of the aquifer, above the piezometric level (Jiménez de Cisneros and Caballero 2013). In it, aerial and subaerial speleothems can be differentiated, such as stalagmites, stalactites, columns, eccentrics, and gours, among others (Fig. 2). In general, they are formed by calcite and/or aragonite and in the case of moonmilk-type speleothems, also by hydromagnesite, huntite, and hydroxyapatite. Other forming minerals are dolomite and magnesite in variable proportions and quartz as traces (Casas et al. 2001; Arrese 2009; Pedrera et al. 2016).

## 3 Methodology

The analyzed speleothem corresponds to a stalagmite (NE) with a length of 17.9 cm and a diameter of ~ 5 cm and presents a blackened crust that partially covers it (Fig. 3). It was cut along its growth axis, obtaining two halves. Seven thin slices were made from one of them, thus covering the entire growth of the speleothem, to carry out the petrographic study.

For the petrographic study, a polarizing microscope "LEITZ WETZLAR standard WL" and an "OLIMPUS DP-20" equipped with microphotographic equipment "OLIMPUS BX60" were used.

The description and characterization of the textures of the speleothem, as well as the typology of the fluid inclusions, have been carried out following the methodology of López-Elorza (2019) and López-Elorza et al. (2021) based on Frisia (2015).

## 4 Results

The study by optical microscopy allows us to distinguish the type of present fabric and its characteristics. In general, the stalagmite is a holocrystalline fine-grain ($\Phi$ < 1 mm) formed

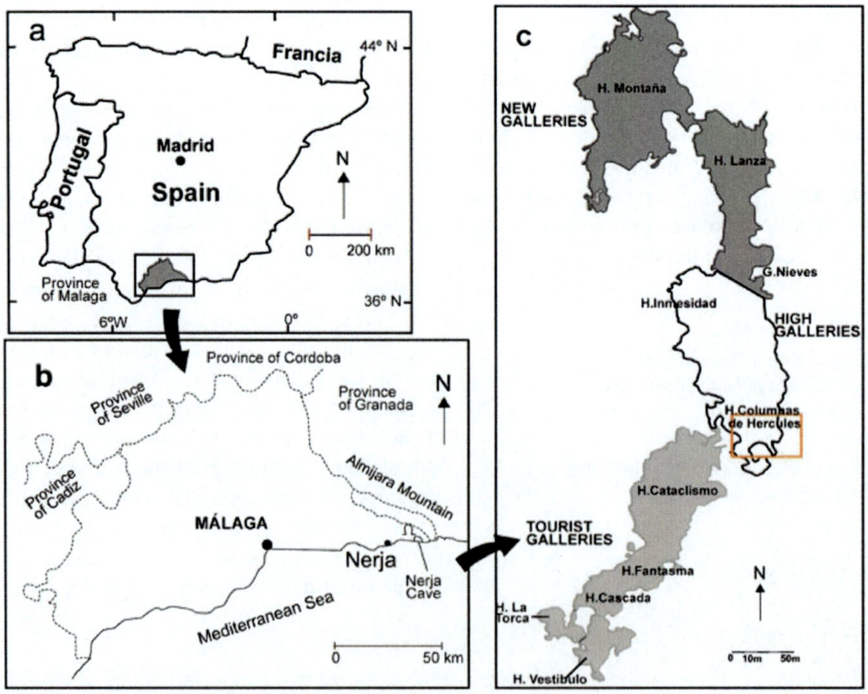


**Fig. 1** a and b Geographical location of the Nerja cave. c Diagram of the Nerja cave with its main halls and galleries. The inset shows the location where the studied stalagmite was found (modified from Arrese 2009)

**Fig. 2** Some types of speleothems from the Nerja cave (Authors: M. Ibáñez and A. Sanchidrián)

**Fig. 3** Stalagmite NE of the Nerja cave (Columnas de Hercules hall, high galleries)

mainly by aragonite and calcite. The blackened crust corresponds to an irregular contact and consists of a thin film with brown minerals (iron oxides and clay) forming a kind of tuffaceous-stromatolitic-styrolitic structure (Fig. 4a). In some areas, there is a replacement of aragonite by calcite (Fig. 4b).

In general, the lamination is continuous but not very distinctive and of small thickness. The aragonite occurs as acicular crystals that branch out and converge when they reach the aragonite–calcite intergrowth zone, where they end up with a crustal texture (Fig. 4c). In some areas, where only aragonite is observed, porosity is more evident. The calcite crystals show columnar open (Co) (Fig. 4d), columnar compact (C) (Fig. 4e), and columnar elongated (Ce) (Fig. 4f) textures. To a lesser extent, micrite (M) appears, generally with small thickness and located in the boundary zones or near the calcite–aragonite intergrowth. In the calcite, the lateral changes appear as a step of texture, from columnar open to columnar elongated and columnar closed. The aragonite is porous, and its acicular crystals also characterize the texture as columnar open.

Figure 5 shows a graph reflecting the microstratigraphic analysis of the fabrics observed in the analyzed speleothem. The figure shows the vertical evolution of the textures present in the speleothem from the base to the top, in general, predominantly consisting of columnar, columnar open, and columnar elongated fabrics.

Fluid inclusions are closely related to the type of texture observed. In general, inclusions have a tendency to be located in voids and between crystal filaments. Both elongated and spherical primary inclusions have been distinguished, as well as macro-inclusions that are usually filled with air and water (Fig. 6a). The fluid inclusions observed (Kendall and Brougthon 1978; López-Elorza et al. 2021) are inter- and intra-crystalline of elongated, boudin, pyriform, thorn-shaped, and down arrow type (Fig. 6b, c, and d), the latter two are restricted to calcite crystals. It is observed that inclusions prevail in the columnar texture.

The textural variations of speleothem fabrics are related to the availability of drip water (Frisia et al. 2000),

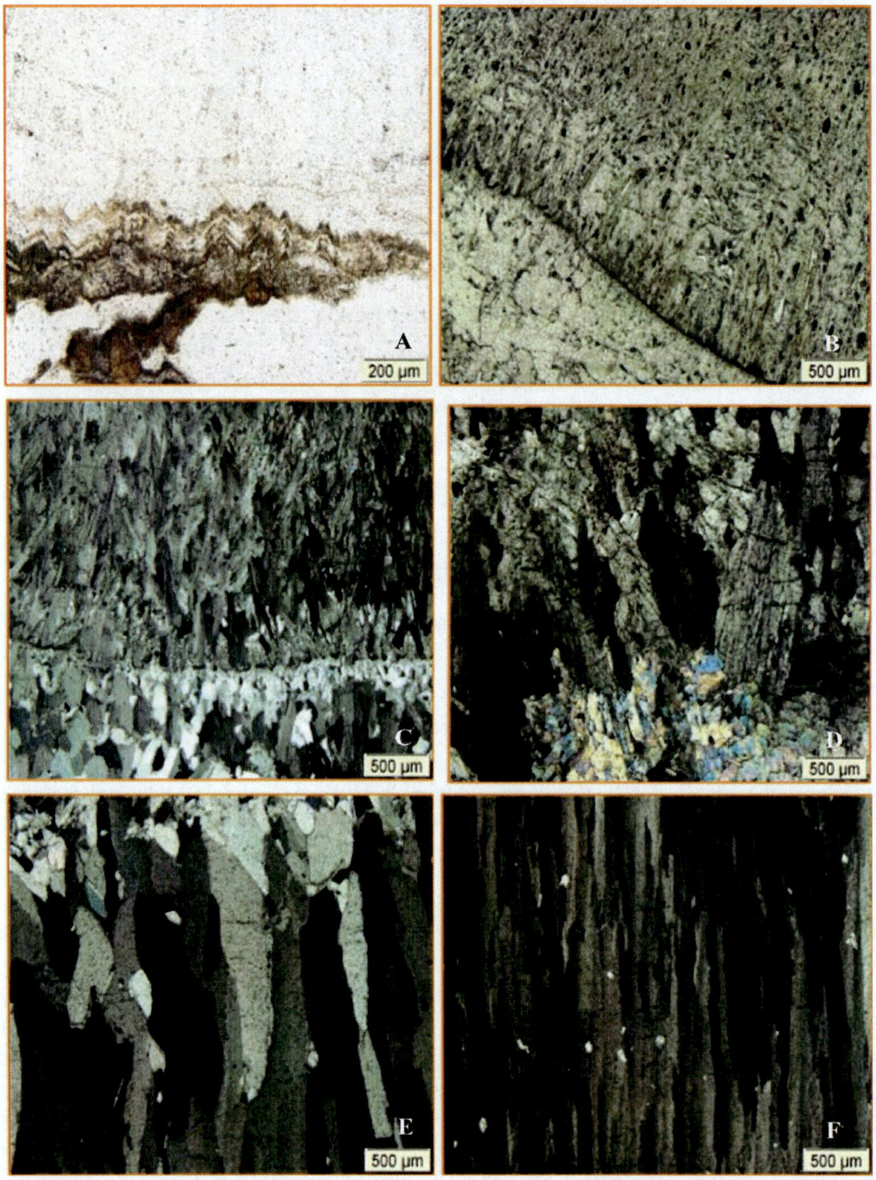


**Fig. 4** Thin sections photomicrographs of stalagmite NE. **a** Tobaceous-stromatolitic structure on the edges of the lamina. **b** Aragonite–calcite rearrangement zone **c** Calcite–aragonite intergrowth zone. The aragonite has a closed texture at the contact. **d** Columnar open texture in the calcite. Primary calcite is also observed. **e** Calcite with columnar compact texture. **f** Calcite with columnar elongated texture

evaporative processes, supersaturation of drip water, and increased magnesium and impurities, as well as diagenetic processes (Jiménez de Cisneros et al. 2020). Each texture is associated with a specific growth period of the speleothem.

The recognized columnar, compact, and fibrous textures are considered as equilibrium textures and show that the formation of the speleothem took place under conditions of isotopic quasi-equilibrium with low supersaturation (Frisia et al. 2000, 2002). The presence of the columnar elongate fabric is common when the speleothem develops from parent waters flowing through dolomitic or $Mg^{2+}$-rich rocks (Jiménez de Cisneros et al. 2020). It is produced from a fast flow and high drip rate, greater than columnar compact or columnar open type (Frisia 2015). This type of texture is indicative of interglacial periods with high flow circulation.

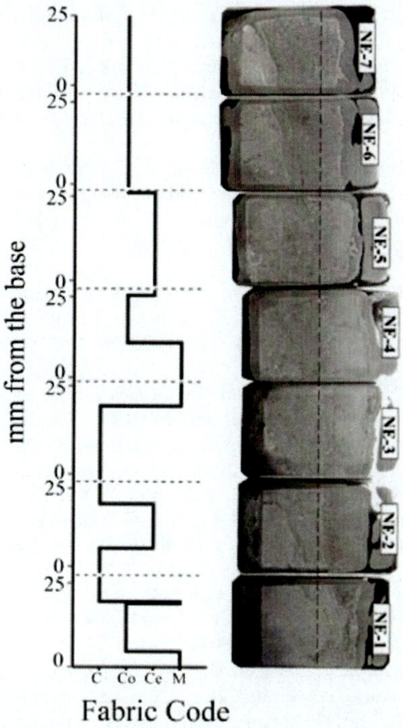


**Fig. 5** Microstratigraphic log of a thin sections from the base of the speleothem. Fabric code: C (columnar compact), Co (columnar open), Ce (columnar elongate), and M (micrite)

## 5 Discussion

The fluid inclusions observed are mainly associated with columnar textures and were formed during the speleothem growth phases, so they are considered primary, and the water they contain comes from the original dripping. The macro-inclusions observed, which would serve to obtain a larger amount of water, are secondary, i.e., not coeval with the formation of the speleothem. However, the primary water-releasing primary inclusions, which are located in the calcite crystals and which are inter- and intra-crystalline in the columnar texture, can be used as paleoclimatic archives.

## 6 Conclusions

Microstratigraphic characterization of speleothems allows better planning of geochemical sampling to obtain better paleoclimatic results. The textural analysis of the stalagmite provides information on the possible dripping conditions that existed during the stages of its formation and allows us to recognize the most favorable zones to obtain more reliable paleoclimatic data since they are related to equilibrium conditions.

It can be deduced that the formation of this stalagmite took place in changing environmental conditions, with alternating warm and humid episodes, reflected in the

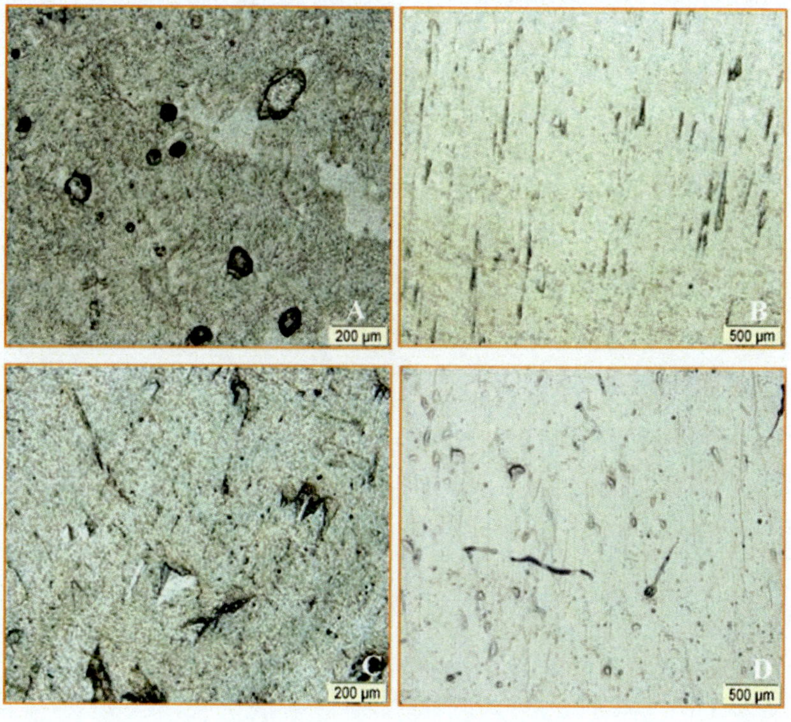


**Fig. 6** **a** Intergrowth inclusions filled with air and/or water in aragonite crystals. **b** Elongated and thorn-shaped fluid inclusions in calcite crystals. **c** Fluid inclusions of thorn-shaped and down arrow types in calcite crystals. **d** Boudin, elongated, and pyriform fluid inclusions in calcite crystals

columnar calcite fabrics and other colder and drier episodes related to the presence of micritic fabrics.

**Acknowledgements** This work has been carried out within the project Contract Technological Support between the CSIC (IACT) and the Nerja Cave Foundation "Speleothems and archaeological records of the Nerja Cave as indicators of paleoclimatic and paleoenvironmental change" (2018–2021) and PID20231-125619OB-C22 project.

## References

Arrese B (2009) Geomorfología, evolución geológica y condiciones ambientales actuales de la Cueva de Nerja, Málaga. (Tesis Doctoral). Universidad Autónoma de Madrid. http://hdl.handle.net/10486/149

Carrasco F, Durán JJ, Andreo B, Liñán C, Vadillo I (1998) Consideraciones sobre el karst de Nerja. Karst en Andalucía 173–181

Casas J, Vidales JL, Durán Valsero JJ, López MJ, Barea J (2001) Mineralogía de depósitos tipo moonmilk en la cueva de Nerja (Málaga, España). Geogaceta 29:29–32

Frisia S (2015) Microstratigraphic logging of calcite fabrics in speleothems as tool for paleoclimate studies. Int J Speleol 44 (1):1–16. https://doi.org/10.5038/1827806x.44.1.1

Frisia S, Borsato A, Fairchild IJ, McDermott F (2000) Calcite fabrics, growth mechanisms, and environments of formation in speleothems from the Italian Alps and Southwestern Ireland. J Sediment Res 70 (5):1183–1196. https://doi.org/10.1306/022900701183

Frisia S, Borsato A, Fairchild IJ, McDermott F, Selmo EM (2002) Aragonite-calcite relationships in speleothems (Grotte de Clamouse, France): environment, fabrics, and carbonate geochemistry. J Sediment Res 72(5):687–699. https://doi.org/10.1306/020702720687

Jiménez de Cisneros C, Caballero E (2013) Paleoclimate reconstruction during MIS5a based on a speleothem from Nerja Cave, Málaga South Spain. Nat Sci 05(05):533–540. https://doi.org/10.4236/ns.2013.55067

Jiménez de Cisneros C, Caballero E, Vera J, Durán J, Juliá R (2003) A record of Pleistocene climate from a stalactite, Nerja Cave, southern Spain. Palaeogeogr Palaeoclimatol Palaeoecol 189(1–2):1–10. https://doi.org/10.1016/s003101820(02)00589-8

Jiménez de Cisneros C, González-Ramón A, Sequero C, Andreo B, Fairchild IJ (2020) Isotopic and petrographic evidence as a proxy in paleoclimatic reconstructions from flowstones in Southern Spain. Open J Geol 10(06):597–611. https://doi.org/10.4236/ojg.2020.106027

Kendall AC, Broughton PL (1978) Origin of fabrics in speleothems composed of columnar calcite crystals. J Sediment Petrol 48 (2):550–552. https://doi.org/10.1306/212F74C3-2B24-11D7-8648000102C1865D

López-Elorza M (2019). Microestratigrafía de inclusiones fluidas en espeleotemas pleistocenos de la Cueva del Tortero (Valencia) y su uso en paleoclimatología (Tesis Doctoral). Universidad Complutense de Madrid

López-Elorza M, Muñoz-García MB, González Acebrón L, Martín-Chivelet J (2021) Fluid-inclusion petrography in calcite stalagmites: implications for entrapment processes. J Sediment Petrol 91 (11):1206–1227. https://doi.org/10.2110/jsr.2021.016

McDermott F (2004) Palaeoclimate reconstruction from stable isotope variations in speleothems: a review. Quat Sci Rev 23(7–8):901–918. https://doi.org/10.1016/j.quascirev.2003.06.021

Pedrera A, Mérida Rodríguez M, Reyes Corredera S, Liñán Baena C, Carrasco F, Jorda JF, Aura Tortosa JE (2016) Cueva de Nerja y Litoral Oriental de Málaga. Excursiones de la XIV Reunión Nacional de Geomorfología. Sociedad Española de Geomorfología, Instituto Geológico y Minero de España, Universidad de Málaga, & Fundación Cueva de Nerja

Scheidegger Y, Baur H, Brennwald MS, Fleitmann D, Wieler R, Kipfer R (2010) Accurate analysis of noble gas concentrations in small water samples and its application to fluid inclusions in stalagmites. Chem Geol 272(1–4):31–39. https://doi.org/10.1016/j.chemgeo.2010.01.010

# Gypsum Dissolution Rate, New Data and Insights

A. Busetti, C. Calligaris, and L. Zini

**Abstract**

Sinkholes linked to covered evaporite karst in urban environments still represent a challenge in hazard and risk assessment. The Quinis hamlet, located in Friuli Venezia Giulia region (NE Italy), is heavily affected by sinkhole phenomena (linked to an evaporitic bedrock), which deeply interested infrastructures and houses. In order to understand the evolution of the sinking phenomena, a field experiment started on the dissolution rate of the gypsum. In 17 existing piezometers, at different depths, 51 evaporitic rock samples were exposed to the naturally occurring variation of relative humidity, air flow and hydrodynamics. The rock samples were placed respectively in the aeration, in the fluctuation and in the phreatic section of the piezometric tubes. Data related to groundwater level fluctuations, temperature and electrical conductivity were collected. After four months, rock samples were removed, weighted and the volume loss evaluated. The obtained results indicate that rock sample reduction is not only dependent on the groundwater level fluctuations and on the number of days during which the samples are immersed in the groundwaters but also on the mineralization of the latter. Some of the rock samples have been almost completely dissolved, with dissolution rate values almost eight times bigger than expected if compared to the available literature data. The proposed approach had as aim to evaluate the quickness of the dissolution process, which is dependant on several causes (groundwater level fluctuations, type of rocks, chemical characteristics of the groundwaters, etc.) and represents a novel contribution to the overall knowledge of karst processes with noticeable impacts on human-built construction.

A. Busetti (✉) · C. Calligaris · L. Zini
Mathematics and Geosciences Department (DMG), University of Trieste, Via Weiss 2, 34128 Trieste, Italy
e-mail: abusetti@units.it

**Keywords**

Sinkhole • Evaporite karst • Dissolution rate • Risk assessment

## 1 Introduction

Sinkhole phenomena, linked to the presence of an evaporitic bedrock, developed in urban areas are very dangerous if considering the damage that they can cause on man-made structures. The results of recent investigations (Calligaris et al. 2020) have shown that Friuli Venezia Giulia region (here after noted as FVG) is one of the most affected areas in northern Italy with 1199 sinkholes inventoried.

In FVG, only 1% of the karstifiable lithologies are represented by evaporites (Calligaris et al. 2017), which can be identified mainly along two E–W alignments: one in correspondence with the Tagliamento valley to the south, and the other along Pesarina and Pontaiba valleys (Fig. 1) to the north.

The main problem with these phenomena is the occurrence speed with which they take place and develop over time due to the extremely high karstifiability and solution rate of the evaporites (0.68–1.14 mm/y, Klimchouk et al. 1996 and 0.4–1.0 mm/y, Cucchi et al. 1998). The latter is very high if compared for example to limestone one (0.009–0.14 mm/y, Furlani et al. 2009) allowing thus to understand the extreme vulnerability linked to territories where these lithologies are present.

The hazard assessment jointly with the presence of element at risk took Calligaris et al. (2019) to try to quantify, thanks to an on field experiment, the solution rate of the evaporite bedrock in the test site area of Quinis (Enemonzo—UD). The field experiment, carried out for the first time, consisted of installing core rock evaporitic samples, drawn from the drilled boreholes, in 7 piezometers at different depths for a period of one year. The aim was to recreate for


B. Andreo et al. (eds.), *EuroKarst 2022, Málaga*, Advances in Karst Science,
https://doi.org/10.1007/978-3-031-16879-6_30

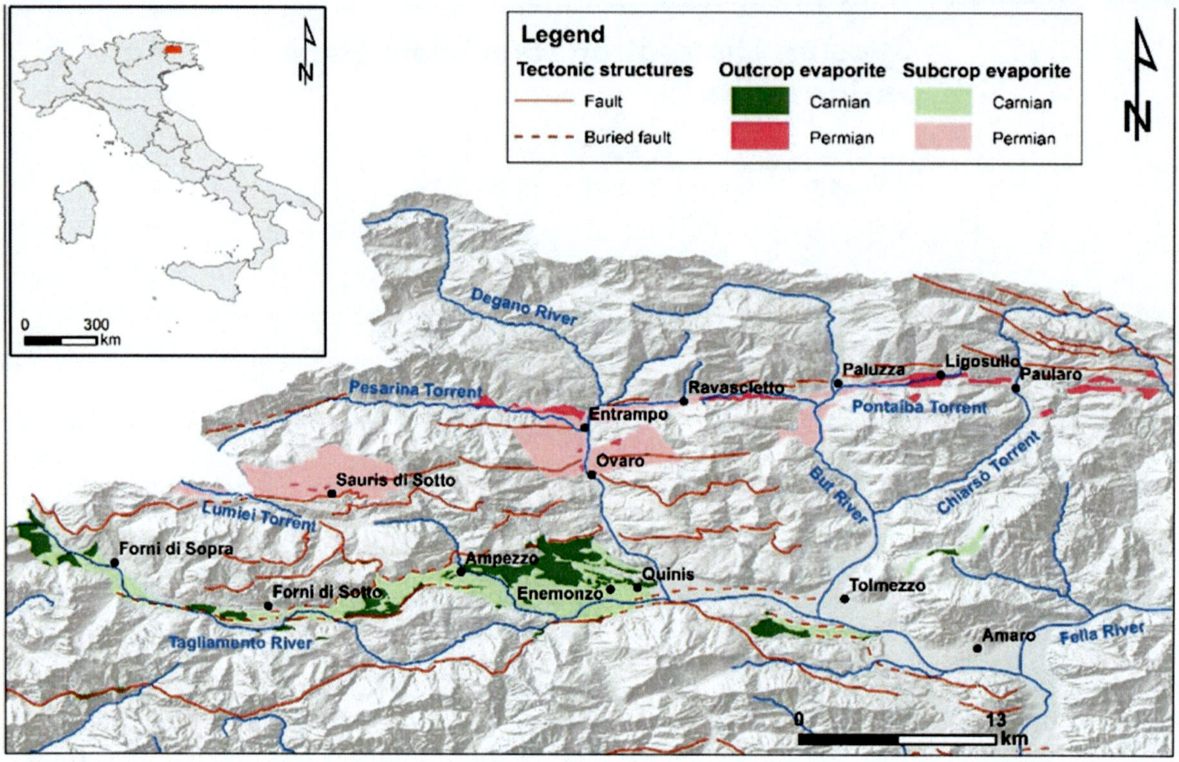


**Fig. 1** Sketch of the evaporite outcrops (dark green and dark pink) and of the mantled or overlaid by a non-karst rock evaporites (light green and light pink) in the NW sector of the Friuli Venezia Giulia region

the rock samples the original conditions in order to assess their loss of weight and volume. The obtained results witnessed a range of solubility in between 0 and 8.1 mm/y with an average value of 2.1 mm/y, double the known maximum one. The high dissolution rate demonstrated and its spatial variability encouraged to investigate more in detail the particular situation in the inhabited area of Quinis.

In the present paper, we propose a new field experiment by placing rock samples in piezometric tubes at different depths to better understand the causes of the fast dissolution rate.

## 2 Study Area

The sinkhole evolution at Quinis is related to the presence of the Raibl Formation (RBA—Carnian in age) in the bedrock, which is subdivided into three different members (Venturini et al. 2009): red shales member (RBA1), gypsum and gray dolostones member (RBA2) and marls and dolostones member (RBA3). The RBA2 is the one present in the study area, and according to Venturini et al. (2009) is composed mainly by gray and white saccharoid gypsum with marl inclusions at the top, yellowish dolomitic marls, and to a lesser extent, blackish or greenish clays and dark limestone in thin layers.

The evaporitic bedrock is mantled by loose quaternary deposits such as glacial till, alluvial and colluvial deposits having a variable thickness, increasing from north to south, from a few meters up to more than 50 in correspondence of the Tagliamento riverbed. These materials are extremely heterogeneous due to complex and articulated depositional patterns conditioned by the tectonical setting, the alternation of glacial and interglacial periods, the recent depositional events due to the Tagliamento River and the overlapping alluvial fan present northern of Quinis.

In this framework, the high solubility of the gypsum and gray dolomites, scarcely outcropping in the investigated area, together with important oscillations of the groundwater level (Zini et al. 2015), has led over time to the formation of important sinkhole phenomena (Gutiérrez et al. 2008), which in turn caused extensive damages to local infrastructures and houses.

In the Enemonzo municipality, since the end of the nineteenth century, 208 sinkholes were inventoried, of which

46 are cover suffosion, 40 are cover collapse, and the remaining have an undefined typology. The latter was detected by desk activities and were not surveyed or not anymore identifiable in the field. In the Quinis village, actually 32 are the recognized sinkholes and most of them are active. Since the first time that these phenomena have been identified in the area (Marinelli 1898), some buildings have been demolished (Fig. 2) and others are actually damaged.

## 3 Materials and Methods

In the Quinis village, over time, 24 boreholes were drilled and later equipped with different instruments: 2 assestimeters, 2 Casagrande piezometers, 1 borehole drilled for geophysical proposes, and 19 piezometric tubes. 17 of the 19 piezometers (Fig. 2) have been used in the ongoing field experiment. The piezometers are well spread, covering the whole study area.

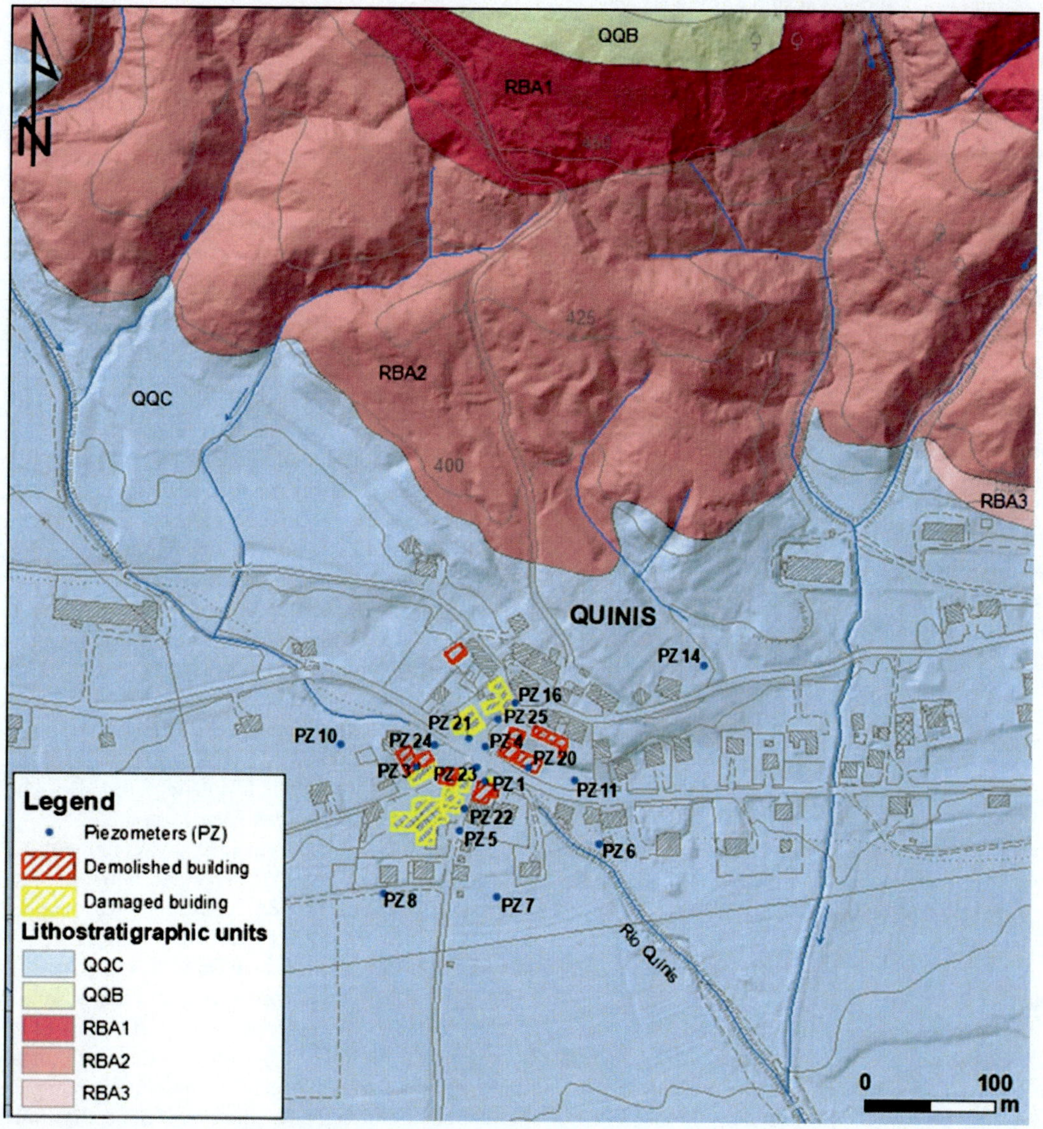


**Fig. 2** Geological map of the study area with the evidence of the demolished and damaged building from 1800s to nowadays. Piezometers are identified as PZ. Lithostratigraphic units: (QQC) fluvial gravel and sand and (QQB) glacial till, Pleistocene-Holocene in age; (RBA1) red shales member, (RBA2) gypsum and gray dolostones member and (RBA3) marls and dolostone member belong to the Raibl Formation, Triassic in age (modified after Venturini et al. 2009)

In each piezometer, 3 rock samples have been installed placed at different depths:

- the shallower one at about 2 m below the ground level (in the vadose zone never reached by the groundwater);
- the middle one in the epiphreatic zone, where the groundwater level fluctuates;
- the third one in the phreatic zone (always immersed in the groundwater).

The positioning depth has been decided according to the level of the groundwater and the position of the screens.

Before the installation, a preparatory phase took place. In order to prepare rock samples as uniform as possible not only from a dimensional point of view, but also from a mineralogical one. 12 boulders were collected in the Entrampo (Ovaro) quarry and later shaped as 115 parallelepiped having 8 cm of height and a square base of 4 cm with an area ($A$) of 160 $cm^2$.

All samples were weighed and 78 of these were selected choosing the most similar in weight (range from 273.4 to 300 g) and dimension. Once over-dried in the oven at 50 °C for 48 h, a second weight ($W_1$) was done.

Before the installation, volume and density ($\rho$) were also measured. The first as volume difference by using a graduated cylinder containing water. Density was calculated as the quotient between dry weight and volume. The whole process is summarized in Fig. 3.

The 51 rock samples chosen were then installed in the piezometers wrapping each single parallelepiped in a plastic net and later fixed to a rope.

After the first 4 months (June 18, 2021–October 1, 2021), the samples were collected, over-dried for 48 h at 50 °C, weighted ($W_2$) and weight loss calculated ($W_1$−$W_2$).

The weight loss is functional to the dissolution rate evaluation according to the formula proposed by Plan (2005):

$$R = \left[\frac{(W_1 - W_2)}{A\rho}\right] * 10(\mathrm{mm}/4\ \mathrm{month})$$

Data-logger devices, able to record in continuous groundwater level, temperature and electrical conductivity variations, were installed in 7 piezometers.

Two data-loggers WLT-Diver Eijekelkamp (pressure range 10 m, accuracy ± 0.5 cm, resolution 0.2 cm; temperature range −20 °C to +80 °C, accuracy ± 0.1 °C, resolution 0.01 °C) were installed in PZ10 and PZ11; five CTD-Diver Eijekelkamp (pressure range 10 m, accuracy ± 0.5 cm, resolution 0.2 cm; temperature range −20 ° C to +80 °C, accuracy ± 0.1 °C, resolution 0.01 °C, Electrical Conductivity range 0–120 mS/cm, accuracy ± 1%, resolution ± 0.1%) were installed in PZ7, PZ22, PZ23, PZ24 and PZ25. The compensation of the atmospheric pressure variability was ensured by a Baro-Diver from Eijekelkamp (pressure range 150 cm, accuracy ± 0.5 cm, resolution 0.2 cm) installed in the PZ1, in the core of the study area. All instruments were synchronized and had a recording range interval of 30 min.

Rainfall data were downloaded from the OSMER FVG network (https://www.osmer.fvg.it/archivio.php?ln=&p=dati) for the Enemonzo rainfall station.

## 4 Results

Within the 1-year project, during the first check, after four months after the installation, data-logger devices were downloaded, and rock samples were collected. The monitored period coincided with a summer season characterized by a maximum daily precipitation of 60.5 mm (June 18, 2021–October 1, 2021). Six are the main rainfall events that had a meaningful response on the groundwater levels recorded by the data-logger devices (Fig. 4).

In accordance with the bedrock deepening from north of Quinis where it is seldom exposed to the south where it is mantled by even more than 50 m of deposits, the groundwater level fluctuation is lower in the northern part (PZ11 and PZ25, a few centimeters) and higher (PZ7 and PZ8, 13 m) in the southern of the study area.

During the four-month experiment, 27 rock samples always remained always above the groundwater level. Twenty-three of them recorded a weight loss of up to 6%; three (PZ1, PZ10 and PZ23) showed a weight loss of about 20%. The remaining PZ11 had a higher weight loss of about 42%.

The complete dissolution of the sample in PZ3 requires a separate explanation. In fact, a later inflow is present (at about 6.5 m b.g.l.) within the piezometer due to a local perched aquifer above the rock sample placed at 9.7 m b.g.l. The sample is thus always wet due to the water leaching from above.

Only 4 rock samples (PZ6, PZ7, PZ8, PZ23) in the epiphreatic zone got wet by the fluctuating groundwaters. They recorded a loss of weight between 15 and 25%.

Eighteen rock samples remained always immersed. PZ14, PZ16, PZ20, PZ21, and PZ25 recorded a loss of weight of less than 1%. PZ6, PZ8, PZ11, PZ23 and PZ24 are between 4 and 9%. PZ3, PZ4, PZ5, PZ7, PZ10 and PZ22 are between 30 and 50%. PZ1 had a loss of 96%.

In general, all the samples in the vadose zone never reached by the groundwater show a very low loss of weight. Instead, when the rock samples are occasionally or always immersed in the groundwater, the dissolution is higher. The dissolution rate strongly depends on the mineralization of the groundwaters and so on their saturation index. It comes out (Table 1) that when the groundwaters are characterized by low electrical conductivity values up to 2 mS/cm (PZ1, PZ3,

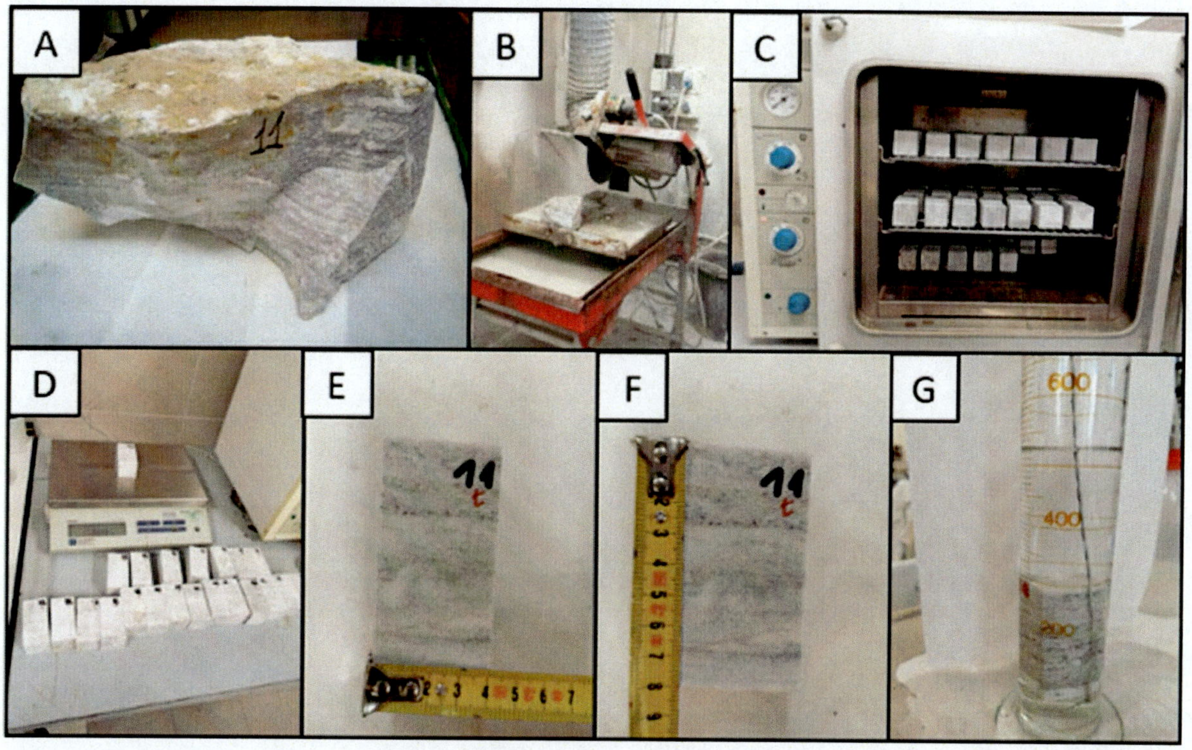


**Fig. 3** Rock samples preparation process: **a** a boulder selected at the Entrampo quarry, **b** cutting and shaping phase, **c** over dried samples in the oven at 50 °C for 48 h, **d** weight procedure, **e–f** measurement of the dimensions **g** volume calculation

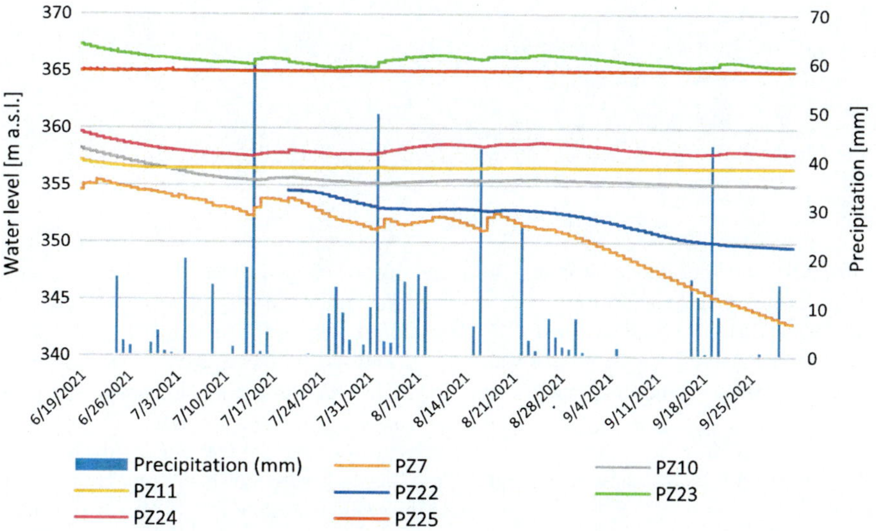


**Fig. 4** Groundwater level fluctuations recorded in the piezometers of the Quinis village in the period June 18, 2021–October 1, 2021. Daily rainfall registered in Enemonzo station (OSMER FVG)

**Table 1** Weight loss, dissolution rate (*R*) and electrical conductivity (EC) calculated for the rock samples used in the field experiment.

| PZ | Depth (m b.g.l.) | Weight loss (%) | *R* (mm/4 months) | EC (mS/cm) | PZ | Depth (m b.g.l.) | Weight loss (%) | *R* (mm/4 months) | EC (mS/cm) |
|---|---|---|---|---|---|---|---|---|---|
| PZ1 | −1.5 | 1.7 | 0.13 | 0.51 | PZ14 | −2 | 0.1 | 0.01 | 2.87 |
| | −8 | 19.5 | 1.65 | | | −12.7 | 0.1 | 0.01 | |
| | −15$^{*}$ | 95.9 | 7.68 | | | −19.4$^{*}$ | 0.2 | 0.02 | |
| PZ3 | −2 | 2.2 | 0.18 | 1.41 | PZ16 | −2 | 0.1 | 0.01 | 3.10 |
| | −9.7$^{**}$ | 100.0 | > 8.125 | | | −16.3$^{*}$ | 0.2 | 0.01 | |
| | −21.2$^{*}$ | 49.4 | 3.77 | | PZ20 | −2 | 1.3 | 0.10 | 10.70 |
| PZ4 | −2 | 2.1 | 0.16 | 1.75 | | −8.15 | 3.1 | 0.25 | |
| | −7.7 | 0.9 | 0.07 | | | −15.15$^{*}$ | 0.3 | 0.02 | |
| | −15.6$^{*}$ | 49.7 | 3.85 | | PZ21 | −2 | 0.5 | 0.03 | 3.28 |
| PZ5 | −0.5 | 0.9 | 0.07 | 1.40 | | −7.7 | 0.5 | 0.03 | |
| | −13.7 | 6.1 | 0.50 | | | −16.8$^{*}$ | 0.7 | 0.05 | |
| | −24.7$^{*}$ | 40.1 | 3.21 | | PZ22 | −2 | 0.2 | 0.02 | 2.18 |
| PZ6 | −2 | 0.2 | 0.01 | 1.80 | | −13 | 0.3 | 0.02 | |
| | −15.6$^{**}$ | 25.1 | 2.01 | | | −28.1$^{*}$ | 46.6 | 3.73 | |
| | −36.7$^{*}$ | 6.9 | 0.55 | | PZ23 | −2 | 22.5 | 1.83 | 2.40 |
| PZ7 | −2 | 0.4 | 0.04 | | | −10$^{**}$ | 20.4 | 1.63 | |
| | −19.6$^{**}$ | 15.0 | 1.17 | 1.36 | | −30$^{*}$ | 4.9 | 0.39 | |
| | −37$^{*}$ | 55.3 | 4.32 | 1.99 | PZ24 | −2 | 0.1 | 0.01 | 4.40 |
| PZ8 | −2 | 0.6 | 0.05 | | | −11.6 | 0.2 | 0.01 | |
| | −20$^{**}$ | 20.6 | 1.65 | 1.60 | | −29$^{*}$ | 4.2 | 0.34 | |
| | −37.5$^{*}$ | 4.5 | 0.35 | 2.50 | PZ25 | −2 | 0.4 | 0.03 | |
| PZ10 | −2 | 0.0 | 0.00 | 1.65 | | −12.6$^{*}$ | 0.6 | 0.05 | 7.06 |
| | −15 | 20.6 | 1.57 | | | −17.4$^{*}$ | 1.0 | 0.07 | 8.73 |
| | −29$^{*}$ | 32.0 | 2.50 | | | | | | |
| PZ11 | −2 | 0.1 | 0.01 | 2.50 | | | | | |
| | −11.6 | 42.0 | 3.20 | | | | | | |
| | −18.9$^{*}$ | 8.8 | 0.69 | | | | | | |

Depth (b.g.l.): rock samples depths
$^{*}$when always inmmersed
$^{**}$if inmersed only occasionally

PZ4, PZ5, PZ7, PZ8, PZ10, and PZ22), the rock samples dissolve more (in the range between 30 and 50%).

In the piezometers, PZ14, PZ16, PZ20, PZ21, and PZ25 instead, it have been observed stability of the groundwater levels, which indicates a low water replacement and a consequent high mineralization (EC > 2.5 mS/cm). In this situation, the weight loss is quite low, less than 1%. The remaining (PZ6, PZ11, PZ23 ,and PZ24) showed a weight loss between 2 and 9%.

## 5 Conclusion

Quinis is one of the most prone areas to sinkhole phenomena in the FVG (NE Italy). Known since the 1800s, they are still active and involve built-up structures. Despite the intense activity developed over the years, researchers are still working in the area, trying to understand the evolution rate in order to assess the jointed risk.

During the summer of 2021, 51 evaporite rock samples were placed in 17 piezometers at different depths in order to evaluate the loss of weight and the relative dissolution rate. From data analysis emerged that the samples always above groundwater level have a minimum weight loss. The ones always immersed in the groundwaters have a weight loss related to the mineralization and the water replacement. The result is a non-homogenous dissolution within the area and depth mainly linked to the permeability and the groundwater circulation.

Regarding the dissolution rate, in the four months investigated, some values can reach rates of 8 mm, which if compared with literature (Cucchi et al. 1998; Klimchouk et al. 1996), yearly value is already much higher (at least eight times).

The experiment is still ongoing and future months will be used to evaluate the dissolution of rock samples according to seasonal variations.

## References

Calligaris C, Devoto S, Zini L (2017) Evaporite sinkholes of the Friuli Venezia Giulia region (NE Italy). J Maps 13:406–414. https://doi.org/10.1080/17445647.2017.1316321

Calligaris C, Ghezzi L, Petrini R, Lenaz D, Zini L (2019) Evaporite dissolution rate through an on-site experiment into piezometric tubes applied to the real case-study of Quinis (NE Italy). Geosciences 9:298. https://doi.org/10.3390/geosciences9070298

Calligaris C, Zini L, Nisio S, Piano C (2020) Sinkholes in the Friuli Venezia Giulia Region focus on the evaporites. J Appl Geol 5. https://doi.org/10.1007/978-3-030-43953-8_5

Cucchi F, Forti P, Finocchiaro F (1998) Gypsum degradation in Italy with respect to climatic textural and erosional conditions. Geogr Fis Din Quat 3:41–49

Furlani S, Cucchi F, Forti F, Rossi A (2009) Comparison between coastal and inland karst limestone lowering rates in the northeastern Adriatic Region (Italy and Croatia). Geomorphology 104:73–81. https://doi.org/10.1016/j.geomorph.2008.05.015

Gutiérrez F, Cooper AH, Johnson KS (2008) Identification, prediction and mitigation of sinkhole hazards in evaporite karst areas. Environ Geol 53:1007–1022. https://doi.org/10.1007/s00254-007-0728-4

Klimchouk A, Cucchi F, Calaforra JM, Aksem S, Finocchiaro F, Forti P (1996) Dissolution of gypsum from field observations. Int J Speleol 25:37–48. https://doi.org/10.5038/1827-806X.25.3.2

Marinelli O (1898) Fenomeni di tipo carsico nei terrazzi alluvionali della Valle del Tagliamento. Studi orografici nelle Alpi orientali. Mem Soc Geol Ital 8(2):415–419

Plan L (2005) Factors controlling carbonate dissolution rates quantified in a field test in the Austrian alps. Geomorphology 68:201–212. https://doi.org/10.1016/j.geomorph.2004.11.014

Venturini C, Spalletta C, Vai GB, Pondrelli M, Delzotto S, Fontana C, Longo Salvador G, Carulli GB (2009) Note illustrative carta geologica d'Italia alla scala 1:50.000 foglio 031 ampezzo. ISPRA: Rome, Italy, pp 7–222 (In Italian)

Zini L, Calligaris C, Forte E, Petronio L, Zavagno E, Boccali C, Cucchi F (2015) A multidisciplinary approach in sinkhole analysis: the Quinis village case study (NE-Italy). Eng Geol 197:132–144. https://doi.org/10.1016/j.enggeo.2015.07.004

# A Multidisciplinary Investigation of Karstic Subsidence in a Madrid Urbanization

E. Sanz Pérez, C. Sanz Riaguas, and J. Sanz de Ojeda

**Abstract**

A multidisciplinary investigation was carried out in a karstic depression in a housing development under construction in Madrid to assess its stability. It was found that it is a small basin within a larger depression as a result of subsidence accumulated during the Quaternary. Subsidence has built up progressively in the Miocene clay cap and bedrock due to the underlying dissolution of gypsum-rich intercalations. The groundwater flow has flowed preferentially through a major fault that crossed the Miocene. The progressive dissolution over time along the fault caused the formation of an elongated syncline by subsidence. This subsidence is also conditioned to the formation of a small alluvial basin. It seems that under the alluvial deposits, a slow and diffuse dissolution is taking place at the shallower clayey gypsiferous levels; this is somewhat more intense in the fault zone, which is more active hydrodynamically. Microgravimetry surveys indicate that only 5% of the area under the alluvium shows anomalies, interpreted as residual soft clayey masses, or anomalous alluvial fillings of old dissolution voids. These pockets have no more dimensions than 20 × 20 m and a maximum vertical dimension of 20 m. These measurements have been confirmed by boreholes and are the only points that would require special attention in the future construction of the urbanization. The urbanization work, in full development, is implementing solutions aimed at the stability of the road in the strips of alluvial studied. The affected areas by this difficulty are located in areas classified as green zones in the planning (parks and open spaces). However, despite obtaining non-relevant subsidence data, exhaustive research works have been carried out, based on extensive field prospecting (trial pits, boreholes, microgravimetry, etc.). These works have led to satisfactory conclusions for the proper development of urbanization works.

**Keywords**

Sinkholes • Evaporitic karst • Karstic subsidence • Hazard • Risk • Microgravimetry survey

## 1 Introduction and Objectives

The nature of karst in evaporite rocks presents a variety of risks associated with construction or civil engineering activities (Mancebo et al. 2014). In the case of the construction of a housing development, the most important factor is the stability of the foundations, referring above all to the existence of possible areas of subsidence and still active karstification. One of the natural risks of karst origin associated with subsidence areas is sinkholes (Cooper and Waltham 1999; Gutierrez et al. 2008).

Early identification of subsidence areas is one of the most effective prevention strategies in areas of future urbanization. Hazard assessment involves not only the identification of existing sinkholes, but also their characterization, the prediction of future subsidence phenomena, probabilities of formation, the mechanism of subsidence, expected subsidence rates, etc. These predictions are based on the assumption that future activity will have a similar effect to that of the past. The implementation of these preventive measures requires accurate mapping of the areas affected by subsidence and the delimitation of the most dangerous

E. Sanz Pérez
Laboratorio de Geología, Departamento de Ingeniería y Morfología del Terreno, Universidad Politécnica de Madrid, 28040 Madrid, Spain
e-mail: eugenio.sanz@upm.es

C. Sanz Riaguas
Desarrollos Logísticos y Fomento de Suelo S.L. (DELFOS), C. de Narvaez, 15, 28009 Madrid, Spain
e-mail: csanz@grupodelfos.net

J. S. de Ojeda (✉)
ETSI Minas y Energía, Universidad Politécnica de Madrid, 28040 Madrid, Spain
e-mail: joaquin.sanzdeojed@upm.es


B. Andreo et al. (eds.), *EuroKarst 2022, Málaga*, Advances in Karst Science,
https://doi.org/10.1007/978-3-031-16879-6_31

sectors. It is intended to exclude dangerous sectors if they are close to construction sites, in order to avoid significant economic losses.

When dynamic processes develop in the ground are difficult to quantify and predict, conventional geotechnical criteria are not useful, and it becomes necessary to consider and understand geological processes as such. It is no longer an exclusively "technical" problem; it is now a "scientific" problem. A scientific approach is thus required to fully understand a natural phenomenon that involves interconnected processes controlled by the lithology, structure, morphology, hydrology, hydrochemistry, and rheology of the terrain.

It is, therefore, a complex phenomenon that must be approached from a multidisciplinary perspective. A specific investigation is required to detect underground cavities and their fillings. This should include geological and geomorphological explorations, geophysical prospecting, and mechanical soundings that allow sufficient detail to be obtained on the anomalies of the subsoil. The most relevant geological and geotechnical aspect in the southeast area of Madrid is the karstification processes that take place in the gypsiferous rocky substratum (Escolano 2005). In this sense, and as part of previous geological, geotechnical, and hydrogeological work (Sanz 2019; Sanz de Ojeda et al. 2021), a small basin of 500 by 300 m was revealed on the southern edge of "Los Ahijones" housing estate, in a sector ("Estevillas" alluvial) that rests on a Miocene clayey and gypsiferous substrate, covered by Pleistocene and Holocene alluvial deposits that are interpreted as the filling of a dissolution depression currently located in a fluvial valley. Project soundings in this area have detected signs of dissolution at the contact of the gypsum layers and the overlying clays, as well as the loss of penetration resistance to considerable depths. The alluvial basin is located in a relatively lower area, which favours the concentration of surface and subsurface flow and thus dissolution, as found in the work of Bárez del Cueto et al. (2016).

The study area is located in the southeast of Madrid, specifically in the urbanized area of "Los Ahijones", which is located in the Vicálvaro district of Madrid. The research focuses on several aspects relevant to assessing hazards related to subsidence phenomena, such as the origin, mechanisms, and rates of subsidence, as well as the possible presence of hypersoluble salts.

## 2 Methodology

The application of different combined methods is also an objective for the achievement of this work. Among these methods, geophysical prospecting stands out. This type of prospecting did not try to cover the entire karst depression (except in the foundation of a viaduct as the most important structure) but rather to know the type and frequency of these karstified zones and the degree of advance of this karstification. This technique, however, is not conclusive, and it was necessary to check by means of surveys (soundings).

In this type of study, it is essential to investigate as many surface and subsurface sources of information as possible to provide data on current or past subsidence activity in the study area. The Prediction will depend to a large extent on an understanding of the local geology and hydrogeology. A phased sequence of investigation has been applied, using, for example, geophysical prospecting prior to drilling boreholes. The methodology is explained and justified below:

(1) Literature review and compilation, and analysis of all previous geotechnical studies of the area and its surroundings (2) Detailed topographic survey of the basin with equidistance of 0.25 m in order to identify closed depressions. (3) Detailed geological mapping of the subsidence depression, including slopes and surroundings of the alluvial sector studied, at a scale of 1/2000 (Fig. 1). (4) Geophysical prospecting. The strategy consisted of carrying out a preliminary study by means of refraction seismic and electrical tomography. On these last geophysical surveys, another method of higher resolution, microgravimetry, was applied. (5) The drilling of two research boreholes in the geophysical anomalies detected in the two aforementioned alluvial sectors (SN-17 and SN-18) (Fig. 1). Detailed test drilling has been carried out, accompanied by X-ray diffraction analysis to identify the possible presence of hypersoluble minerals. (6) Retrodeformation analysis. The documentation of 35 boreholes and 90 research trial pits distributed throughout the development from previous works (Sanz 2019; Sanz de Ojeda et al. 2021) made it possible to reconstruct the geological structure of the site in great detail: synclines, fault planes, collapse zones, etc., and gave information on the mechanisms and cumulative displacement on subsidence. This has allowed for the retrodeformation analysis through the progressive restoration of the sedimentary layers and the interpretation of subsidence episodes over time, as well as finding out the somewhat uncertain origin of the "Los Ahijones" depression. (7) Direct inspection of road excavation, earthworks, and trenches. (8) Hydrogeological investigations. Understanding the hydrogeology (including hydrogeochemistry) of a study area is a crucial aspect of subsidence hazard analysis. Groundwater flow is the geological agent responsible for karstification in evaporite rocks and is often one of the most important conditioning and triggering factors of the generation of sinkholes. We also carried out a hydrogeological study of the area, including flow modeling (Sanz de Ojeda et al. 2021).

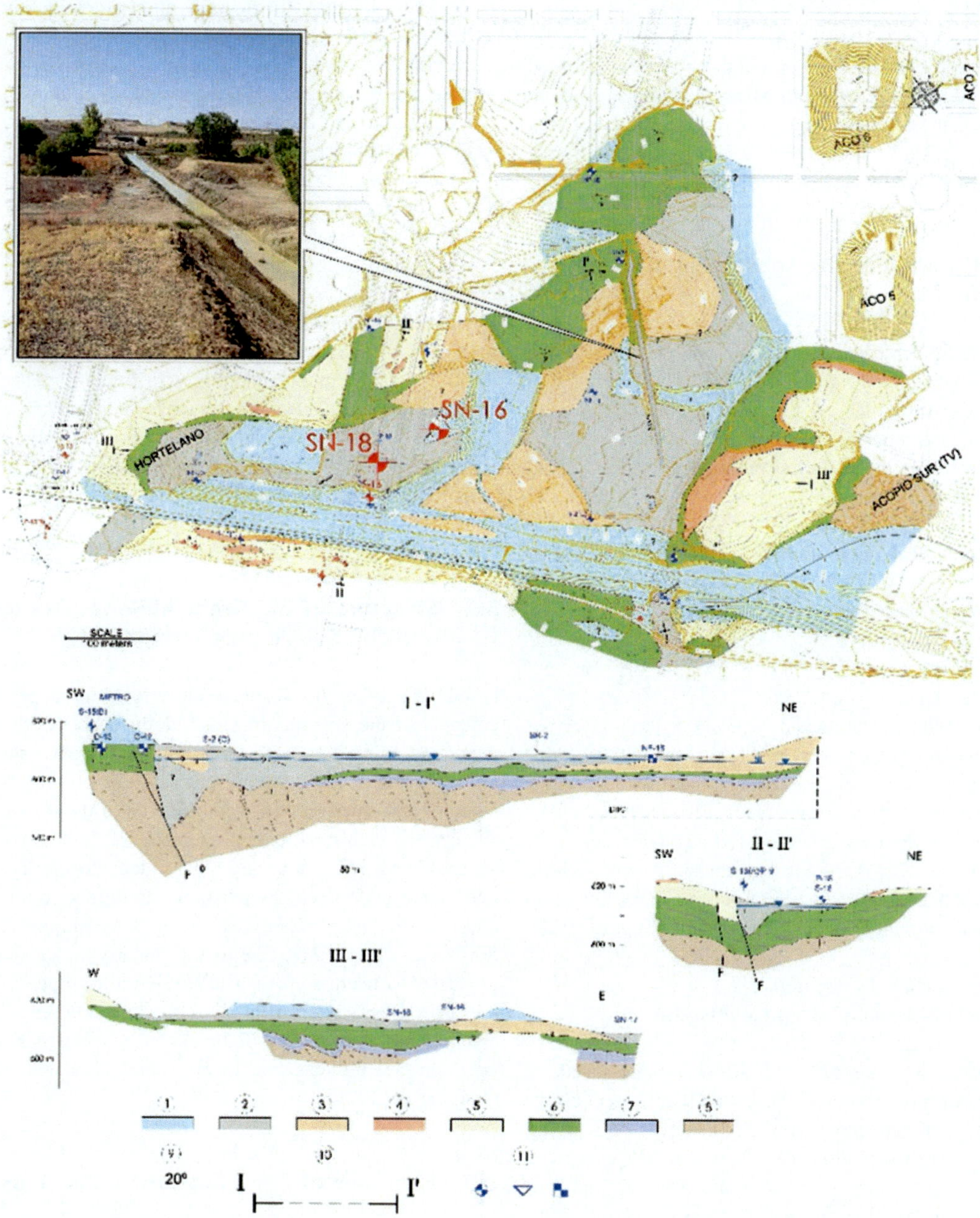


**Fig. 1** Geological map of the subsidence basin. (1) Anthropic landfill (Quaternary). (2) Alluvial (Holocene). (3) Terrace (Pleistocene). (4) Calcareous tufa (Miocene). (5) Sands, chert nodules, and dolomites. (6) Green clays and silts ("peñuela"). (7) Clays and green silts with gypsum intersperses. (8) Massive gypsum. (9) Direction and dip. (10) Geological cross-sections. (11) Surveys: boreholes, penetrometers, test pits, and trenching. The photo shows the northern edge of the subsidence basin with the geological contact in the trench between the green rock (7) and the Holocene alluvial overburden (2). In the foreground, sandy deposits corresponding to terrace level T4 (3), which is overhanging in this sector

## 3 Results

As a result of a detailed topographic survey, a rounded topographic depression of about 80 m in diameter and no more than 0.5 m of depth was revealed in the southeast area of the basin. It can be interpreted as a very discrete doline that tends to become waterlogged when it rains.

### 3.1 Stratigraphy of the Subsidence Alluvial Basin

The stratigraphy described in detail here refers to the small subsidence basin under study, although the stratigraphy of the Miocene substratum of the rest of "Los Ahijones" depression is also reported. In the geological map in Fig. 1, we can differentiate the following geological units, ordered from oldest to most recent:

**Miocene Geological Formations:**

- More or less massive lower Miocene–Middle Miocene lower unit.
- Clays with intercalated layers of gypsum or gypsum with intercalated layers of clay.
- Clay and green silt laminated with intercalated layers of fine to medium-grained, very micaceous sands ("Peñuela" in the area). According to soundings, the thickness varies between 5 and almost 10 m.
- Sandstones, dolomites, limestones, flint levels, and clay intercalations. This is the most permeable "layer" within the intermediate unit and here it has a small thickness, no more than 3 m; under the alluvial it hardly appears at all.

**Quaternary Surface Formations**

We distinguish, from older to more recent:

- Limestone calcareous tufa. Remnants are preserved on the edges of this depression, between 10 and 11 m above the bottom of the depression, in the form of hanging terraces towards elevations of 620–622 m asl, except on the southeastern edge, where they are scattered and broken all along the slope up to 612 m asl, probably indicating a resubmergence towards the interior of the depression. According to the geomorphological maps of Bárez del Cueto et al. (2016), these tufa seem to be synchronous with those of the T2 terraces (PC-4), also located at an elevation of 622 m asl and dating from the Middle Pleistocene, and have been dated to 115,000 years ago. This is confirmed by the fact that within the tufa, there are prehistoric tools of carved flint from the Mousterian (Bárez del Cueto et al. 2016).
- Terrace R4. Thanks to the excavation of a trench in a N–S direction, and slightly more than 200 m long with a depth varying between 8 and 1.5 m, the northern edge of the basin has been revealed. The schematic geological section along the trench is shown in Fig. 1. From north to south, there is a first section excavated in "peñuelas clays", on which are supported, in the area closest to the depression and for a length of 35 m, the deposits of hanging terrace T2, which reveals indicates an age of 115,000 BP, corresponding to the Middle Pleistocene (Bárez del Cueto et al. 2016). It is a 4-m-thick detrital deposit.
- Holocene Alluvial Deposits. The Holocene deposits have an overall thickness of less than 4 m and consist of dark brown clayey muds and soils with organic matter in the upper part and feldspar and quartz sands and some gravel in the remaining portion. However, in the southern edge fault zone, there are average thicknesses of more than 10 m that are considered Holocene.

### 3.2 Structure of the Great Ahijones Depression and the Subsidence Trough

Isoline maps of the absolute elevation and depth of the contacts of the substratum layers defined above have been generated using data from the numerous boreholes and test pits available. With these maps, it has been possible to make a geological cross-section representative of the urbanization. We can see in this geological section (Fig. 2c) that the layers are subhorizontal, although they adapt to a dissolution depression that affects the gypsum, giving rise to a syncline and a fault. The great depression of "Los Ahijones", some 500 ha in size, where, due to the underlying karstification processes concentrated in a NW–SE fault coinciding with the Migueles stream (Fig. 1), an elongated and narrow syncline has been created in the clayey cover, which has had an important influence on underground drainage (Sanz de Ojeda et al. 2021).

### 3.3 Existence of Karstification in the Gypsum

The microgravimetry survey identified five small-amplitude gravity minima, two of which were anthropic fills. The other three should be associated with the three possible cases mentioned above, or a combination of these. In an anomaly, a borehole (SN-18) was drilled, which allowed the recognition that it was a soft clayey gypsiferous zone from the Miocene up to 16 m. This is probably the same as with the other anomaly B1, where there is little alluvium thickness. Other anomaly data from another borehole (S-2 in Fig. 1)

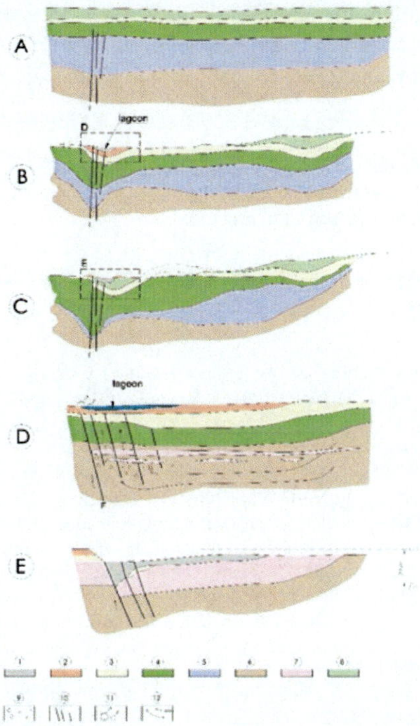


**Fig. 2** Retrodeformation analysis to the investigation of subsidence evolution in Ahijones karst. **a** Undeformed deposits in the Miocene. **b** Gradual subsidence and synsedimentary gradual subsidence in the Quaternary. Along the Pleistocene, where the depression caused by preferential dissolution in the gypsum in the discharge zone of the subsurface flow created a lagoon with tufa on the shores (Detail in **d**). This deformed the Miocene clay cover subsided into a syncline **c**. The process continues today, with the over accumulation of anomalous thicknesses in the fault zone (Detail in **e**). 22 m is the cumulative displacement). (1) Holocene alluvial. (2) Calcareous tuffs. (3) Miocene sandstones (aquifer). (4) Miocene green clays ("peñuelas"). (5) Clays and green silts with gypsum intersperses. (6) Massive gypsum. (7) Residual clays of (4). (8) Clays and sepiolites. (9) Cavities due to karstification. (10) Fault zone. (11) Springs in the lagoon. (12) Groundwater flow lines

were also available, which confirmed the thickening of the alluvium on this side close to the 14 m fault zone.

### 3.4 Hydrogeology

The piezometric control network shows that the water table is well adjusted to the topography, with a fairly epidermal flow associated with the clayey formations of the Miocene. The role played by the elongated faulted syncline of a NW–SE direction and higher permeability, which conditions and redirects the regional flow to a great extent, is highlighted. The hydrogeochemical conceptual model is well explained by the geology of the site: the predominance of calcium–magnesium bicarbonate facies in the Miocene clayey layers, and calcium–sulphate facies when there is the influence of the gypsum if they are close.

Modeling of the aquifer (Sanz de Ojeda et al. 2021) has revealed that natural recharge in the clay aquiclude is minimal and does not reach 1% of the useful water of the hydraulic balance (0.16 mm/year). It has also revealed the general hydrogeological behaviour of the lower gypsum layer. This layer has a general permeability similar to that of the clay aquiclude (despite possible local karstification) and is recharged in an induced manner through this layer. It discharges into the lower part of the alluvial part of "Los Ahijones" stream (Fig. 1) via an upward flow and longer streamlines, which explains the high sulphate content of the groundwater in this area. This flow has been quantified by MODFLOW simulation to be no more than 2 L/s (Sanz de Ojeda et al. 2021).

### 3.5 Presence of Water-Soluble Salts?

In the visual inspection of the evaporite samples obtained from the two boreholes mentioned above, only gypsum was recognized, and no hypersoluble salts were clearly identified. Semiquantitative analyses of samples by X-ray diffraction and thin sheet petrographic microscopy have not detected significant contents of highly soluble mineral species beneath the Quaternary alluvium. There is no halite, polyhalite, or epsomite, which could be suspected to occur because of the high Mg content in the water; for example, it is seen that the high content of this ion is due to the magnesian clays in the "peñuelas".

## 4 Discussion and Conclusions

Thanks to the application of different combined methods, it has been possible to determine that the small subsidence basin studied corresponds to the most active sector of a large depression formed by underlying karstification during the Quaternary. This clay-covered doline has been produced by slow flexural subsidence due to the removal of the bedrock and cover by the dissolution of the interstratified layers of gypsum-rich clays. It conforms to a fault and axis of an atectonic syncline that has concentrated the dissolving subsurface flow, accentuating the underlying karstification, and topographic subsidence until the formation of a microlake basin during the Pleistocene. The karstification progressively enlarges the syncline, and in a sector of the fault and bottom of the syncline, there is collapse. As we can see, the dissolution has created the syncline, and the syncline is the cause of the concentration of the flow: these have a cause-and-effect relationship with each other over time. The result is a basin or doline filled with alluvial formations of mainly Holocene age. Three streams flow into it. As it is a low-lying area, there are also underground water discharges.

The current average decrease of subsidence rate could be evaluated at about 0.4 mm/year and concerns the alluvial zone, and the maximum is 1.4 mm/year in the fault zone. The possible karstified areas, karstified residues, soft patches, and anomalous alluvial fills in former hollows only represent between 5 and 12% of the hidden area beneath the alluvium. As it was mentioned before, these pockets have no more dimension than 20 × 20 m and a maximum vertical dimension of 20 m. In the future, no acceleration in dissolution is expected because the flow will be substantially reduced by the draining action of a drainage ditch in the process of urbanization, and because there are no significant quantities of highly soluble mineral species.

The urbanization works, which are in full progress, are undertaking complementary measures aimed at the stability of the road by removing and cleaning up the alluvial deposits studied. In those alluvial areas with a shallow water table, where a road infrastructure will be located, alluvial material has been completely removed (whose depths have not exceeded 5 m) and have been filled in with a reinforced geotextile protection riprap. This construction procedure has allowed to work in that wet alluvial, guaranteeing a good mechanical behaviour of the upper embankment.

There is only one area where a road structure is going to be built, where the foundation avoids karstic fillings, and it is reinforced with piles; this guaranteed its stability and functionality. To this end, geotechnical soundings will be taken at the site where the foundation piles are to be driven to guarantee the exact depth of the stable support stratum at the tip of the pile. The area most affected by the aforementioned subsidence refers to the area designated by the territorial planning as an urban park, where there are no relevant infrastructures that need to be treated with special construction procedures.

**Acknowledgements** The authors would like to make heartfelt thanks to Dr. Augusto Auler for his review of the manuscript, suggestions, and comments that have helped to improve it.

## References

Bárez del Cueto S, Preysler JB, Perez-Gonzalez A, Torres C, Rus Perez I, Vega de Miguel J (2016) Acheulian flint quarries in the Madrid Tertiary basin, central Iberian Peninsula: first data obtained from geoarchaeological studies. Quat Int 441:329–348

Cooper AH, Waltham AC (1999) Subsidence caused by gypsum dissolution at Ripon, North Yorkshire. Q J Eng Geol 32:305–310

Escolano F (2005) Caracterización geológico-geotécnica de las facies de transición y evaporíticas de la cuenca de madrid en el interfluvio Manzanares-Jarama. PhD Thesis, Universidad Politecnica de Madrid, Madrid, Spain

Gutiérrez F, Cooper AH, Johnson KS (2008) Identification, prediction and mitigation of sinkhole hazards in evaporite karst areas. Environ Geol 53:1007–1022

Mancebo JA, Sanz E, Menéndez-Pidal I (2014) Karstología de terrenos salinos aplicada a la ingeniería civil; editorial garceta. Madrid, Spain

Sanz E (2019) Modelación matemática del flujo subterráneo en el acuicludo de la Urbanización Uzpp.02.03 Los Ahijones Distrito 19. Vicálvaro (Madrid) y su Entorno; Fundación Agustín de Betancourt Escuela Técnica Superior de Ingenieros de Caminos: Madrid, Spain. Informe inédito

Sanz de Ojeda J, Sanz E, Elorza FJ, Riaguas CS, Liaño MdP (2021) Simulation of groundwater flow in an aquiclude for designing a drainage system during urban construction: a case study in Madrid, Spain. Sustainability 13:1526

# Morphometric Comparison of Dolines in Three Karst Landscapes Developed on Different Lithologies

E. Pardo-Igúzquiza, J. M. Gil-Márquez, M. Mudarra, B. Andreo, and J. J. Durán-Valsero

**Abstract**

The present contribution compares the geomorphological characteristics of dolines developed in three different geological contexts in southern Spain: an evaporite tectonic melánge, a bare-carbonate massif, and an outcrop of inter-stratified sedimentary gypsum sequence. In the three cases, the closed depressions have been identified and mapped using digital elevation models with a spatial resolution of 5 m, and the results were validated in the field. To reveal similitudes and differences due to distinctive geological and climatological settings between the three sites, morphometric and spatial analyses were applied to the threes sites: size-distributions, relationships between area and depth, preferential directions, and point field analysis, among others. The results show apparent differences between the carbonate and the two evaporitic settings. Carbonate outcrop displays larger dolines, implying a lower doline density, while the density of depressions rises in the evaporite sites. The results also show a dominant ENE-WSW alignment of dolines in two areas, in agreement with the main fault families of the Betic Cordillera, which evidences the tectonic influence on the karst development. However, in the gypsum karst, other directions appear related to the surficial drainage network connected to dolines. Finally, in the tectonic melange, the uneven distribution of lithologies affects karstification distribution. Geological differences (both lithology and structure) imply changes in solubility and karstification dynamics that, together with other factors such as the climatic conditions and the exposure time, affect the genesis of the depression and, ultimately, its size and shape. All that combined explains the differences in morphometric parameters and spatial.

**Keywords**

Karst • Morphometric analysis • Digital elevation models • Sinkholes

## 1 Introduction

Dolines are the most typical landform of the karst landscape on the metric scale (Sauro 2003). In karst terrains, most dolines have been formed by the solution of the bedrock (limestone, dolostone, gypsum, and halite), although some dolines form by terrain subsidence or collapse (Ford and Williams 2007), which ultimately are also related to the solution of the bedrock. Dolines have been intensively studied since the times of Cvijić (White 1988) and the first karst geomorphologists. Doline mapping has been an important task in karst geomorphology, not only by its scientific interest but also by its importance in sinkhole hazards (Gutiérrez et al. 2008), preferential recharge in hydrogeology (Somaratne 2014), land use management (Hughes et al. 1994), and biodiversity studies (Bátori et al. 2019). Bondesan et al. (1992) recognized that measuring morphometric parameters through field surveys or over maps and aerial photographs requires hard work and is time-consuming. However, they also pointed out that the morphometric analysis has been facilitated by the availability of computers and the incipient, at the time, geographical information

E. Pardo-Igúzquiza (✉) · J. J. Durán-Valsero
Instituto Geológico y Minero de España-CSIC, Ríos Rosas 23, 28003 Madrid, Spain
e-mail: e.pardo@igme.es

J. J. Durán-Valsero
e-mail: jj.duran@igme.es

J. M. Gil-Márquez · M. Mudarra · B. Andreo
Edificio de Investigación Ada Byron, Centro de Hidrogeología de la Universidad de Málaga (CEHIUMA), C/Arquitecto Francisco Peñalosa, 18, 29071 Málaga, Spain
e-mail: josemgil@uma.es

M. Mudarra
e-mail: mmudarra@uma.es

B. Andreo
e-mail: andreo@uma.es


B. Andreo et al. (eds.), *EuroKarst 2022, Málaga*, Advances in Karst Science,
https://doi.org/10.1007/978-3-031-16879-6_32

systems. Also, they highlighted that the altimetric parameters of dolines had been very often neglected while the analysis has focused on planimetric attributes.

In the last few decades, the availability of digital elevation models (DEMs) has improved quantitative terrain analysis in general and the morphometric analysis of dolines in particular (Lyew-Ayee et al. 2007). Based on the availability of high-resolution DEMs, Pardo-Igúzquiza et al. (2013) propose for the automatic detection and delineation of depressions a simple idea: a digital map of depressions can be easily obtained by the map algebra operation of subtracting the depression-free DEM from the original DEM. The DEM can be combined with satellite images and morphometric analysis to discriminate different types of depressions. Improvements in the acquisition of lidar data and the use of crewless aerial vehicles have improved the spatial resolution of the DEMs, and there is an increase in the detail that can be observed of individual dolines in order to extend their morphometric analysis. All the previous references are morphometric and spatial analyses of dolines ensembles in carbonate karst systems. Fewer studies have been done on morphometric analysis of karst depressions in other types of karst systems, which is related to their smaller world abundance. Nevertheless, morphometric analysis of gypsum karst is not rare (Keskin and Yilmaz 2016).

This contribution analyses the geomorphological characteristics of dolines developed in three different geological contexts in Southern Spain: an evaporite tectonic mélange, a bare-carbonate massif, and an outcrop of inter-stratified sedimentary gypsum sequence. The method described by Pardo-Igúzquiza et al. (2013) has been used for the identification and delineation of the closed depressions, and the results are shown and discussed next. By comparing the geomorphological difference between them, this work aims to advance in understanding their genesis and assess the role of different factors, including geological ones, on the evolution of depression landforms in karst.

## 2 Study Area

The location of the three different karst areas selected in southern Spain is shown in Fig. 1. The carbonate karst is the Sierra Gorda karst massif (SG), a medium relief karst (with maximum altitudes lower than 1700 m a.s.l.) with a NW–SE oriented elliptic shape that is developed on Jurassic limestones and dolostones (López-Chicano 1992). The karst in an evaporitic mélange is part of the so-called "Trías de Antequera" (Sanz de Galdeano et al. 2008), a Triassic tectonic mélange and olistostrome with diapiric structure that encloses blocks of gypsum and halite (among other lithologies) embedded in a fine-grained matrix of clays and marls. The gypsum karst of Sorbas is formed by 60-m thick Neogen gypsum deposits related to the Messinian Salinity Crisis event (Braga et al. 2006).

Figure 2 shows the digital elevation models (DEMs) of the three karst systems. The three DEMs have a spatial resolution of 5 m and are freely available from the Internet site of the Instituto Geográfico Nacional (www.ign.es). The southern limit of the Sierra Gorda DEM has been arbitrarily chosen in order to exclude the Zafarraya polje, a huge karst depression that can be considered as an exception if compared with a typical doline of the highest part of the Sierra Gorda karst system. The area of each of the three karst systems considered is 267, 69, and 31 km$^2$ for the Sierra Gorda carbonate karst, Meliones evaporitic mélange karst, and Sorbas gypsum karst, respectively (Table 1). The method described by Pardo-Igúzquiza et al. (2013) has been used for the identification and delineation of the closed depressions, and the results are shown and discussed next.

## 3 Results

The maps of karst depressions are shown in Fig. 3a–c. Results were validated in the field. In the Sorbas gypsum karst, three large depressions corresponding with gypsum quarries were filtered out from the map. It may be seen how the maximum depth of a depression from its rim is −38.44 m in SG, then −18.08 m in ME and −14.16 m in the SO (Fig. 3). Dolines in the Sierra Gorda carbonate karst appear in a doline field with the individual dolines smoothly rounded and aligned, most likely along fractures. The dolines in the Meliones evaporitic mélange karst seem more asymmetrical with an almost triangular outline. On the other hand, the dolines in the Sorbas gypsum karst are the most

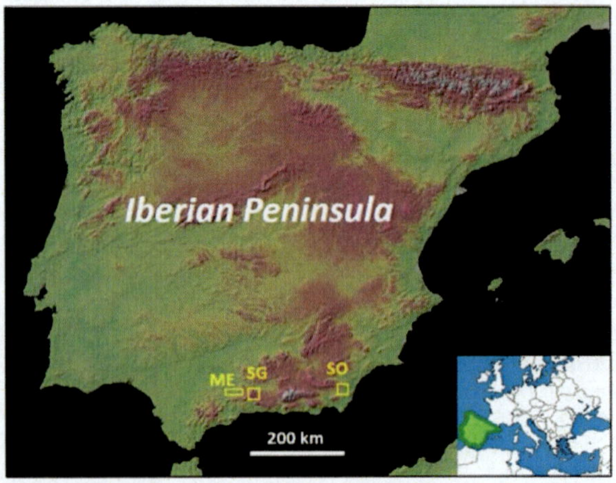


**Fig. 1** Geographical location of the three karst systems in Southern Spain. SG: Sierra Gorda carbonate karst; ME: Meliones evaporitic mélange karst; SO: Sorbas gypsum karst

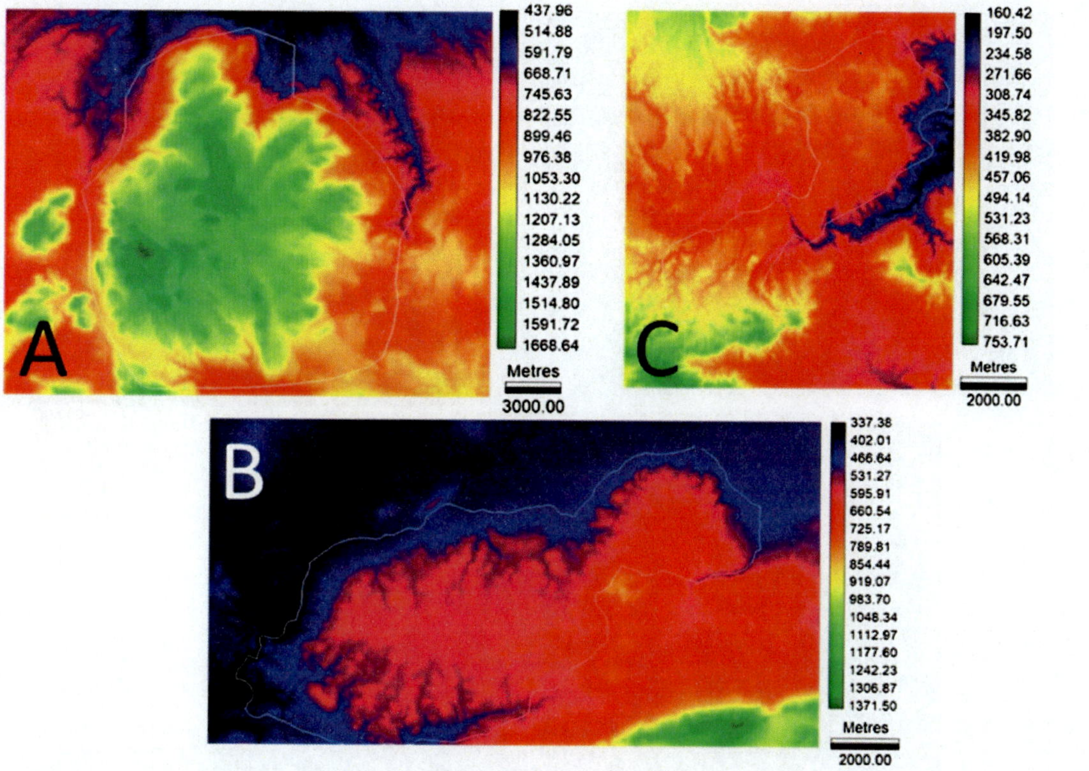


**Fig. 2** Digital elevation models (DEMs) of the three karst systems in Southern Spain. A: Sierra Gorda carbonate karst; B: Meliones, evaporitic mélange; C: Sorbas gypsum karst. The colour legend is the altitude in metres above the mean sea level. The spatial resolution of each DEM is 5 m. The white lines represent the borders of the karst systems

**Table 1** Surface of the study areas (first column) and statistics of the morphometric characteristics of dolines. N: number of dolines. D: density of dolines in dolines per km$^2$. 1 N: number of one-cell dolines (dolines < 25 m$^2$). N > 250: number of dolines larger than 250 m$^2$. D: Median of the diameter assuming the shape of the doline is a circle with the same area. D/R: the ratio diameter/depth

| | Area (Km$^2$) | N | D Density N/Area | 1 N | N > 250 m$^2$ | Mean area (m$^2$) | Median area (m$^2$) | Median depth R (m) | Median equivalent diameter D (m) | Ratio D/R |
|---|---|---|---|---|---|---|---|---|---|---|
| Sierra | 267 | 3100 | 12 | 370 | 1931 | 1510 | 400 | 0.7 | 22 | 31:1 |
| Gorda | | | | 29% | 62% | | | | | |
| Meliones | 69 | 1273 | 18 | 370 | 267 | 885 | 75 | 0.36 | 10 | 28:1 |
| | | | | 29% | 21% | | | | | |
| Sorbas | 31 | 1145 | 37 | 366 | 161 | 309 | 75 | 0.31 | 10 | 32:1 |
| | | | | 32% | 14% | | | | | |

asymmetrical and seem to have been developed along a gully. The identified closed depressions have been 3100, 1273, and 1145 for the SG carbonate karst, ME mélange karst, and SO gypsum karst, respectively (Table 1). Thus, considering that the surface of the analysed outcrops is 267 km$^2$, 69 km$^2$, and 31 km$^2$ for SG, ME, and SO, respectively, their mean density of kart depressions per km$^2$ is 12, 18, and 37. However, the mean size of the depression progresses in the reverse order, and is 1510 m$^2$ (SG), 885 m$^2$ (ME), and 309 m$^2$ (SO). Other statistics of the number and area of the dolines are shown in Table 1.

Figure 4 shows the histograms of the areas of the karst dolines in the three study areas. It may be seen how the depressions developed in evaporitic systems (gypsum and the evaporitic mélange) have a similar size distribution, which is markedly different from the distribution of doline size in the carbonate karst. The evaporitic karsts have a more significant proportion of small depressions.

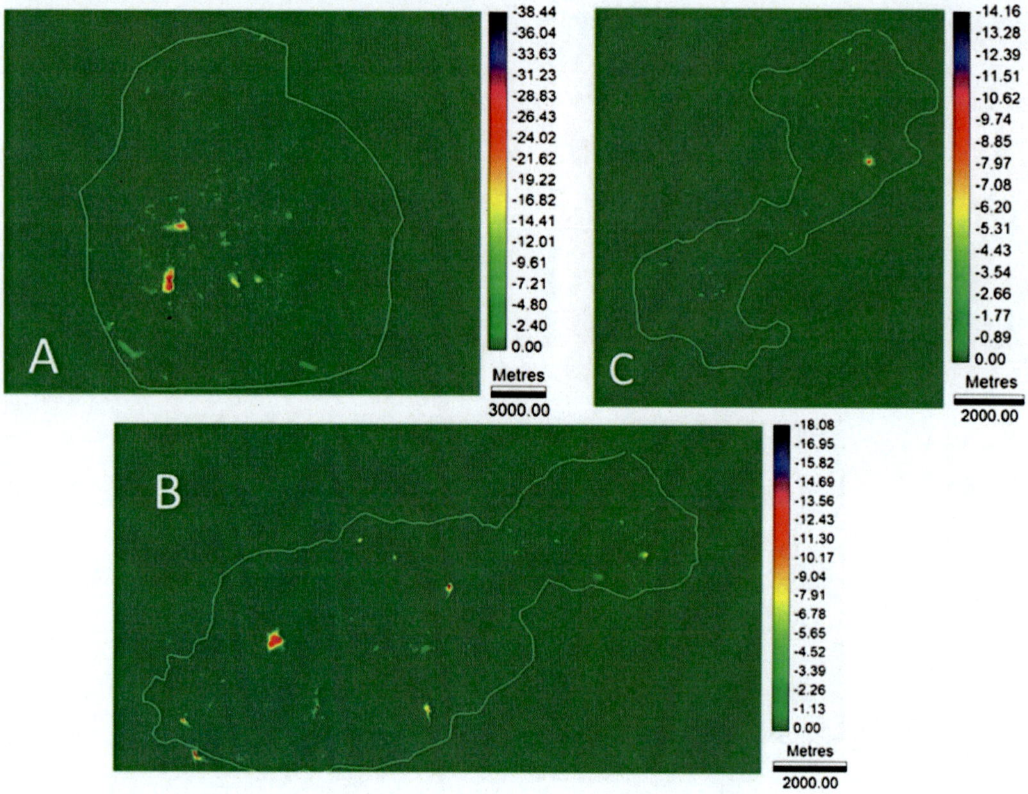


**Fig. 3** Maps of identified and delineated karst depressions in A: Sierra Gorda carbonate karst (SG); B: Meliones evaporitic mélange (ME); C: Sorbas gypsum karst (SO). The colour legend bar is the depth of each doline from its lowest edge in metres

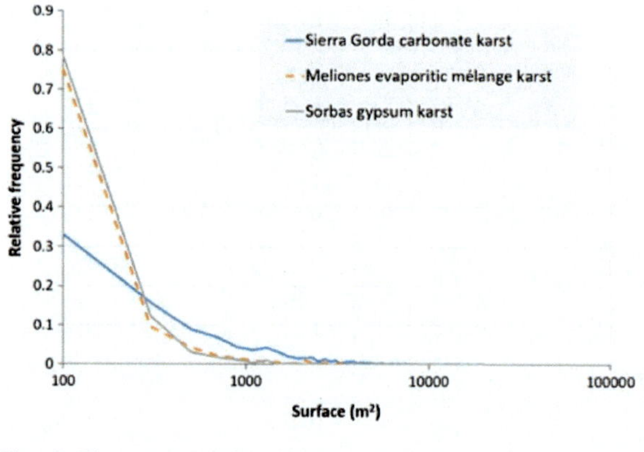


**Fig. 4** Histogram (relative frequencies) for the area of the karst depressions in the three study sites

Figure 5 shows the relationships between the maximum depth of the dolines and their area. In all the cases, there is a positive correlation between both parameters. A power law has been fitted to all the scatterplots (Fig. 5a–c), with exponent values of 0.62 (SG), 0.51 (ME), and 0.64 (SO). Thus, there is a clear similarity between the carbonate outcrop (SG) and the gypsum karst systems (SO), which differ from the evaporitic melange system (ME). Thus, dolines of the same surface developed in carbonates and gypsum outcrops will have a higher maximum and mean depth than if formed in evaporitic melange. More frequent collapse processes may explain that in the later karst system.

Figure 6 shows the power law fitted to the log–log plots of the size distribution of depressions for the SG, ME, and SO karst systems, respectively. The exponent of the power law can be considered the fractal dimension of the size distribution and has values of −1.04 (SG), −0.833 (ME), and −0.627 (SO). The fitting of straight lines to the log–log plot is qualitative proof of the fractal character of the sizes of dolines in karst systems independently of their lithology.

When each doline is substituted by its centroid, the fields of points shown in Fig. 7a–c are obtained for the SG, ME, and SO karst systems, respectively. It may be seen how the points in the SG carbonate karst cluster in given areas around the centre of the karst massif, while for the evaporitic karst systems (ME and SO), the dolines appear more isolated or concentrated in linear clusters along with drainage

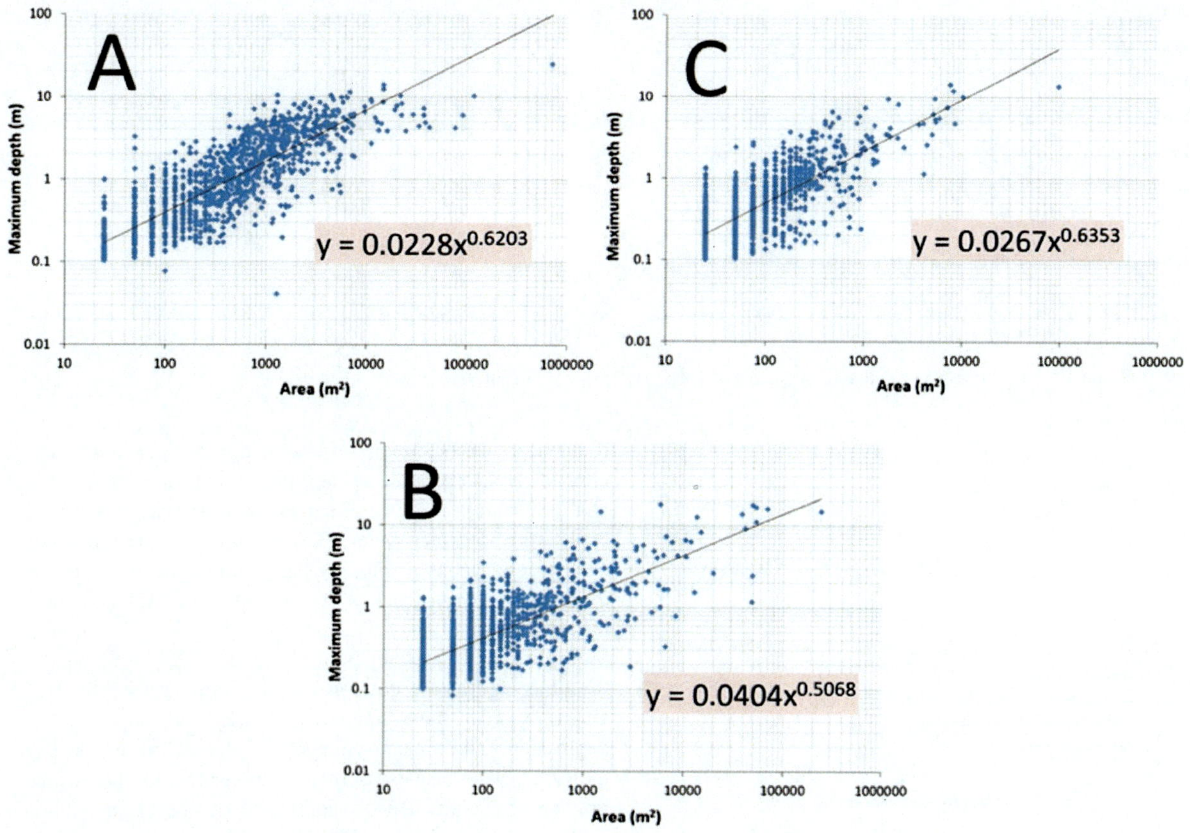


**Fig. 5** Relationship between the maximum depth of each depression and its area. A: Sierra Gorda carbonate karst; B: Meliones evaporitic mélange; C: Sorbas gypsum karst

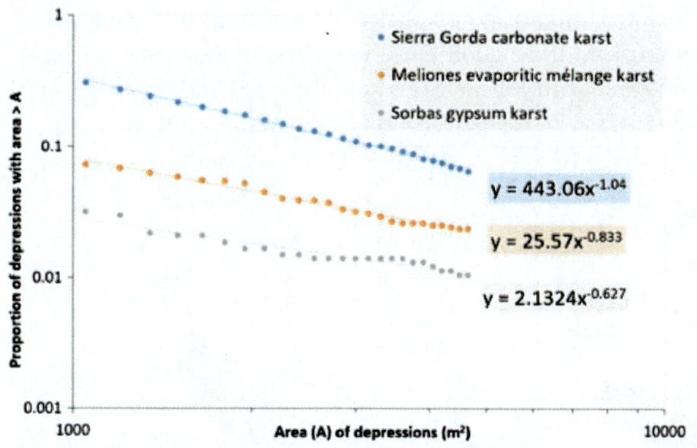


**Fig. 6** Power law fitted to the size distribution of dolines in a log–log plot for in Sierra Gorda (A), Meliones (B) and Sorbas (C)

patterns. Nevertheless, the SO system has larger clusters than the ME system. This can be reflected in the fractal dimension of the spatial distribution of points shown in Fig. 8, which has a higher value for the SO system: 1.14, 1.06, and 1.40 for the SG, ME, and SO karst systems, respectively.

Figure 9a–c shows the rose diagram (or angular histogram) of the main directions of the alignments of points in Fig. 7a–c for the SG, ME, and SO karst systems, respectively. The main modes are for azimuths N70E and N150E for the SG system, N65E for the ME system, and N35E for the SO system.

## 4 Discussion

Dolines in a carbonate karst area have a slow evolution so that the landscape looks invariable from year to year. This is because of the relatively low solubility of carbonates and their high mechanical strength. In general, the probability of occurrence of a new doline, from year to year, in carbonate karst is small, and thus, there is little risk of subsidence or collapse. In gypsum karst terrains with a thick geological formation of massive gypsum, the behaviour may be similar although the rock is more soluble. On the other hand, in a

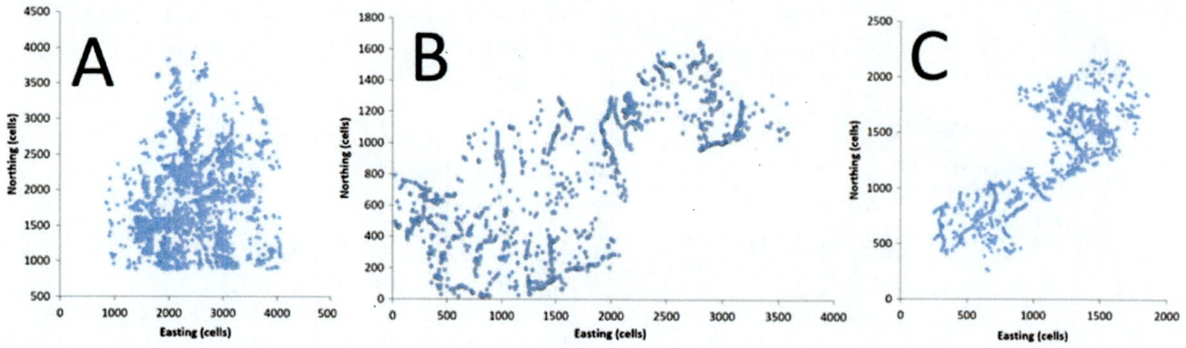


**Fig. 7** Location of centroid points of doline in Sierra Gorda (SG), Meliones (ME), and Sorbas (SO)

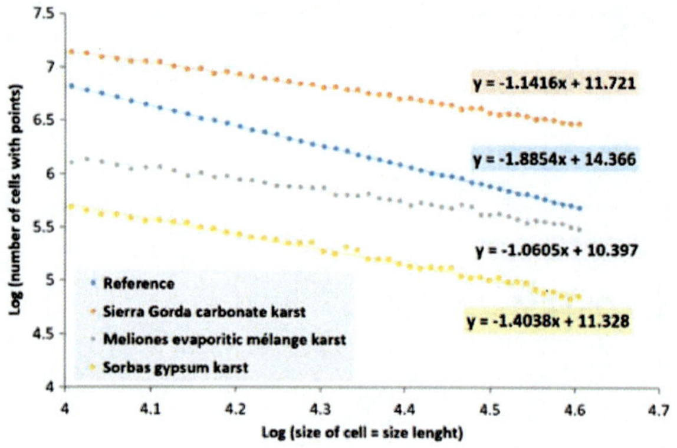


**Fig. 8** Box counting fractal dimension of point fields of karst depressions for Sierra Gorda (A), Meliones (B), and Sorbas (C)

mélange with gypsum blocks, the karst evolution is more dynamic because there is more concentration of flow crossing the marl and clay landscape and focusing towards the dolines developed in the gypsum blocks. Thus, in the karst developed in an evaporitic málange, the high dynamical development of karst depressions and their changes are perceptible at the human scale with noticeable changes from year to year. That has been detected in the field trips to the Meliones system. The histogram of the surface of closed depressions in Fig. 4 clearly shows the similarity of the two evaporitic rock systems (ME and SO) and their difference from the carbonate system (SG). The greater proportion of large dolines in the latter could be related to the older age of the outcrop and the more wet conditions, which would have favoured further development of depression. The clear asymmetry of the doline in the SO gypsum karst is because of their relationship with the surface drainage network. This is confirmed by the alignment of such zones along with gullies that differ from Sierra Gorda (SG) and Meliones (ME), where the dolines align along with the main fault families in the Betic Cordillera that is ENE-WSW (Fig. 9).

A distribution of points that will tend to fill the space will have a fractal dimension close to 2. The higher value of the fractal dimension of the SO system implies that the field of points in Fig. 7c tends to fill the plane more uniformly than the SG and ME karst systems. From a morphodynamic point

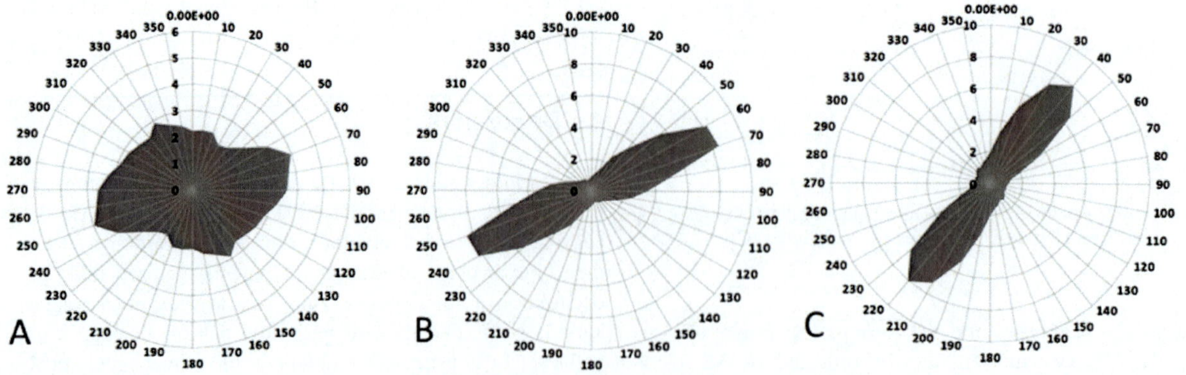


**Fig. 9** Rose diagram (or angular histogram) of the main directions of the alignments of centroid points of dolines in Sierra Gorda (A), Meliones (B) and Sorbas (C)

of view, dolines constitute elementary hydrographic units (Bondesan et al. 1992), which with their systems of slopes, covey water to a central point where recharge occurs (Ford and Williams 2007). Thus, where the bedrock is more soluble, and the outcrops are extensive, the distribution of dolines would be more uniform and they would be more abundant. That explains the differences between the gypsum system (SO) and the mélange systems (ME), where the gypsum outcrops are patched.

## 5 Conclusions

Morphometric and spatial analysis provides a mean for studying the differences between karst landscapes. In this study, the results of the morphometric and spatial analysis of karst closed depressions (dolines) developed in a carbonate, an evaporitic mélange, and a gypsum karst system have been compared. The geologic differences are clear, but the purpose has been to reveal intrinsic differences between the karst depressions in relation to their lithology and its implication in the karst surface dynamics. It is obvious that even between three different carbonate karst, there will be differences because there are many factors affecting the development of dolines, but in this study, in addition to those factors, there is the fact of the differences between landscapes being developed on carbonates, gypsum, and an evaporitic mélange. The comparison between such three different karst landscapes is also a novel aspect of this work. The differences in rock solubility, lithology, and structure of the outcrops, which imply different karst dynamics, explain the differences between the morphometric parameters of the three karst systems. However, other factors such as the climatic conditions and the exposure time cannot be neglected.

**Acknowledgements** The present work was supported by research projects PID2019-106435GB-I00 of the Ministerio de Ciencia e Innovación of Spain and PI20-01118 of the Junta de Andalucía

## References

Bátori Z, Vojtkó A, Maák IE, Lőrinczi G, Farkas T, Kántor N, Tanács E, Kiss PJ, Juhász O, Módra G, Tölgyesi C, Erdős L, Aguilon DJ, Keppe G (2019) Karst dolines provide diverse microhabitats for different functional groups in multiple phyla. Sci Rep 9(7176):1–13

Braga JC, Bassi D, Martín JM, Riding R, Aguirre J, Sánchez-Almazo IM, Dinarées-Turell J (2006) Testing models for the Messinian salinity crisis: the Messinian record in Almería, SE Spain. Sed Geol 188–189:131–154

Bondesan A, Meneghel M, Sauro U (1992) Morphometric analysis of dolines. Int J Speleol 21(1–4):1–55

Ford D, Williams P (2007) Karst hydrogeology and geomorphology. John Wiley & Sons, Chichester

Gutiérrez F, Cooper AH Johnson KS (2008) Identification, prediction, and mitigation of sinkhole hazards in evaporate karst areas. Environ Geol 53:1007–1022

Hughes TH, Memon BA Lamoreaux PE (1994) Landfills in karst terrains. Environ Engin Geosci 31(2):203–208

Keskin I, Yilmaz I (2016) Morphometric and geological features of karstic depressions in gypsum (Sivas, Turkey). Environ Earth Sci 75 (1040):1–14

López-Chicano M (1992) Contribución al conocimiento del Sistema hidrogeológico kárstico de Sierra Gorda y su entorno (Granada y Málaga). PhD thesis, University Granada, p 492

Lyew-Ayee P, Viles HA, Tucker GE (2007) The use of GIS-based digital morphometric techniques in the study of cockpit karst. Earth Surf Proc Landforms 32(2):165–179

Pardo-Igúzquiza E, Durán JJ, Dowd PA (2013) Automatic detection and delineation of karst terrain depressions and its application in geomorphological mapping and morphometric analysis. Acta Carsologica 42(1):17–24

Sanz de Galdeano C, Lozano JA, Puga E (2008) El «Trias de Antequera»: Naturaleza, origen y estructura. Revista De La Sociedad Geológica De España 21(3–4):111–124

Sauro U (2003) Dolines and sinkholes: aspects of evolution and problems of classification. Acta Carsologica 32:41–52

Somaratne N (2014) Characteristics of point recharge in karst aquifers. Water 6:2782–2807

White WB (1988) Geomorphology and hydrology of karst terrains. Oxford University Press, p 480

# The Importance of Snow in the Hydrogeology of a High Relief Karst System: Sierra De Tendeñera, in the Pyrenees Mountain Range (Huesca, Northern Spain)

J. J. Durán-Valsero, E. Pardo-Igúzquiza, R. Morales-García, J. A. Luque-Espinar, S. R. Durán-Laforet, D. Balard, E. Quiroga, J. Borrás, and J. Ferreres

**Abstract**

The Sierra de Tendeñera is a mountain located in the Western Pyrenees, in the province of Huesca (Northern Spain). The maximum altitude is Pico Tendeñera, with 2845 m a.s.l., while the snow line is around 2600 m a.s.l. Both the landscape and a well-developed network of deep karst conduits are conditioned by the structural arrangement of a carbonate sequence with almost vertical strata, modeled by glacial and karst processes. The highland landscape is characterized by numerous closed depressions that connect with an extensive subvertical network of karst conduits. During the winter, the accumulation of snow in the high parts implies the filling of sinkholes and potholes by snow and ice, which can remain until the end of August, which has some hydrogeological implications. First, snow and ice at the bottom of sinkholes and at the entrance to potholes and caves act as a secondary reservoir of water in a suspended frozen aquifer. This implies that the base flow during the summer is higher than it would be in the absence of snow, and the minimum discharge is reached in October instead of the end of August. Second, the hydrochemistry of the water discharged from the springs has a very low content of dissolved chemicals and an unusually low electrical conductivity in karst aquifers.

J. J. Durán-Valsero · E. Pardo-Igúzquiza · R. Morales-García · J. A. Luque-Espinar
Instituto Geológico Y Minero de España-CSIC, Ríos Rosas 23, 28003 Madrid, Spain
e-mail: jj.duran@igme.es

E. Pardo-Igúzquiza
e-mail: e.pardo@igme.es

R. Morales-García
e-mail: r.morales@igme.es

J. A. Luque-Espinar
e-mail: ja.luque@igme.es

S. R. Durán-Laforet (✉)
Ed. de Investigación Ada Byron, Centro de Hidrogeología de la Universidad de Málaga (CEHIUMA), C/Arquitecto Francisco Peñalosa, 18, Málaga, Spain
e-mail: sduran@uma.es

D. Balard · E. Quiroga · J. Borrás · J. Ferreres
Espeleo Club Muntanyenc Barcelonès (ECMB), Carrer de Sant Pere Més Alt, 25, 08003 Barcelona, Spain
e-mail: espeleo@evaristoquiroga.com

J. Borrás
e-mail: jborras@ecmbarcelones.com

**Keywords**

High relief karst • Snow • Karst springs • Dolines • Perched frozen aquifer

## 1 Introduction

Alpine, high-mountain or high relief karst systems (Sauro et al. 2013) are karst massifs that reach altitudes above the snow line and that host a well-developed network of karst conduits (caves) that reach a considerable vertical development (i.e., a deep karst system in the sense of a thick vadose zone (Sauro et al. 2013)). From a hydrogeological point of view, these high relief karst systems are important karst aquifers (Ballesteros et al. 2015) a with high density of fractures and karstic conduits that affect the porosity. The presence of a well-developed network of karst conduits introduces a high anisotropy and heterogeneity in the structure of these aquifers (Fandel et al. 2022). Another important difference with between classical hydrogeology in detritic aquifers and in low-relief karst aquifers is the scarcity of experimental data (Pardo-Igúzquiza et al. 2016b, a). In fact, the high altitude and the abrupt terrain imply important problems in accessibility. Moreover, boreholes are absent, and no pumping tests have been performed. Thus, most of the hydrogeological information is provided by the discharges at karst springs, where flow, temperature and


B. Andreo et al. (eds.), *EuroKarst 2022, Málaga*, Advances in Karst Science,
https://doi.org/10.1007/978-3-031-16879-6_33

hydrochemistry can be measured (Bonacci 1987). Nevertheless, further knowledge can be gained by integrating different sources of information like speleological, climatological, and geomorphological information and geophysical and remote sensing data (Pardo-Igúzquiza et al. 2012). In this work, the preliminary results from the research of the deep karst system of Sierra de Tendeñera are provided. The methods used for the first approximation in this study are the hydrochemistry of the spring waters, and the geomorphologic analysis of the exokarst and the speleological network explored until now.

## 2 Study Area

### 2.1 Geographical Location

The Sierra de Tendeñera is located in the central part of the Pyrenees Mountain Range in the province of Huesca in Northern Spain, close to the border of Spain and France (Fig. 1). Tendeñera has a preferential development in the E–W direction (Fig. 2) with 15 km of length and 3 km of extension in the N–S direction. In the E–W direction, the Sierra is limited by the valleys of the rivers Ara in the East and Gallego in the West. Sierra de Tendeñera reaches the highest altitude at Tendeñera peak with 2855 m a.s.l. The shape of the Sierra has been sculpted by glaciers and there is a notable lineation of glaciers cirques in the E–W direction that isolate horns that give the following highest altitudes (peaks) from East to West (in meters above the sea level): Tozal del Cebollar (2177), Pico Fenés (2538), Peña de Otal (2705), Pico Tendeñera (2845), Punta de la Ripera (2814), Peña Forato (2711), Mallo de las Peñas (2657), Peña del Verde (2618), Peña Sabocos (2754), Peña Roya (2573), Peña Blanca (2557) and La Muralla (2365). The slopes have high grades, and thus, the topography is very abrupt.

### 2.2 Geology and Geomorphology

The structure of Sierra de Tendeñera is part of a thrust fault (Teixell 1992) that overlays rocks from the Upper Cretaceous, Paleocene, and Early Eocene over Paleozoic rocks (Fig. 3). The stratigraphic succession is highly dipping by a large anticline fold that corresponds to the anti-formal stack of the Axial Zone of the Pyrenees (Izquierdo-Llavall et al. 2018). The main karstic part of the Sierra is a sequence of carbonate rocks, with a width of 2 km in the N–S directions, formed by sandy limestone, dolostone in thin layers, massive limestone, and limestone with chert, reaching a thickness of around 800 m with E–W strike and high dip, almost vertical in some zones. This high dip is very important in making the vertical magnitude of karstic rocks much larger than the real thickness of the sedimentary sequence.

Another important characteristic of the sequence is that it is highly fractured and faulted, with directions N20W and N130W. These fracturing directions, together with the bedding direction, impose an important structural control in the geomorphology of Sierra de Tendeñera (Serrano Cañadas 1995). The geomorphology is dominated by important scarps and slope processes (e.g., talus) associated with the high angles of the slopes and the production of rock fragments by freezing and thawing of water in the rock cracks. An important part of the geomorphology is caused by glacial processes, as a heritage from the past glacial events that affected the area. Glacier cirques, horns, and glacial valleys dominate the sculpting of the mountain range. Additionally, there is well-developed karstic geomorphology caused by the soluble character of the carbonate rocks (Ford and Williams 2007). On the surface, there is karren at a small scale and closed depressions (dolines) at a larger scale. Underground, there is an important network of karst conduits (caves) (Cardona Oliván 1992).

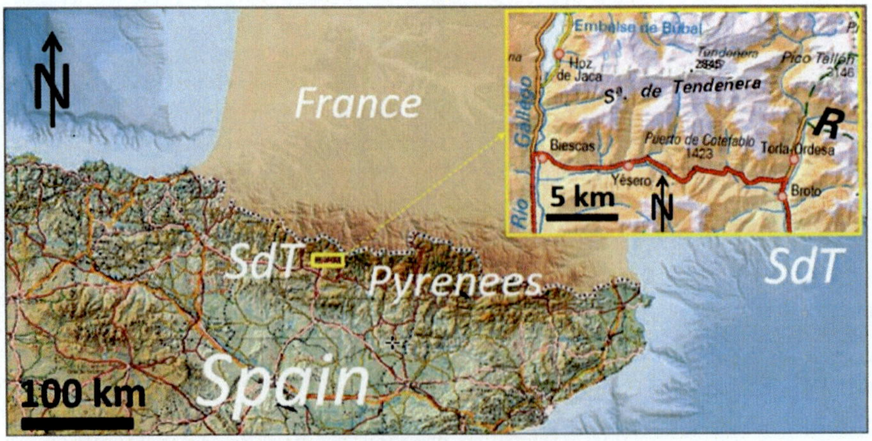


**Fig. 1** Geographical location of the high relief karst system of Sierra de Tendeñera (SdT) in the Pyrenees Mountain Range in Northeastern Spain, close to the border between Spain and France. The inset shows a detail of the SdT topography with light colors representing higher altitudes

**Fig. 2** Satellite image, from Copernicus Sentinel satellite, taken in the summer of 2019. The snow patches can be seen as turquoise blue color (white arrows), in the center of the image, mainly in glacial cirques on the north-facing side of the Sierra de Tendeñera mountain range. Water appears as black patches representing a reservoir (large polygon on the left-hand side of the image) and glacial lakes (the two small rounded polygons). The E–W development of the Sierra de Tendeñera is evident in the image

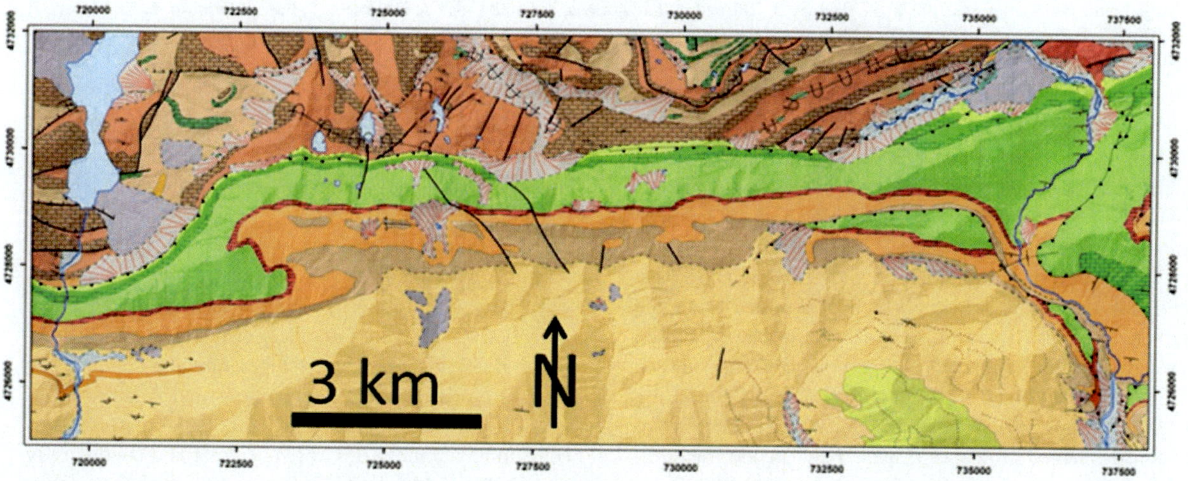


**Fig. 3** Geological map of the karst massif of Sierra de Tendeñera. The east–west band of green, red, orange, and dark brown colors represent the upper Cretaceous (greens), Paleocene (red and orange), and early Eocene (dark brown) calcareous sandstones and carbonate materials of Sierra de Tendeñera. To the north, there are Paleozoic materials, and to the south, there are Cenozoic flysch materials (light brown)

## 3 Results

In this study area, there are some geological and geomorphological studies (Serrano Cañadas 1995); Ford and Williams 2007) as well as studies published by speleologists (Cardona Oliván 1992). However, there are few published studies on the hydrogeology of Sierra de Tendeñera and geomorphological studies (Pardo-Igúzquiza et al. 2016b, a; Durán et al. 2016) that can be considering supporting works. Figure 4 shows the digital elevation model (DEM) of Sierra de Tendeñera, where the white dots represent the karst springs studied. A group of karst springs are located to the East (1, 2, 3, and 4 in Fig. 4) and discharge to the Ara River, while other karst springs are located to the West (6, 7 and 8 in Fig. 4) and discharge to the Gallego River. Spring 5 in Fig. 4 drains a perched aquifer which eventually will end in the Ara River in the East.

Geomorphological observations (Figs. 6 and 7) show significant water storage in high altitudes due to snow/ice in depressions and potholes. Also, it may be seen in Fig. 6a how the density of closed depressions is higher in the eastern part of the aquifer. This storage provides prolonged discharge during summer/autumn with a shift of the hydraulic regime. Spring measurements show low and differing electrical conductivities. This confirms significant snowmelt contribution and suggest two different sub-systems with diverging groundwater flow. This hypothesis is supported by speleological observations that indicate a higher influence of meltwater and higher karstification in the Eastern part. Thus, there were two remarkable features at the springs: an important flow of water despite being summer and the

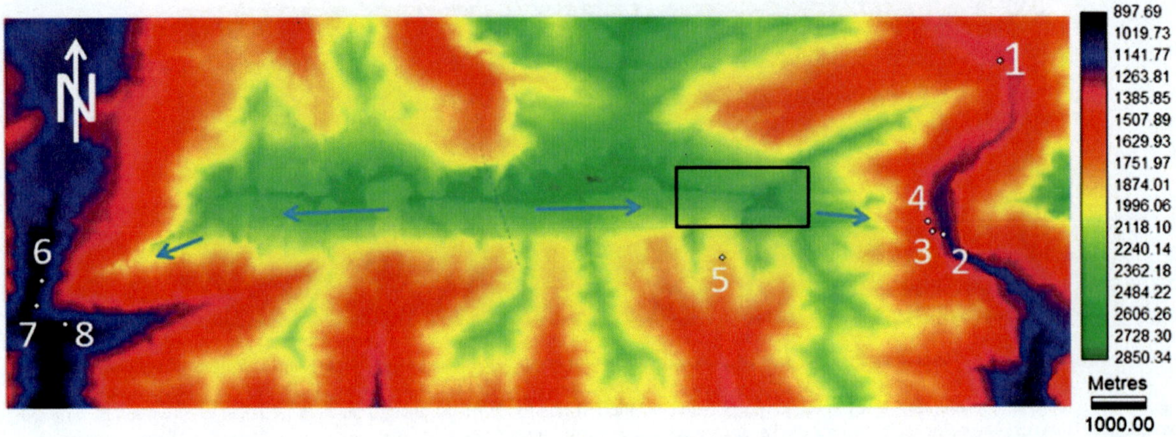


**Fig. 4** Digital elevation model of the Sierra de Tendeñera. The color bar represents altitude in meters above sea level. The white dots represent the discharge points (karstic springs): 1: Fuente Pantoja, 2: El Bozo, 3: La Gloriosa, 4: Santa Elena de Bujaruelo, 5: Fuente Carduso, 6: Espumoso, 7: Santa Elena de Biescas and 8: Batanes. The blue arrows represent the general direction of groundwater flow and details of the zone inside the black rectangle are shown in Fig. 5

chemical composition of the spring water (Table 1) shows low mineralization, especially in the karst springs of the eastern part, where there is an important development of a network of karst conduits (Fig. 6). Both facts can be explained by the effect of snow (Fig. 7a). During winter, there is important precipitation in the form of snow that accumulates forming a frozen perched aquifer during the summer. That is because of the abundance of closed depressions (dolines (Fig. 7b) and potholes), as may be seen in Fig. 5 in a map obtained by an automatic procedure using the DEM (Pardo-Igúzquiza et al. 2013). Water in the form of ice is also retained in taluses (Fig. 7c) and inside fractures. This ice stays in the summer and supplies water to the karst springs by melting during the daytime. Certainly, this effect is stronger in the eastern part, where there are more dolines and potholes storing snow, and the karst network is much more developed.

The working hypothesis in this work is that the Sierra de Tendeñera has two sub-systems, maybe hydrogeological basins (Fig. 6), one that drains toward the East and the other that drains toward the West. As no data of flow, temperature or hydrochemistry of the karst springs have been recorded so far, in July 2021 a hydrogeological campaign took place in the Sierra de Tendeñera where temperature and conductivity loggers were placed in the karst springs and water samples were taken for chemical analysis in the laboratory. It is expected to have time series of temperature and conductivity data available in August 2022. The results of the analytical hydrochemistry of water samples taken on July 14, 2021, are shown in Table 1.

**Table 1** Hydrogeochemistry of the main karst springs of Sierra de Tendeñera (July 14, 2021). Fuente Pantoja, Santa Elena de Bujaruelo (Santa E. Bujaruelo), La Gloriosa, el Bozo and Fuente Carduso are karst springs in the East part (blue color) draining to the Ara river. Espumoso, Santa Elena de Biescas (Santa E. Biescas) and Batanes 2 are karst springs in the West part (green color) draining to the Gallego river. The low mineralization of the karst springs of the eastern part can be related to the high development of the underground karst conduits in this part of the Sierra (Fig. 6b)

| Karst spring | pH | Conductivity 20° (μS/cm) | Ca (mg/L) | $Mg^{2+}$ (mg/L) | $SO_4^{2-}$ (mg/L) | $HCO_3^-$ (mg/L) |
|---|---|---|---|---|---|---|
| Fuente Pantoja | 7.47 | 230 | 22 | 13 | 29 | 96 |
| Santa E. Bujaruelo | 7.46 | 171 | 26 | 4 | 21 | 73 |
| La Gloriosa | 7.44 | 191 | 30 | 5 | 13 | 96 |
| El Bozo | 7.53 | 181 | 27 | 4 | 14 | 82 |
| Fuente Carduso | 7.56 | 208 | 34 | 5 | 19 | 103 |
| Espumoso | 7.25 | 430 | 59 | 25 | 154 | 103 |
| Santa E. Biescas | 7.20 | 434 | 58 | 27 | 119 | 154 |
| Batanes 2 | 7.40 | 195 | 25 | 7 | 31 | 71 |

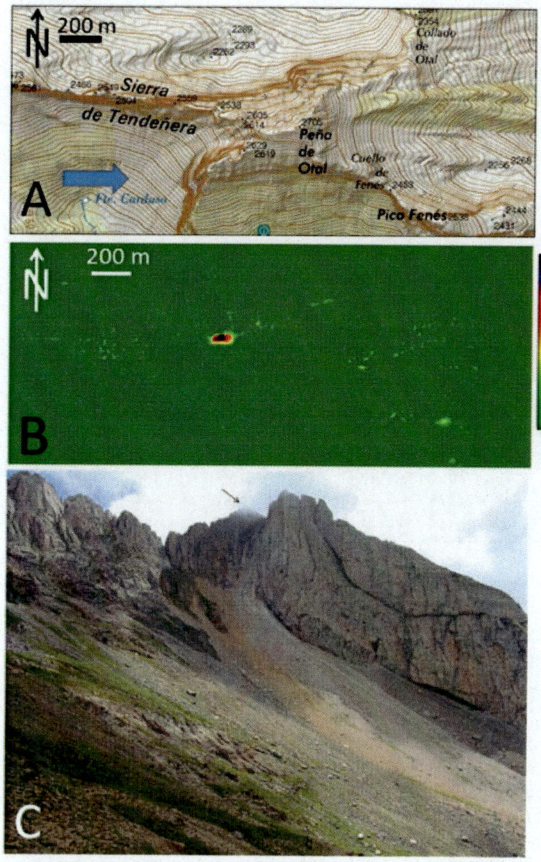


**Fig. 5** **a** Detail of the topographic map of the area inside the black square in Fig. 4. **b** Map of closed depressions (showing its abundance) of the area in **a** and calculated from a digital elevation model with a spatial resolution of 5 m. **c** Photograph of Peña de Otal (black arrow) shown in **a** and with the photograph taken from the perspective of the blue arrow in **a**

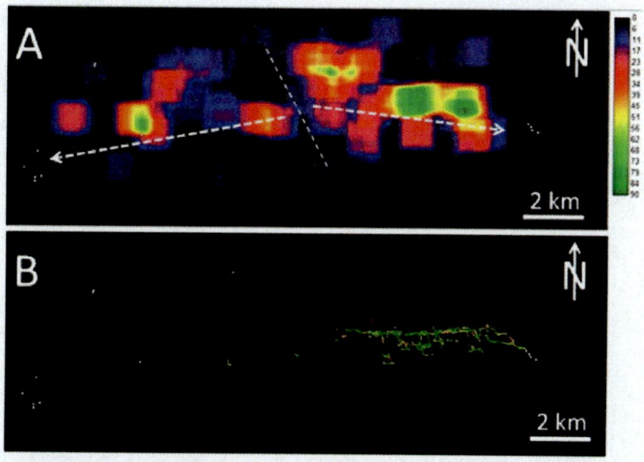


**Fig. 6** **a** Map of density of closed depressions (dolines) in number per square kilometer. This map can be used for a tentative hypothesis of establishing the divide between the two hydrogeological basins of Sierra de Tendeñera. **b** Plan view of the network of karst conduits that has been mapped by the speleologists. The eastern part of the Sierra de Tendeñera has a deep karst system that reaches −1400 m in the vertical direction

## 4 Conclusions

The Sierra de Tendeñera karst massif is a high-relief karst system with two clear sub-systems (hydrogeological basins). The exact location of the groundwater divide is under study. The recharge is autogenic; that is, recharge is derived from precipitation directly onto the karst surface. However, the precipitation is not only in the form of rainfall but snow (solid precipitation) plays an important role in the hydrogeology of the system. During winter, snow accumulates in the summit of the mountain range and penetrates into the underground along fractures, potholes and other closed depressions such as dolines and talus. With time, this snow in the underground is transformed into ice, which constitutes a reservoir of water. After the snow season and until late summer (September) the melting of the underground ice implies a contribution of fresh water that has a double effect: firstly, the recession of the karst springs is delayed and the discharge is maintained during the summer; and secondly, the fast travel time of the melting water across the network of fractures and karst conduits that connect the recharge area with the karst springs causes the chemical composition of the spring water to have very low mineralization and electric conductivity. With increasing climate change, it is expected that the snow season will decrease by increasing temperatures and decreasing rainfall, and that will imply a modification of the discharge regime that can become closer to high relief karst systems with no snow. Research on these high-relief karst systems is currently still under development.

**Fig. 7** **a** Central part of Sierra de Tendeñera in winter and covered by snow. **b** Snow (in the photo as ice) that was accumulated in a closed depression. The photo was taken in the middle of July 2021. **c** Different large taluses were observed on the steep terrain at the foot of the upper cliffs at the head of a glacier valley. The photo, taken in the middle of July 2021, shows the persistence of patches of snow at the bottom of the valley, which is facing North. It is expected that in the underground of the talus, the high porosity will store water in the form of ice

**Acknowledgements** The present work was supported by research project PID2019-106435GB-I00 of the Ministerio de Ciencia e Innovación of Spain.

## References

Ballesteros D, Malard A, Jeanning P-Y, Jiménez-Sánchez M, García-Sansegundo J, Meléndez-Asensio M, Sendra G (2015) KARSYS hydrogeological 3D modeling of alpine karst aquifers developed in geologically complex areas: Picos de Europa National Park (Spain). Environ Earth Sci 74:7699–7714

Bonacci O (1987) Karst Hydrology, Springer Series in Physical Environment, 2, Springer Verlag, p 184

Cardona Oliván F (1992) La aventura de Arañonera: al descubrimiento del Pirineo subterráneo. Espleo Club de Gràcia, p 160

Durán JJ, Balart D, Pardo-Igúzquiza E, Borrás J, Ferreres J, Quiroga E, Guillén J, Garza X, Marqués F, Robador A, Cabrera A, Robledo-Ardila P (2016) Geomorfometría y control geológico del endokarst del sector oriental de la Sierra de Tendeñera. Geo-Temas 16(1):307–310

Fandel C, Miville F, Ferré T, Goldscheider N, Renard P (2022) The stochastic simulation of karst conduit network structure using anisotropic fast marching, and its application to a geologically complex alpine karst system. Hydrogeol J. https://doi.org/10.1007/s10040-022-02464-x

Ford D, Williams P (2007) Karst hydrogeology and geomorphology. John Wiley & Sons, Chichester, p 562

Izquierdo-Llavall E, Casas-Sainz AM, Oliva-Urcia B, Villalaín JJ, Pueyo E, Scholger R (2018) Rotational kinematics of basement antiformal stacks: paleomagnetic study of the Western Nogueras Zone (Central Pyrenees). Tectonics 37(10):3456–3478

Pardo-Igúzquiza E, Durán-Valsero JJ, Dowd PA, Guardiola-Albert C, Liñan-Baena C, Robledo-Ardila PA (2012) Estimation of spatio-temporal recharge of aquifers in mountainous karst terrains: application to Sierra de las Nieves (Spain). J Hydrol 470–471:124–137

Pardo-Igúzquiza E, Durán JJ, Dowd PA (2013) Automatic detection and delineation of karst terrain depressions and its application in geomorphological mapping and morphometric analysis. Acta Carsologica 42:17–24

Pardo-Igúzquiza E, Durán JJ, Robledo-Ardila P, Luque-Espinar JA, Martos-Rosillo S, Guardiola-Albert C, Pedrera A (2016b) The karst network system of the Sierra de las Nieves (Málaga, Spain). An example of a high relief Mediterranean karst. Boletín Geológico y Minero 127(1):193–204

Igúzquiza EP, Durán JJ, Balart D, Borrás J, Ferreres J, Quiroga E, Guillén J, Garza X, Marqués A, Robador A, Robledo-Ardila C (2016a) Morfometría y control geológico de las depresiones kársticas de la Sierra de Tendeñera. Geo-Temas 16(1):303–306

Sauro F, Zampieri D, Filipponi M (2013) Development of a Deep Karst System within a Transpressional Structure of the Dolomites in North-East Italy 184:51–63

Cañadas ES (1995) Geomorfología de la Sierra de Tendeñera (Pirineo aragonés). Eria 143–158

Teixell R (1992) Estructura alpina en la transversal de la terminación occidental de la Zona Axial Pirenaica. University de Barcelona, Tesis doctoral, p 252

# The Link Between Man and Water in Karst, Through Examples From Apulia (S Italy)

M. Parise and I. S. Liso

**Abstract**

In this contribution three different sites from the Apulian karst of southern Italy are illustrated about the issue of water flowing out in caves, and how these latter represented remarkable sites for the historical development of ancient local communities. The goal of the work is to highlight the strong link between water and human presence, and the related cults, in the karst settings of Apulia. The study areas cover all main karst sub-regions of Apulia, from the Gargano promontory to the Murge plateau, and Salento peninsula, and they are examined through integration of detailed scrutiny of the historical and archaeological research and field work consisting of geological and hydrogeological surveys. Scaloria Cave, in Gargano, preserves testimonies of the water cult during Neolithic times, as documented by vessels and containers, located in remote places of the cave system. At Minervino Murge, the cave dedicated to Saint Michael Archangel, which played an important role in the pilgrimages along the main religious routes of southern Italy, also hosts places to collect water, dripping from both cave ceilings and speleothems. Eventually, at Manduria, in the Taranto province, *Fonte Pliniano* is the spring originally documented by Pliny the Elder in his work *Naturalis Historia;* given the archeological relevance of the site, the spring has always been a remarkable place of cult. Describing these sites provides interesting hints for further analysis about the connection of water with human settlements, in Apulia and in other karst areas of the Mediterranean Basin.

**Keywords**

Springs • Water • History • Apulia

M. Parise (✉) · I. S. Liso
Department of Earth and Environmental Sciences, University Aldo Moro, Via Edoardo Orabona, 4, 70125 Bari, Italy
e-mail: mario.parise@uniba.it

## 1 Introduction

The availability of water in karst environments, and the chance to use this natural resource, have always played a prominent role in the choice of sites for human settlements (Aley 2000; Laureano 2001; Bakalowicz 2005; Parise and Sammarco 2015). Especially in arid or semi-arid climatic conditions, water can be very rare at the surface, so that the underground becomes the place where searching for it, and for establishing sites for religious and/or pagan cults related to water as well. As documented by a large number of research worldwide, and by the remarkable underground works performed for tapping, transporting, and distributing the hydric resource (Ashby 1935; De Feo et al. 2012; Parise et al. 2013, 2015a; Voudouris et al. 2013), the history of man is therefore definitely interrelated with water (Valipour et al. 2020).

In this contribution three different sites from the Apulian karst of southern Italy are presented. At many sites in this karst region, water flows out in caves, which have therefore represented since long times remarkable sites for the local communities, and their historical evolution as well. The study areas cover the three main karst sub-sectors of Apulia, namely, the Gargano promontory to the North, the Murge plateau in the central part of the region, and the Salento peninsula farther South (Fig. 1).

## 2 Case Studies

The first site is Scaloria cave (for location, see Fig. 1), in the Gargano promontory, where testimonies of the cult of water during Neolithic times have been documented by archeological sources (Tinè and Isetti 1980; Elster et al. 2016). Among the hundreds of karst caves of Gargano, some of the most remarkable, from an archaeological standpoint, are located at its western and southern rims, the closest to the routes followed by communities to reach the Adriatic Sea, in order to travel toward the east.


B. Andreo et al. (eds.), *EuroKarst 2022, Málaga*, Advances in Karst Science,
https://doi.org/10.1007/978-3-031-16879-6_34

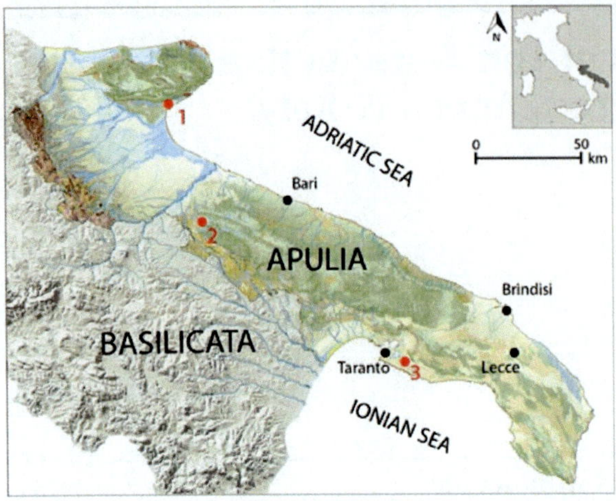


**Fig. 1** Location map of Apulia, showing (red circles) the three case studies: (1) Scaloria Cave (Manfredonia, Gargano), (2) Saint Michael Archangel Cave (Minervino Murge, Murge plateau), and (3) *Fonte Pliniano* (Manduria, Salento)

Scaloria Cave (registered as PU 266 in the regional cadastre of caves; see http://www.catasto.fspuglia.it) opens in the Manfredonia municipality. It is actually part of a larger karst complex also consisting of Occhiopinto Cave (PU 265), linked to Scaloria through a single passage (Fig. 2). The overall system is about 700 m long, and more than 30 m deep; it develops through a series of typically narrow passages, until reaching the sea level. In the lowest sector of the system, many archeological surveys have been carried out in different phases of the last century (Tinè 1972; Tinè and Isetti 1980; Firpo et al. 1990; Whitehouse 1990), documenting the presence of pottery (vessels and other containers) in sites close to the cave lakes in the lower sector of the system, or below dripping stalactites; given the difficult paths to reach such sites, these were interpreted as linked to the cult of water rather than to the daily need to provide with drinkable water.

Moving to the south, in the widest karst sub-sector of Apulia, in the High Murge plateau there is the cave dedicated to Saint Michael Archangel (PU 30 in the regional cadastre of caves; see http://www.catasto.fspuglia.it; location shown in Fig. 1), at the outskirts of the town of Minervino Murge. Today a religious site (Fig. 3), it is one of the many caves in Italy dedicated to the cult of Saint Michael Archangel; among these, it is worth to cite also the cave at Monte Sant'Angelo, in Gargano, which was included since 2011 in the UNESCO World Heritage List. The cave at Minervino Murge has been adapted by man in its first sector, by building a deep stairway to reach, at the base of the main room in the cave, the altar made by beautiful red breccias (Fig. 3), the same ornamental stone that was used for the main portal at the nearby Castel del Monte, another UNESCO world heritage site.

Behind this altar, a natural unmodified cave represents what is left of the original karst system (Fig. 4), extending for some additional tens of meters.

Geologically, the whole cave, and its surroundings as well, are within the Altamura Limestone Fm., a Cretaceous micritic limestone belonging to the Apulian carbonate platform, and now making up the backbone of Murge (Ciaranfi et al. 1988). With a sub-horizontal setting, the up-to-30 m

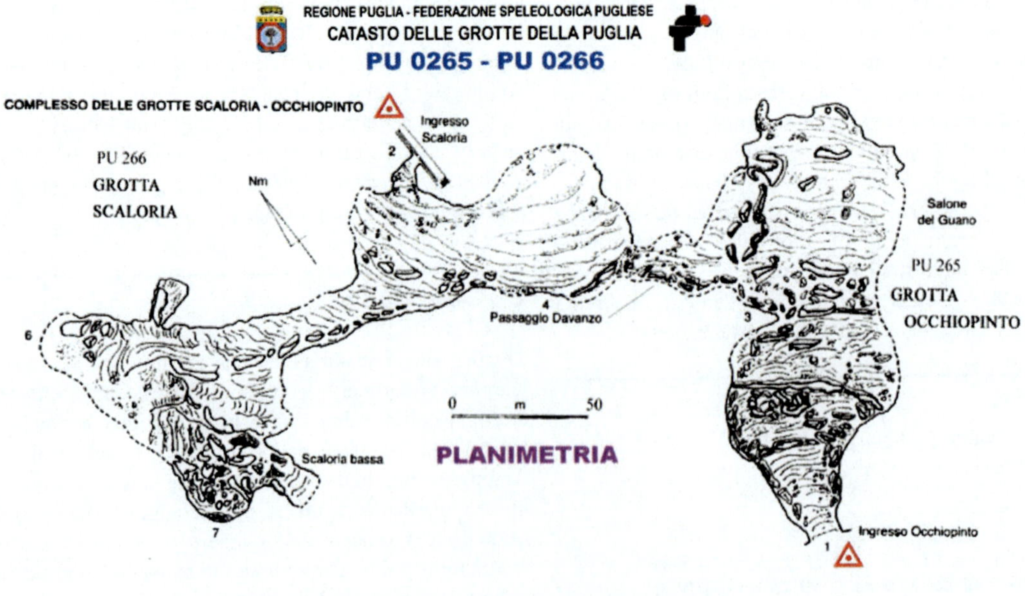


**Fig. 2** The karst system Scaloria–Occhiopinto, in Gargano (survey by the Apulian Speleological Federation; after http://www.catasto.fspuglia.it)

**Fig. 3** The church/cave of Saint Michael Archangel at Minervino Murge: left, view from the outside; right, the beautiful altar, built with a carbonate breccia in red matrix

**Fig. 4** Saint Michael Archangel cave: left, view of the entrance of the natural cave behind the altar, also showing the holy water font; right, detail of the water font, largely covered in calcite

high walls of the cave are strongly interested by rock failures, with the possibility of detachment of blocks of variable size, due to several sets of primary (bedding) and secondary (joints, fractures, and faults) discontinuities.

Besides the religious importance, the Saint Michael Archangel is a cave of interest for the cult of water: it actually hosts several sites to collect the water, dripping down from both the cave ceiling and the speleothems. A holy water font, entirely covered by calcite, is for instance visible at the entrance of the natural cave, just behind the altar (Fig. 4). The Saint Michael Archangel cave has played an important role in history and in the pilgrimages along the main religious routes of southern Italy (Russo 2016).

Eventually, the third case study is *Fonte Pliniano* (Plinian Source; Fig. 5), located at Manduria municipality, in the Taranto province (for location, see Fig. 1). It is a spring originally documented by Pliny the Elder in his work *Naturalis Historia.* Today the spring is within the Archeological Park of Manduria, where remains of the ancient city walls are preserved. These are named *Mura Messapiche* (Messapian Walls), after *Messapi*, the indigenous population living in southern Apulia before the Romans. They funded *Mandyrion* (the old Manduria) between fifth and third century B.C.

Several swallow holes and caves of limited size develop within the area, due to the specific geological characteristics

**Fig. 5** General view of the *Fonte Pliniano* at Manduria

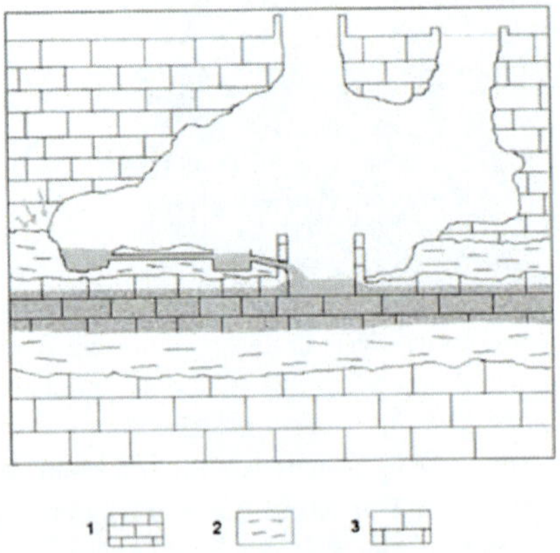

**Fig. 6** Simplified sketch of *Fonte Pliniano* at Manduria (after Maggiore and Salerno 2007). Key: (1) calcarenite; (2) clay; (3) limestone

(low strength of the calcarenite deposits), dominated by karst processes. Generally, falls and failures tend to close the voids, originating the many cover collapse sinkholes (*sensu* Gutierrez et al. 2014; Parise 2019, 2022) recognizable in the territory of Manduria and its surroundings (Lacaita 1967; Giaccari 2001).

From a hydrogeological standpoint, the main groundwater is contained within the Cretaceous limestones bedrock (Ricchetti 1972; Ciaranfi et al. 1988); above this, interlayering among the Plio-Pleistocene calcarenite and sand deposits, with less permeable materials such as silts and clays, originates several minor perched aquifer, as typical of many sectors in southern Apulia (Cotecchia 2014). The *Fonte Pliniano* spring comes out in the overlying Upper Pliocene–Lower Pleistocene calcarenite deposits (Fig. 6); within this latter it is possible to recognize a series of water venues, due to the presence of marly-clay interlayers (Fig. 7).

More than a spring, *Fonte Pliniano* should be more correctly described as a bell-shaped well, dug by man within the Gravina Calcarenite Fm., down to a depth of 9 m. The overall cave is of circular shape, about 15 m in diameter, and takes light from a wide skylight in the vault. In its specific case, water at *Fonte Pliniano* comes out in correspondence of the shallower aquifer, and namely at the contact between the calcarenites and an underlying clay layer, and, at the bottom of the cave, in a calcarenite layer contained between two clay layers. Overall, the total discharge is 0.5 l/s.

The first hydrogeological studies in the Manduria area (De Giorgi 1884, 1922) pointed out to the presence of three

**Fig. 7** Details of the springs at *Fonte Pliniano*: left, water dripping from the wall at the contact between calcarenites and the silty and clay interlayers; right, checking the water parameters with a field probe

shallow aquifers, at a depth ranging from 7–9 m, 10–12 m, and 20–25 m. Since the main groundwater surface floats on the seawater, the content of chlorides tends to increase moving from the surficial to the deeper aquifers (the coastline is about 10 km far). The hydro-geochemical analysis performed at the water from *Fonte Pliniano* pointed out to its character of calcium bicarbonate water, while, based upon the total dissolved solid values (<0.2 g/l), they are considered as soft water (Giaccari 2001). This indicates (1) the brief path and short resident time of rainwater within the ground before coming out, and (2) the absence of any mixing with brackish or seawater at greater depth (Drew and Goldscheider 2007; Ford and Williams 2007).

## 3 Discussion

Apulia, an elongated region of south-eastern Italy, making the heel of the Italian boot, has always been a bridge toward the East, and a site of passage for many ancient populations. These latter inevitably came in contact with the indigenous people of Apulia (*Daunii, Peucetii, Japigii, Messapi*) and with their ability and expertise in finding and collecting water in the semi-arid environment of the region. The long history of the sustainable use of water resources is worth to be further deepened, by adding other interesting sites such as Grotta della Poesia Grande, in a district where still today a dialect very close to ancient Greek is spoken (Delle Rose and Parise 2005), or the many villages where wells and cisterns were built in the past to allow conservation of water near the surface during the hot season (Costantini 1988; Lopez et al. 2009; Parise 2009). A further development of the research could be the comparison of the Apulian case studies with those from karst areas in other countries in the Mediterranean basin.

## 4 Conclusions

Through illustrations of the three described cases, the goal of this work is to highlight the strong connection between water and human presence, and related cults, in karst settings. The history of man, and the development of cultures and communities, have been always and everywhere linked with availability of water. The need to use it in a sustainable way pushed communities to be creative in finding the best solution for the conservation of the precious freshwater resource (Bakalowicz 2005; Brinkmann and Parise 2012; Valipour et al. 2020). For these reasons, looking for ancient springs, or for site in karst systems where water was object of cult, provides remarkable data and information about hydrogeology, also in the context of climate changes we are nowadays living, as an effect of many human actions (Parise et al. 2015b). As documented by the many still functioning ancient aqueducts, some of these old structures could be still used also as a water supply for drinkable, agricultural or industrial purposes, and this could help to face some challenges we are today facing (Howard 2015; Stevanovic 2015), for example the scarcity of water, especially in karst environments.

## References

Aley T (2000) Water and land-use problems in areas of conduit aquifers. In: Klimchouk AB, Ford DC, Palmer AN, Dreybrodt W

(eds) Speleogenesis. National Speleological Society, Huntsville, Alabama, Evolution of karst aquifers, pp 481–484
Ashby T (1935) The Aqueducts of Ancient Rome. Clarendon Press, Oxford, UK
Bakalowicz M (2005) Karst groundwater: a challenge for new resources. Hydrogeol J 13:148–160
Brinkmann R, Parise M (2012) Karst environments: problems, management, human impacts, and sustainability. J Cave Karst St 74(2):135–136
Ciaranfi N, Pieri P, Ricchetti G (1988) Note alla carta geologica delle Murge e del Salento (Puglia centro-meridionale). Mem Società Geol Italiana 41:449–460
Costantini A (1988) Del modo di conservare le acque e la neve. Pozzelle e neviere del Salento leccese. Sallentum 1–2:71–88
Cotecchia V (2014) Le acque sotterranee e l'intrusione marina in Puglia: dalla ricerca all'emergenza nella salvaguardia della risorsa. Mem Descrittive Carta Geol d'Italia 92:416
De Giorgi C (1884) Cenni di geografia fisica della provincia di Lecce, Lecce
De Giorgi C (1922) Descrizione geologica e idrografica della provincia di Lecce. Tip. Salentina F.lli Spacciante, Lecce
Delle Rose M, Parise M (2005) Speleogenesi e geomorfologia del sistema carsico delle Grotte della Poesia nell'ambito dell'evoluzione quaternaria della costa Adriatica Salentina. Atti Mem Comm Grotte "E. Boegan" 40:153–173
Drew D, Goldscheider N (eds) (2007) Methods in Karst Hydrogeology. Taylor & Francis/ Balkema, London, p 264
De Feo G, Laureano P, Mays LW, Angelakis AN (2012) Water Supply Management Technologies in the Greek and Roman Civilizations. Evolution of Water Supply throughout Millennia. IWA Publishing, London, pp 351–382
Elster ES, Isetti E, Robb J, Traverso A (2016) The archaeology of Grotta Scaloria: ritual in Neolithic southeast Italy. Monumenta Archeologica 38
Firpo M, Rellini I, Ciampalini A (1990) Geoarchaeological and micromorphological studies of Scaloria Cave. In: Scaloria Cave, a Neolithic world in the 6th millennium B.C., Los Angeles
Ford DC, Williams PW (2007) Karst Geomorphology and Hydrology. Wiley, Chichester
Giaccari E (2001) Studio idrogeologico delle falde idriche superficiali del territorio di Manduria, in provincia di Taranto. Memorie Società Geologica Italiana 56:213–218
Gutierrez F, Parise M, De Waele J, Jourde H (2014) A review on natural and human-induced geohazards and impacts in karst. Earth Sci Rev 138:61–88
Howard KWF (2015) Sustainable cities and the groundwater governance challenge. Environ Earth Sci 73:2543–2554
Lacaita A (1967) Manifestazioni carsiche e pseudocarsiche nei tufi pliocenici di Manduria. Note Manduriane 1:45–49
Laureano P (2001) Water atlas. In: Traditional knowledge to combat desertification, Bollati Boringhieri, Torino
Lopez N, Spizzico V, Parise M (2009) Geomorphological, pedological, and hydrological characteristics of karst lakes at Conversano (Apulia, southern Italy) as a basis for environmental protection. Environ Geol 58(2):327–337
Maggiore M, Salerno SE (2007) Caratteri idrogeologici del Fonte Pliniano, Manduria (TA). Grotte e Dintorni 14:33–42
Parise M, Galeazzi C, Germani C, Bixio R, Del Prete S, Sammarco M (2015a) The map of ancient underground aqueducts in Italy: updating of the project, and future perspectives. In: Proceedings international congress artificial cavities "Hypogea 2015a", Rome, pp 235–243
Parise M, Ravbar N, Živanovic V, Mikszewski A, Kresic N, Mádl-Szonyi J, Kukuric N (2015b) Hazards in karst and managing water resources quality. In: Stevanovic Z (ed) Karst aquifers: characterization and engineering. Professional practice in earth sciences, Springer, Heidelberg, Germany, pp 601–687
Parise M (2019) Sinkholes. In: White WB, Culver DC, Pipan T (eds) Encyclopedia of caves. Academic Press, Elsevier, 3rd (ed) pp 934–942
Parise M (2022) Sinkholes, subsidence and related mass movements. In: Shroder JJF (ed) Treatise on geomorphology, vol 5. Elsevier, Acad Press, pp 200–220. ISBN 9780128182345
Parise M (2009) Lakes in the Apulian karst (Southern Italy): geology, karst morphology, and their role in the local history. In: Miranda FR, Bernard LM (eds) Lake pollution research progress. Nova Science Publishers Inc, New York, pp 63–80
Parise M, Galeazzi C, Germani C, Sammarco M (2013) Hydraulic works: the map of the ancient underground aqueducts. Opera Ipogea 1:21–28
Parise M, Sammarco M (2015) The historical use of water resources in karst. Environ Earth Sci 74:143–152
Ricchetti G (1972) Osservazioni geologiche e morfologiche preliminari sui depositi quaternari affioranti nel F. 203 Brindisi. Boll Soc Naturalisti Napoli 81:543–566
Russo R (2016) Minervino Murge. Storia, cultura, turismo. Editrice Rotas, Barletta, p 46
Stevanovic Z (ed) (2015) Karst aquifers: characterization and engineering. In: Professional Practice in Earth Sciences, Springer, Heidelberg, Germany
Tinè S (1972) Culto neolitico delle acque nella Grotta Scaloria. In: Actes symposium international Les réligions de la Préhistoire, Capo di Ponte, pp 185–190.
Tinè S, Isetti E (1980) Culto neolitico delle acque e recenti scavi nella Grotta Scaloria. Bollettino Di Paletnologia Italiana 82:31–70
Valipour M, Abdelkader TA, Antoniou GP, Sala R, Parise M, Salgot M, Sanaan Bensi N, Angelakis AN (2020) Sustainability of underground hydro-technologies: from ancient to modern times and toward the future. Sustainability 12:8983
Voudouris K, Christodoulakos Y, Stiakakis M, Angelakis AN (2013) Hydrogeological characteristics of Hellenic Aqueducts-like qanats. Water 5:1326–1345
Whitehouse R (1990) Caves and cult in Neolithic Southern Italy. Accordia Res Pap 1:19–37

# Hydrogeological Setting of Las Loras UNESCO Global Geopark (Palencia-Burgos, Spain): State of Knowledge and Needs for Water Resources Sustainability Research

A. de la Hera-Portillo, M. Llorente, J. Lopez-Gutiérrez, Karmah Salman, Jose Ángel Sánchez, Nicolás Gallego Rojas, D. Ballesteros, L. Moreno, M. M. Corral Lledó, E. Galindo Rodríguez, Marwan Ghanem, Alsharifa Hind Mohammad, Badiaa Chulli, Nour-Eddice Laftouhi, and Fagr Khamis Abdel-Gawad

**Abstract**

Las Loras UNESCO Global Geopark (UGGp) is located in the contact between the southernmost sector of the Basque-Cantabrian Basin and the North of the Cenozoic Duero basin. The bedrock is mainly made up of carbonate aquifers, mainly limestones, dolomites, and marls. Hydrogeologically, these aquifers are shared by the Ebro basin (Páramo de Sedano y Lora groundwater body, ES091MSBT002) and the Duero basin (Quintanilla-Peñahorada-Las Loras groundwater body, ES020MSBT 00400004), being of the order of 80% of the UGGp drained towards the latter. This work aims to present a state-of-the-art overview of the regional hydrogeological knowledge and identifies the main lines of work for a detailed hydrogeological characterization of the Las Loras UGGp. The work methodology, includes a bibliographic review of the materials published in various sources, both official and sources from other non-profit entities, that

A. de la Hera-Portillo (✉) · M. Llorente · J. Lopez-Gutiérrez · L. Moreno · M. M.C. Lledó · E. G. Rodríguez
Consejo Superior de Investigaciones Científicas (IGME-CSIC), Centro Nacional Instituto Geológico Y Minero de España, Ríos Rosas 23, 28003 Madrid, Spain
e-mail: a.delahera@igme.es

M. Llorente
e-mail: m.llorente@igme.es

J. Lopez-Gutiérrez
e-mail: j.lopezgu@igme.es

L. Moreno
e-mail: l.moreno@igme.es

M. M.C. Lledó
e-mail: mm.corral@igme.es

E. G. Rodríguez
e-mail: e.galindo@igme.es

K. Salman · J. Á. Sánchez
Las Loras UNESCO Global Geopark, Av. Cervera, 26, 34800 Aguilar de Campoo Palencia, Spain
e-mail: geoloras@gmail.com

J. Á. Sánchez
e-mail: geoloras@gmail.com

N. G. Rojas
Asociación Argeol, C/ La Fuente 11, San Martín de Perapertú, Palencia, Spain
e-mail: nicolasgallegorojas@gmail.com

D. Ballesteros
Department of Geodynamics, University of Granada, Campus de Fuentenueva s/n, 18071 Granada, Spain
e-mail: dballesteros@ugr.es

M. Ghanem
Birzeit University, Ramallah, Palestine
e-mail: marwan.ghanem2012@gmail.com

A. H. Mohammad
Jordan University, Amman, Jordan
e-mail: alsharifahind@gmail.com

B. Chulli
Water Research and Technologies Centre, University of Carthage, Teknoparc, Carthage, Tunisia
e-mail: bchoulli@yahoo.fr

Nour-E. Laftouhi
Faculty of Sciences Semlalia of Marrakech, Earth Sciences Department Morocco, Cadi Ayyad University, Marrakesh, Morocco
e-mail: laftouhi@gmail.com

F. K. Abdel-Gawad
Water Pollution Research Department, Centre of Excellence for Research and Applied Studies On Climate Change and Sustainable Research, National Research Center, Cairo, Egypt
e-mail: fagrabdlgawad@gmail.com


B. Andreo et al. (eds.), *EuroKarst 2022, Málaga*, Advances in Karst Science,
https://doi.org/10.1007/978-3-031-16879-6_35

have contributed to regional and local knowledge in the region. The capture of data from official sources, mainly Duero River Basin Authority (CHD) and the IGME-CSIC databases, and their treatment, allowed the obtaining of a snapshot of the state of the aquifers on a regional scale. The results lead to the need for a hydrogeological study that includes recent trends and the assessment of the impact of global change on groundwater resources. Some elements of hydrogeological interest are identified as Sites of Geological Interest, and are compiled on Las Loras UGGp website. Other, potential new points are commented in this work. These results are part of the IGCP-730 project funded by UNESCO's International Geosciences Program.

**Keywords**

Groundwater • Las Loras UNESCO Global Geopark • Hydrogeology • Karst • Southern edge of Cantabrian basin

## 1 Introduction, Objectives and Methodology

Las Loras region was incorporated into the UNESCO European Geoparks Network in 2017. Three core aspects are highlighted in the Las Loras UGGp: the prominence of the impressive relief of its extensive limestone moors, crossed by large valleys, the slowness of the transformations in an area of low-intensity land use and built-up areas, and the predominantly agricultural economic background. It presents a close relationship between geological heritage and cultural patrimony. It is an area where conservation activities require prior hydrogeological studies. The project IGCP-730 provides an umbrella to support this approach, carried out through a collaboration between the CN Instituto Geológico y Minero de España (IGME-CSIC) and the Las Loras UGGp. This work aims to present a state-of-the-art overview of the regional hydrogeological knowledge and identifies the main lines of work for a detailed hydrogeological characterization of the Las Loras UGGp. It aims to the work methodology which includes a bibliographic review of the materials published in various sources, both official and sources from other non-profit entities that have contributed to regional and local knowledge in the region. The capture of data from official sources, mainly Duero River Basin Authority (CHD) and the IGME-CSIC database, and their treatment allowed the obtaining of a snapshot of the state of the aquifers on a regional scale. A first review was done in De la Hera-Portillo et al. (2022) for surface water bodies. This work focuses on the groundwater bodies and their state of knowledge concerning groundwater dynamics and quality within the Las Loras UGGp.

## 2 Study Area

The Las Loras UGGp (900 km$^2$, 16 municipalities) is located in the southernmost sector of the Basque–Cantabrian basin and the North sector of the Duero basin (Fig. 1). Most of the population in the territory is settled in the western area, being Aguilar de Campoo the biggest city (6749 inhabitants in 2020; EpData). The climate is Mediterranean with long winters and short summers, with an average annual precipitation in Aguilar de Campoo (1985–2015) of 585 mm.

The rocks appearing in the area are of Mesozoic age, mainly from the Upper Cretaceous. The most remarkable elements of the geological structure of Las Loras UGGp are the Burgalesa platform, the Ubierna Fault System and the Folded Band (Fig. 1).

The Ubierna Fault separates two areas geologically, geomorphologically and structurally, well differentiated (Figs. 1 and 4): to the north, the great unit of the Paramera de la Lora, which corresponds to a syncline of great dimensions and gentle dips, which the intense erosion has highlighted; and to the south, Las Loras, a profusion of perched synclines (Loras) and interspersed valleys (combes) that give rise to a singular broken landscape (Salman et al. 2015). Although both areas present a similar landscape, they are two well-differentiated areas in terms of their altitude and drainage. While the maximum altitude in Las Loras reaches 1362 m in Peña Amaya, and 1200 m in the rest of the Loras, in La Paramera, it is between 1,000–1,100 m. Martínez-Arnáiz (2013) identified 12 Loras (Fig. 2) within the Folded Band, and two moors within the Burgalesa Platform. A Lora is a Mesozoic moorland-hanging syncline widely represented in this area (Fig. 3). In the La Paramera, the drainage of the rivers and groundwater is towards the Ebro River (towards the northeast) with the exception of the western sector of the Las Loras UGGp, this means the Scientific Protected Sites of Covalagua (Fig. 2), where the drainage is towards the Duero basin. Meanwhile, in Las Loras, the drainage of the rivers is to the Duero basin.

The main carbonate formations in the Las Loras UGGp are Upper Cretaceous formations. The Jurassic carbonate formations, underlying Cretaceous carbonate and detrital deposits, are more abundant in the western sector of the Las Loras UGGp and used to be associated with thrusting structures (Fig. 4). The piezometric level in this area is close to the ground level with abundance occurrence of springs. The drainage of these formations gives rise to the sources of the Hurón, Rudrón, Odra, Lucio or de la Hoz, and Urbel rivers among others shorter watercourses.

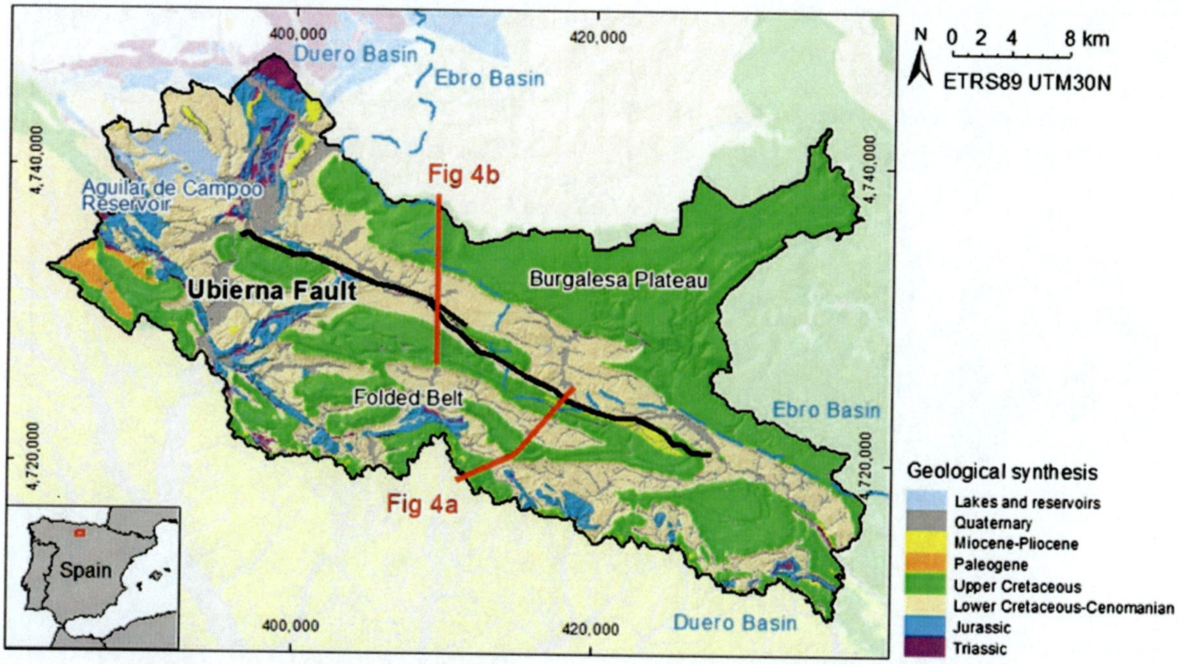


**Fig. 1** Location of the Las Loras UGGp in the north of Spain and geological map with the main structural elements highlighted: Ubierna fault, Folded band, and Burgalesa structural platform. Due to space reasons, the legend of the map can be consulted on the Las Loras Geopark website (http://geoparquelasloras.es/)

## 3 Hydrogeological Setting

The territorial scope of Las Loras UGGp extends over two groundwater bodies (IHP, 2008) that constitute the same permeable geological formation, shared by both basins: Quintanilla-Peñahorada-Las Loras GWB, corresponding to the Duero River Basin; and the Páramo de Sedano y Lora GWB, corresponding to the Ebro River Basin (Fig. 5). Both are carbonate formations that work in a karst regime as evidenced by the existence of sinkholes, siphons and caves.

Las Loras UGGp includes 54 springs (Fig. 5) but their status is unknown. Most are still active today. A review of its condition (quality and quantity) is necessary. The most important spring associated with the Páramo de Sedano y Loras groundwater body (GWB) is the source of the Rudrón River. Within the Quintanilla-Peñaharada-Las Loras GWB, the most prominent spring is Fuente Urbel (Lopez Geta et al. 1998), although there are several sources of rivers, for example, Lucio, Odra and Urbel Rivers. The highest concentration of springs in Las Loras UGGp is located in the southeastern sector of the Las Loras UGGp, associated to the Loras.

The Páramo de Sedano y Lora GWB is made up of sparsely deformed platforms made up mainly of Cretaceous materials, affected by very soft and wide folds that form wide structural tables. This arrangement in platforms perched by rivers translates into the disconnection of the main aquifers of the GWB (CHE). This behaviour requires further research.

The Páramo de Sedano y Lora GWB is recharged through the infiltration of precipitation that is collected in the permeable outcrops of the Cretaceous páramos of the surrounding area. The Ebro River constitutes the natural drainage of the northern limit of the groundwater body. Its elevation decreases throughout its transit through the unit from 660 m asl up to 500 m asl. At the SW end of the Sedano syncline, the Hurón river, which runs from south to north, is completely infiltrated at the level of the bioclastic calcarenites of the upper-middle Santonian, to reappear at a distance of approximately 1200 m downstream in Barrio Panizares at an elevation of 860 masl and give rise to the source of the Rudrón River (CHE), within the Sedano y Lora GWB. This section is known as the blind valley located in a core section of Las Loras UGGp no recognized as GWB (Fig. 5). The groundwater flows converge towards the rivers of the hydrographic network to discharge into the rivers of the Ebro River basin. The marked karst character is shown in the occurrence of important cave complexes among which the Orbaneja del Castillo system (25 km long) and the Basconcillos del Tozo Caves (3.4 km long).

The hydrochemical facies of the Páramo de Sedano y Lora GWB present bicarbonate-calcium facies, with medium

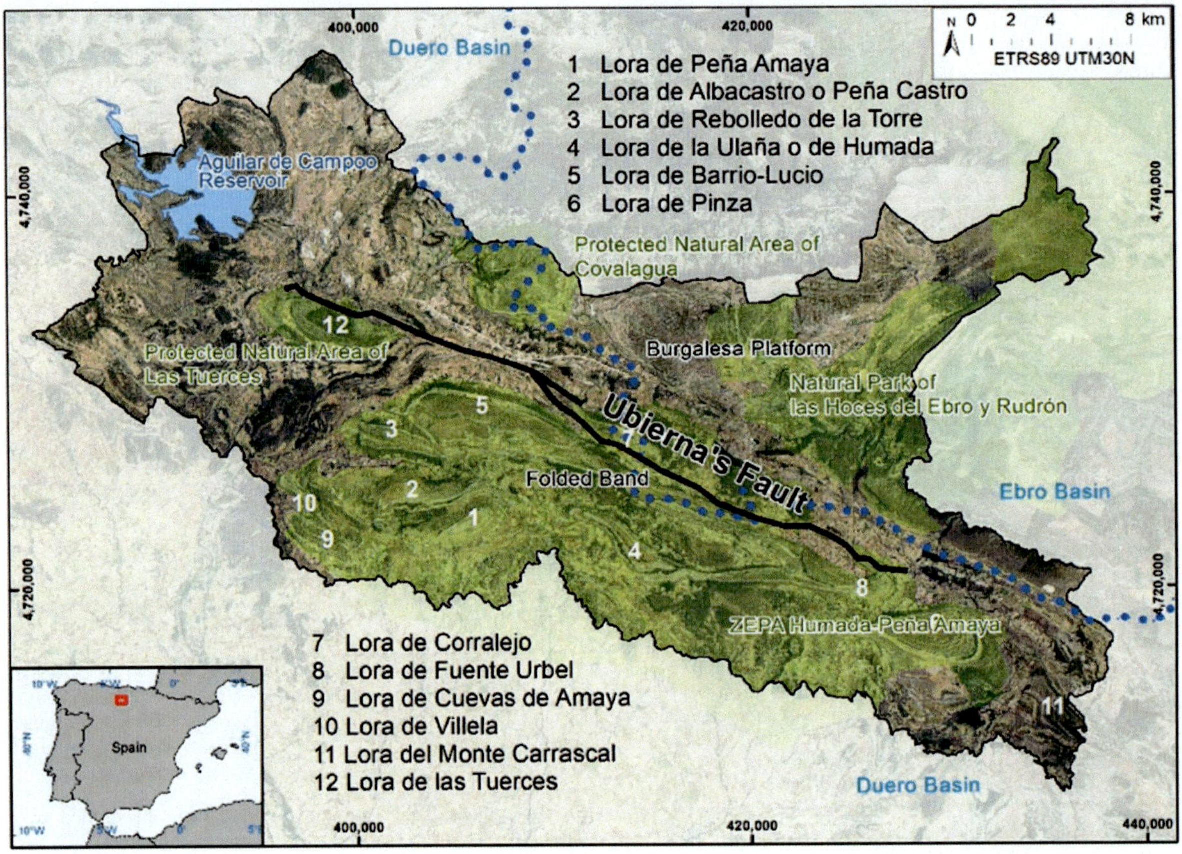


**Fig. 2** Main protected areas within the Las Loras UGGp and identification of the Loras in the southern sector of Ubierna Fault

hardness and moderate to low mineralization. The electrical conductivity values are between 300 and 600 µS/cm (CHE).

The main aquifers of Quintanilla-Peñahorada-Loras GWB are Turoniense, Cenomaniense and Santoniense. The discharge takes places through the main rivers: Urbel, Odra, Bulles and Lucio; and a good number of springs, some of them associated with travertine systems, which generate waterfalls. These sites present a high hydrogeological significance and require special protection; the state of knowledge is still pending of study and is currently under research. In other cases, some springs have disappeared probably caused by the increase of groundwater abstractions in the last decade, as shown by the existing records for Quintanilla-Peñadorada-Las Loras GWB. This GWB presents good quality waters, with calcium bicarbonate facies. The Ubierna Fault is the main geological structure that determines the hydrogeology of the Quintanilla-Peñahorada-Las Loras GWB. There is evidence, which shown the potentially associated hydrothermalism of groundwater. The Ubierna fault is a complex fault system, and this aspect requires a specific study that is being carried out under the IGCP-730 Project.

## 4 Historical Evolution of Water Uses

Las Loras region is characterized by the abundance of water resources. Its proximity to the Cantabrian Sea ensures a rainy regime throughout the year that favours the presence of a water table close to the surface of the land. However, for a decade, certain indications have been observed that lead to changes in the described scenario. More and more people are resorting to wells and boreholes to ensure water resource. In addition, the desiccation of a certain number of wetlands has been observed, as well as a decrease in the levels of the few existing groundwater wells.

**Fig. 3** Ulaña Lora (no. 4 in Fig. 2), from highway from the Villadiego to Humada

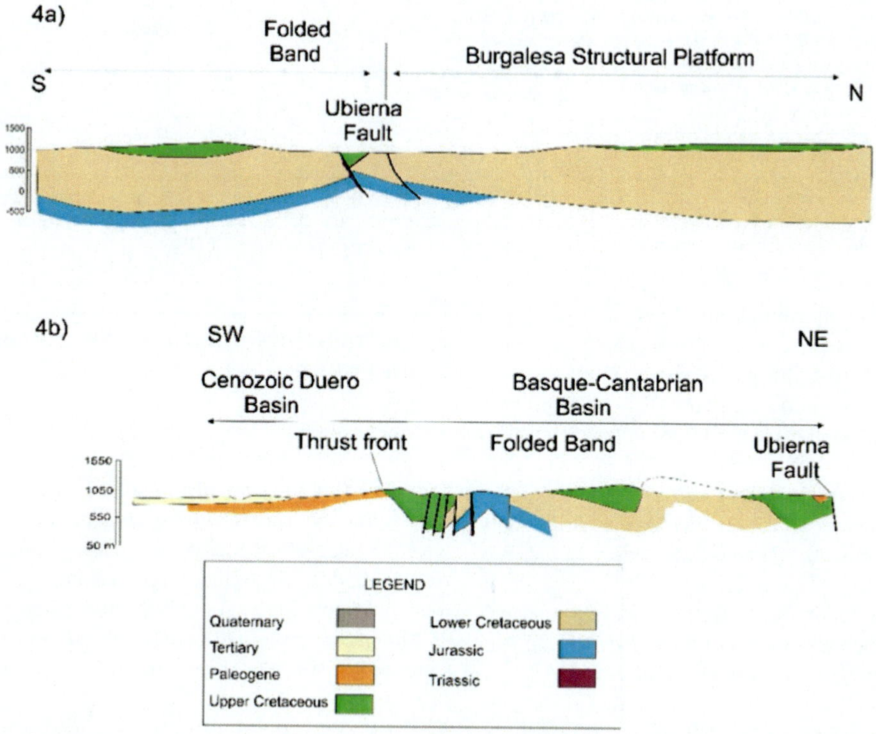


**Fig. 4** Geological cross-sections of the Las Loras UGGp **a** modified from Pendas y Menéndez 1998, and **b** modified from Pineda Velasco 1990. Cross-sections are indicated in Fig. 1

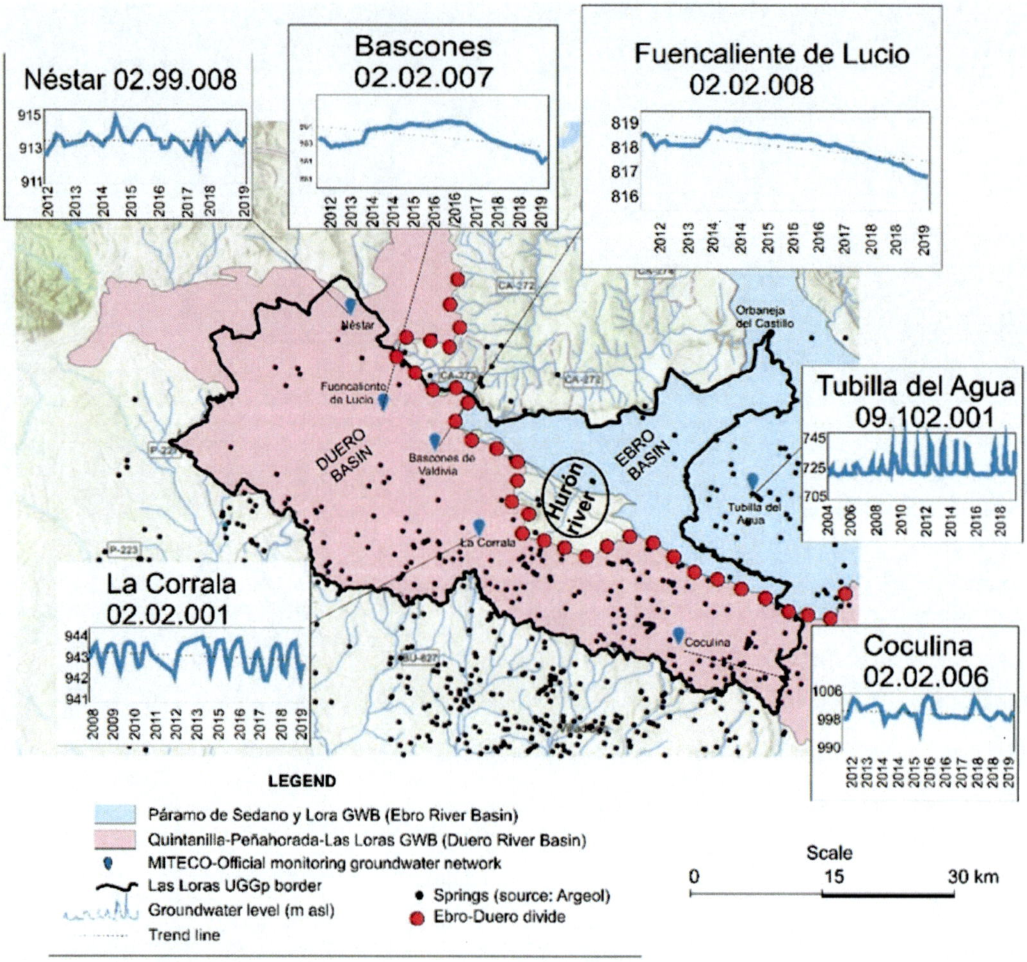


**Fig. 5** Groundwater bodies recognized in the Las Loras UGGp. Groundwater-level monitoring network indicated by their respective codes (data from MITECO) in the Las Loras UGGp. For Quintanilla-Peñahorada-Las Loras GWB, the piezometric levels show a gradual decrease (although they are tenths of a metre) since the beginning of the records with the exception of Néstar, whose evolution shows little variation. For Páramo de Sedano y Lora GWB, there is just one monitoring network that does not allow comparative process (CHE)

The available data on groundwater levels are reduced to eight MITECO observation wells located in well-differentiated sectors of Las Loras Geopark (Fig. 5).

The piezometric evolution diagrams (Fig. 5) show very different behaviours in the recording period, which varies between 15 (in the longest recording, #6) and 8 years (in most of the piezometers) depending on the piezometers. The 09.102.001 piezometer in Tubilla del Agua is the only piezometer that controls the Páramo de Sedano y Lora GWB within the Geopark.

No further information is available. However, such different records for the same period of time (2013–2020) lead to thinking that these are independent aquifers.

## 5 Sustainability of Water Resources in Future

The trend observed is a progressive increase in the demand for groundwater resources in the summer months. In this period of time, the population of the towns in the region triples and the increase in the number of visitors, together with the increase in population, makes it necessary to reinforce supplies. This trend will continue given that rural tourism is growing exponentially in the region.

The most important population nucleus is Aguilar de Campoo (6749 inhabitants in 2020; EpData). Regional economic development is stabilized. The region is characterized by its depopulation and the presence of a small number of companies with the exception of Gullon company. The main pressures that can pose a threat to groundwater resources are mainly slurry, certain livestock uses, and agriculture, in the case of the Quintañilla-Peñahorada-Las Loras GWB, and certain chemical pollutants (nitrogen) in the case of the Páramo de Sedano y Lora GWB, outside the limits of the Geopark.

## 6 Proposal for New Sites of Geological Interest with Hydrogeological Significance

In the light of the study carried out, the Sites of Geological Interest are well identified although they require a deeper hydrogeological understanding. One of the new potential Sites of Geological Interest is the Valdelateja Spa. It is the occurrence of a spring (average 1.8 l/s), which drains the calcarenites of the Cenomanense. It is considered thermal water with calcium bicarbonate facies, low mineralization and neutral pH.

## 7 Conclusions

Las Loras Geopark presents well-differentiated areas both from the geological point of view, including geomorphological, structural and hydrogeological aspects.

The review of the hydrogeological information and the analysis of the existing data suggest that within each of the two GWBs that are within the Geopark, sectors with different hydrogeological behaviour can be differentiated. These sectors would be defined by each one of the moors and Loras that could constitute individualized aquifers, with no connection between them.

Knowledge gap analysis requires a characterization and a detailed hydrogeological investigation of each of the individualized aquifers that allows a characterization of their hydrodynamic functioning and hydrochemical behaviour, as well as an analysis of the main threats to its good condition and state and possible management alternatives.

**Acknowledgements** This work is funded by the International Geosciences Program of UNESCO project IGCP-730 Hydrogeological significance of Mediterranean geoparks.

## References

CHE, Confederación Hidrográfica del Ebro. https://chebro.es. Accessed 12 Nov 2021

De la Hera-Portillo A, Llorente-Isidro M, Moreno L, Sánchez JA, Salman K, Corral MM, Galindo E, López J, Ghanem M, Hind Mohammed A, Chulli B, Laftouhi NE, Abdel-Gawad FKh (2022) Geoheritage and geoconservation in the Las Loras UNESCO Global Geopark. In: Pereira P, Monge-Ganuzas M, Bollati IM, Rouget I (eds.) ProGEO SW Europe regional working group virtual conference on geoconservation abstracts book, 2022. pp 30–31, March: 14–19

EPDATA. URL: https://www.epdata.es/datos/poblacion-inmigrantes-emigrantes-otros-datos-habitantes-cada-municipio/3/aguilar-campoo/444

IGCP-730. Hydrogeological Significance of Mediterranean Geoparks. https://en.unesco.org/international-geoscience-programme/projects/730

IHP (2008). Orden ARM/2656/2008, de 10 de septiembre, por la que se aprueba la instrucción de planificación hidrológica. Boletín Oficial del Estado (BOE) de 22 Sep 2008, Nº 229, pp 38472–38582

Lopez Geta JA, Ballester Rodríguez A, Pinuaga Espejel JI, Bronchalo de la Vega M, Gutiérrez Peña S, Monasterio Pérez S, Cadenas Bernar P (1998) (a). Unidad Hidrogeológica 09.02 Sedano-Lora. (b) Unidad Hidrogeológica 02.02 Quintanilla-Peñahorada-Atapuerca. Diputación provincial de Burgos, Instituto Tecnológico Geominero de España

Martínez-Arnáiz M (2013). Loras y Paramera de la Lora en Burgos. El incierto horizonte del desarrollo rural en un espacio de montaña media en recesión demográfica. Facultad de filosofía y letras. Universidad de Valladolid. https://uvadoc.uva.es/handle/10324/4223.

MITECO. Sistema de Información de Redes de seguimiento del estado e información hidrológica. https://www.miteco.gob.es/es/cartografia-y-sig/visores/visores-agua.aspx

Pendás Fernández F, Menéndez Casares E (1998) MAGNA50 Hoja 134 (18–18) Polientes. IGME, Madrid

Pineda Velasco A (1990) MAGNA50 Hoja 166 (18–9) Villadiego. IGME, Madrid

Salman Monte K, Sánchez Fabián JA, García García F, Basconsillos Arce J, Rodríguez García A, Gallego Rojas N. (2015) Application dossier to be a UNESCO GLOBAL GEOPARK. Diputación de Palencia, Diputación de Burgos, Junta de Castilla y León, ADECO-Camino, Agrupación Comarcal de Desarrollo Montaña Palentina, Páramos y Valles Palentinos, Asociación Argeol. 50 páginas. http://geoparquelasloras.es/wp-content/uploads/2021/06/Application-Dossier-Las-Loras-Geopark-Project-Main-document-b.pdf

# Preliminary Hydrogeological Investigations for Sustainable Development in the Courel Mountains UNESCO Global Geopark (NW Spain)

D. Ballesteros, A. de la Hera-Portillo, M. Llorente, R. Vila, M. M. Corral Lledó, E. Galindo Rodríguez, J. López, L. Moreno, M. Menéndez, P. Caldevilla, M. Ghanem, A. Hind, B. Chulli, N. E. Laftouhi, F. K. Abdel-Gawad, and M. Alemparte


## Abstract

Karst aquifers and, especially, their caves and springs, are singular sites of the geoheritage due to their scientific, touristic and use (water resources) values. Besides, they are recognized as habitats of special protection by the Habitat Directive (92/43/CEE) and would be used for sustainable development in UNESCO Global Geoparks (UGGp). However, their management continues to be a challenge at the present day since karst aquifers and their associated springs and caves are vulnerable environments used by many actors and affected by land uses and changes in surface hydrology and groundwater. Our work aims to implement of a suitable management of aquifers and springs in the Courel Mountains UGGp based on hydrogeological studies. The studies include mainly the characterization of aquifers' functioning, the elaboration



D. Ballesteros (✉) · A. de la Hera-Portillo · M. Llorente · R. Vila · P. Caldevilla · M. Alemparte
Courel Mountains UNESCO Global Geoparks, Rúa do Courel 21, 27320 Quiroga, Spain
e-mail: dballesteros@ugr.es

A. de la Hera-Portillo
e-mail: a.delahera@igme.es

M. Llorente
e-mail: mllorente@igme.es

R. Vila
e-mail: concello.quiroga@eidolocal.es

P. Caldevilla
e-mail: pcald@unileon.es

M. Alemparte
e-mail: info@ribeirasacracourel.es

D. Ballesteros
Department of Geodynamics, University of Granada, Campus de Fuentenueva s/n, 18071 Granada, Spain

A. de la Hera-Portillo · M. Llorente · M. M.C. Lledó · E. G. Rodríguez · J. López · L. Moreno
Centro Nacional Instituto Geológico Y Minero de España (IGME-CSIC), Ríos Rosas 23, 28003 Madrid, Spain
e-mail: mm.corral@igme.es

E. G. Rodríguez
e-mail: e.galindo@igme.es

J. López
e-mail: j.lopezgu@igme.es

L. Moreno
e-mail: l.moreno@igme.es

R. Vila
Museo Xeolóxico de Quiroga. Rúa Do Courel 21, 27320 Quiroga, Spain

M. Menéndez
Centro Nacional Instituto Geológico Y Minero de España (CSIC), C/ Matemático Pedrayes, 25, 33005 Oviedo, Spain
e-mail: m.melendez@igme.es

P. Caldevilla
Escuela Superior Y Técnica de Ingenieros de Minas, Universidad de León, Campus de Vegazana s/n, 24071 León, Spain

M. Ghanem
Birzeit University, Ramallah, Palestine, Israel

A. Hind
Jordan University, Amman, Jordan
e-mail: alsharifahind@gmail.com

B. Chulli
University of Carthage, Water Research and Technologies Centre, Teknoparc, Carthage, Tunisia
e-mail: bchoulli@yahoo.fr

N. E. Laftouhi
Faculty of Sciences Semlalia of Marrakech, Earth Sciences Department Marocco, Cadi Ayyad University, Marrakesh, Marocco

F. K. Abdel-Gawad
Water Pollution Research Department Centre of Research and Applied Studies On Climate Change, NRC (National Research Centre Cairo), Cairo, Egypt
e-mail: fagrabdlgawad@gmail.com

M. Alemparte
Grupo de Desarrollo Rural Ribeira Sacra-Courel. Rúa Doctor López Lallana 6, 1ºD, 27340 Bóveda, Spain




B. Andreo et al. (eds.), *EuroKarst 2022, Málaga*, Advances in Karst Science,
https://doi.org/10.1007/978-3-031-16879-6_36

of the spring inventory, physicochemical analyses and geoconservation actions.

**Keywords**

Cave • UNESCO Global Geopark • Hydrogeology • Karst • Spring

## 1 Introduction and Objectives

Karst aquifers and, especially, their caves and springs, constitute frequently geosites related to habitats of special protection stated in the European Habitat Directive (92/43/CEE). Besides, karst caves and springs show a touristic interest, thus they would be addressed for sustainable development in UNESCO Global Geoparks (UGGp) (e.g., Han et al. 2018; Valente et al. 2021). However, karst cave and spring management for their sustainable use continues to be a challenge at the present day since they are vulnerable environments affected by global processes (e.g., global change), regional factors (e.g., changes in land use and surface hydrology and groundwater in the catchment area), and local drivers (e.g., spring capture or uses of caves by diverse actors as speleologists, tourism companies and scientists).

Courel Mountains UGGp is administered in northwest Spain (Fig. 1) by three municipalities (5 178 inhabitants in 2019) and the Ribeira Sacra-Courel Local Action Group (ES-212; LEADER Program of the European Union) to avoid the economic and demographic depletion of the territory by means of the promotion of the geotourism, science, education, and conservation. The cornerstone of the UGGp is the singular relationship between its noticeable geoheritage, the extraordinary biodiversity, and the abundant cultural patrimony (Ballesteros et al. 2021). In this sense, our work aims to implement a suitable management of watercourses, springs and caves in the sustainable development of the Courel Mountains UGGp in cooperation with the *CN Instituto Geológico y Minero de España* (CN IGME, CSIC). For that, we identified the groundwater bodies, inventoried the springs, and established the water chemical facies.

## 2 Study Area

The Courel Mountains UGGp (578 km$^2$ in extension) was declared in a mountain area (up to 1641 m asl) in 2019 (Fig. 1b). The climate is humid with cool-dry summers; Csb climatic type according to the Köppen-Geiger categorization. (Cunha et al. 2011). The UGGp exhibits Atlantic influence to the north, with 1200 to > 2000 mm of annual precipitation, and Mediterranean conditions to the south, with valleys receiving ca. 800 mm rainfall per year (Fig. 1b). Extreme temperature ranges often from −10 to 40ºC, and snow is usually present in high areas from November to April.

Courel Mountains resulted mainly from the Variscan and Alpine geological cycles (Ballesteros et al. 2022). The bedrock comprises mainly Paleozoic rocks affected by large recumbent folds and low-grade metamorphism (Martínez Catalán et al. 1992), and the landscape is dominated by fluvial incision and slope processes, as well as local karstification and old periglacial and glacial activity (see Pérez Alberti 2018). The UGGp is located within the *Cuenca del Sil* groundwater body (GWB) (011.003), which presents point sources, diffuse source pollution, and abstraction pressures, although it is recognized as having good status according to the Water Framework Directive requirements (2000/60/CE). Its annual water resources are estimated in ca. 5000 MMC according to *Confederación Hidrográfica del Miño-Sil* (Miño-Sil River Basin Authority). The UGGp is drained by the Sil River and its tributary watercourses: Bibei and Lor rivers, and Quiroga, Soldón, Parteme and Castelo streams (Fig. 1b). Selmo stream also enters the Sil River upstream of the UGGp.

## 3 Methodology

The implementation of hydrogeological research in the management of the Courel Mountains UGGp includes (1) the identification of groundwater bodies, (2) the elaboration of the spring inventory (which discharge is higher than 1 l·s$^{-1}$), reporting also the water use, and (3) chemical analysis of principal watercourses and singular springs. In detail, physicochemical analysis was performed in 24 water samples carried out in September 2021. The analysis provided the pH, electrical conductivity, oxidability, and the anions and cation concentrations via ion chromatography (IGME laboratory), allowing the first identification of the dominant hydrochemical facies and a preliminary environmental evaluation of terrestrial and aquatic related ecosystems. All hydrogeological results were integrated into the geographic information system (GIS) database of the UGGp, allowing us to establish recommendations for management.

## 4 Groundwater and Water Use

Courel Mountains UGGp is formed by four main types of aquifers depicted in Fig. 2:

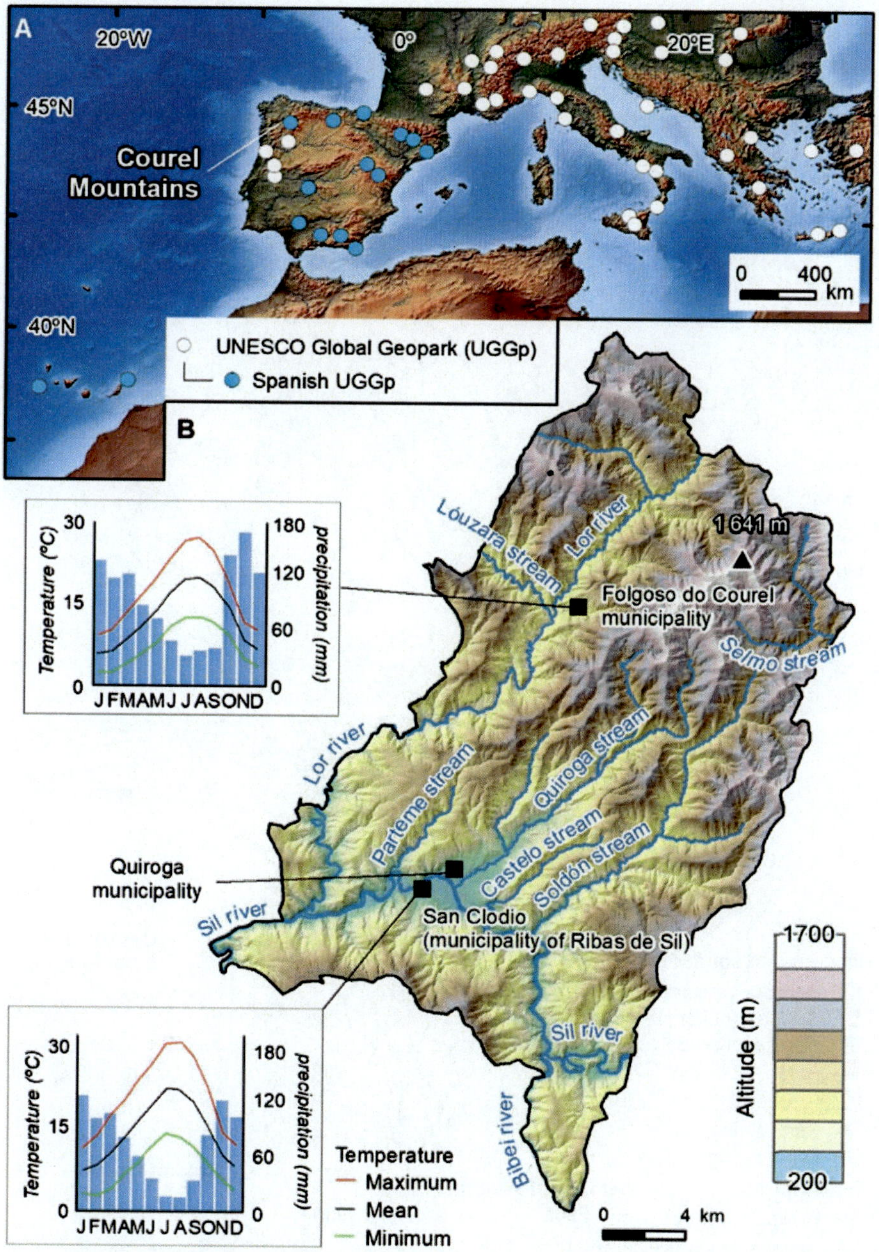


**Fig. 1** **a** Location of the Courel Mountains UGGp in SW Europe. **b** Relief and main watercourses of Courel Mountains. Climographs were carried out using raw data (2000–2020) from *Meteogalicia* (meteorology agency of Galicia; www.meteogalicia.gal)

(1) bedrock aquitards (81% of the UGGp extension) within Paleozoic siliceous metamorphic rocks of low permeability by local fracturing;

(2) karst aquifers (5% of the UGGp) formed by Cambrian to Devonian meta-limestone and marble that crop out in northern UGGp; these aquifers host 8 km of karst conduits documented by speleological groups and are recharged by the infiltration of rain and snow melting water, as well as a local losing stream (Fig. 2); the discharge occurs via springs or directly to the watercourses;

(3) surface porous aquifers (12%) represented mainly by slope deposits (e.g., talus deposits) over the UGGp, alluvial deposits mapped in the south, and periglacial

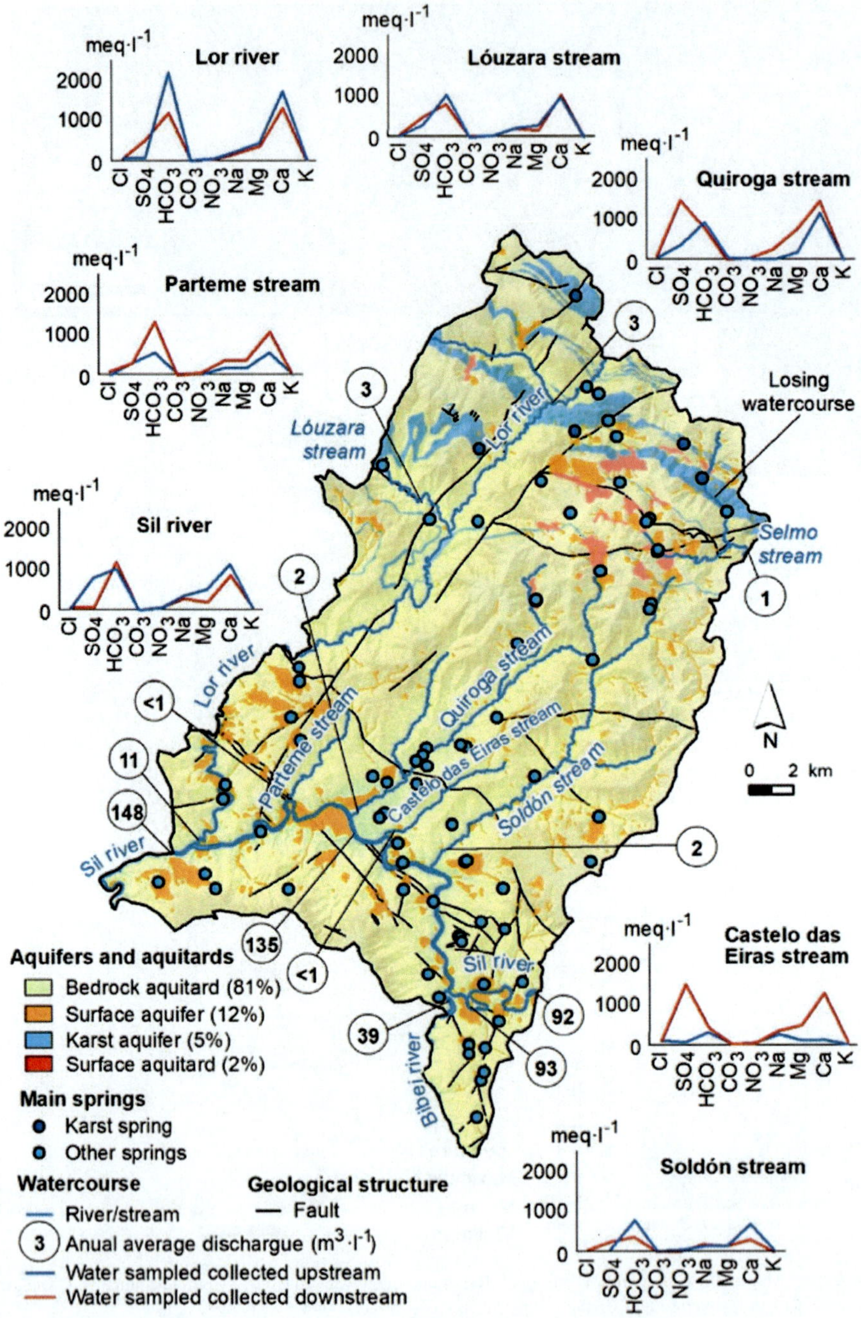


**Fig. 2** Hydrogeological map of the Courel Mountains UGGp showing the chemical analysis of water collected upstream and downstream of the main watercourses. The annual average discharge (1980–2012) of these watercourses is from the water plan 2016–2021 of *Confederación Hidrográfica del Miño-Sil*; stated by the real decretory 1/2016 of Spain)

deposits (e.g., block river deposits, talus rampant) in the north; surface aquifers are recharged mainly by rainwater infiltration; among them, alluvial aquifers supply primarily the influent Sil River while other surface aquifers are drained by springs with a variable flow (0–10 $l \cdot s^{-1}$);

(4) surface aquitards (2%), represented mostly by glacial sediments and other clayey deposits.

The hydrogeological map includes 79 springs: 5 karst springs (up to 20 $l \cdot s^{-1}$ flow) only used for irrigation and other 74 springs linked to surface porous aquifers and locally fractured bedrock aquitards. Springs supply fresh water for rural communities (e.g., irrigation, drinking) albeit minor streams related to slope deposits were captured to provide water to the main villages in the S.

## 5 Water Physicochemical Analysis

The goal of this hydrogeological characterization is also addressed to establish an environmental monitoring network of the environmental conditions of the UGGp for the future and in the long-term, to study its resilience to global change.

The samples were carried out in two reservoirs, 4 springs, 2 caves and 7 rivers and streams named Sil, Bibei, Lor, Parteme, Castelo das Eiras, Soldón, Quiroga and Lóuzara (Fig. 2). The main river basins have been sampled upstream and downstream in order to compare its changes along the course of the river.

(a) 50% of the analyzed samples show a Ca bicarbonate facies (Fig. 3), and the remaining 50% are mostly Ca bicarbonate-sulfate water, although in some cases Na, or with the presence of Ca and Na as dominant ions. The samples that present a sulfate or Ca sulfate facies are: Lóuzara stream; Cervo 1 and springs; Quiroga stream; Soldón stream, San Martiño Reservoir and Parteme stream that present a dominant sulfate facies (Quiroga stream and Fonte do Cervo 1 spring) or very abundant (Lóuzara, Soldón and Parteme streams, San Martino Reservoir and Fonte do Cervo 2 spring).

(b) The samples taken in karst cave springs (Vellos and A Ceza caves) correspond to a Ca bicarbonate facies with similar ionic composition (Fig. 3).

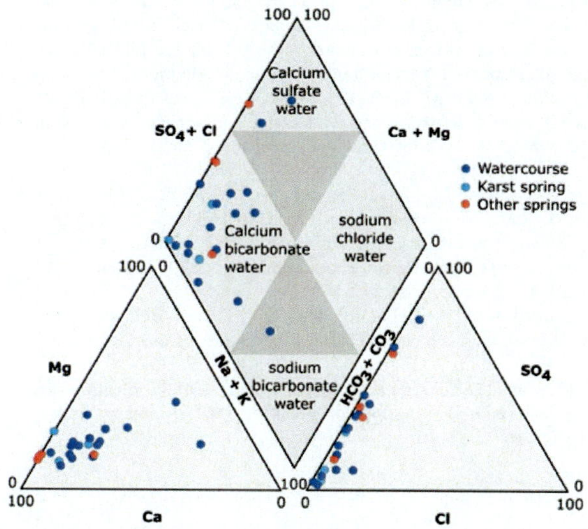


**Fig. 3** Piper diagram of watercourses and springs of the Courel Mountains UNESCO Global Geopark (September 2021) showing a Ca bicarbonate water as the main chemical facies

The comparative analysis of the river sub-basins present in the UGGp has been approached by representing the river channels in independent graphs, but with the same vertical scale and with the same legend (Fig. 2). The comparison between samples taken upstream and downstream of the watercourse points out the following conclusions:

- Sil River: upstream it presents bicarbonate-sulfate Ca water, and downstream it is definitively Ca sulfate with a notable reduction in salinity.
- Lor River: the concentration of bicarbonates decreases and sulfates increase downstream; the water is less saline downstream.
- Parteme stream: the concentration of bicarbonates and Ca increases downstream; the increase in salts is reflected in an increment in electric conductivity.
- Castelo das Eiras stream: it presents a noticeable increase in the sulfate concentrations downstream, as well as in Ca abundance; the increase in salinity is an order of magnitude.
- Soldón stream: interestingly, the water reduces salinity downstream, becoming sweeter; this depletion resulted mainly from the concentrations of $HCO_3^-$ and Ca.
- Quiroga stream: its water is Ca bicarbonate upstream, showing a change in hydrochemical facies downstream, becoming Ca sulfate water with an increase in salinity due to the increment in these two ions.
- Lóuzara stream: it experiences a slight reduction in salinity downstream. Its water is Ca bicarbonate with a slight increase in sulfates downstream.

The preliminary comparative analysis between sub-basins shows that the Lor River exhibits the highest salinity, and in decreasing sequence in the Sil river and Quiroga, Castelo das Eiras, Parteme, Lóuzara and Soldón streams; the latter has the most freshwater.

The analysis of the springs (Rubia, Cervo 1 and 2 springs, O Incio spa) reveals that:

- The Cervo 1 spring presents sulfate-Ca bicarbonate water.
- The Rubia and Cervo 2 springs show a bicarbonate-sulfate facies, and O Incio spa exhibits a Ca bicarbonate facies without significant sulfates.

- Among the studied, Cervo 2 spring is the less mineralized, and the water from Rubia spring and O Incio spa depicts the highest salinity.

The available water analyses (Fig. 3) show that the composition of the sampled waters seems to correspond to two different genetic processes: on the one hand, the dissolution of carbonates that is responsible for the carbonated facies and on the other hand the oxidation of sulfides that is responsible for the sulfate facies. In general, karst springs respond to the first group and other sources as well as some streams to the second. The mixing processes determine how the samples are placed in the Piper diagram, where an almost continuous distribution of the facies is observed between the calcium sulfated and calcium bicarbonate ends, especially in the anionic facies.

## 6 Management of Hydrogeological Systems

Preliminary studies allowed to provide recommendations for the management body of the Courel Mountains UGGp, which stated five geosites of hydrogeological main interest for adding value to four springs and an influent watercourse section. The hydrogeological geosites represent 7% of the totally geosites of the UGGp. In general, the apparent relative adequate conservation of the surface hydrology and groundwater suggests the conduction of surveillance efforts, as well as further studies to increase the hydrogeological knowledge of Courel Mountains. The surveillance reported the occurrence of solid urban waste in streams and karst caves. The UGGp ran four cleaning actions involving the regional Government, local communities and volunteers in 2021, including teenagers for removing residues from a losing stream.

## 7 Conclusions

The Courel Mountains UGGp promoted investigations in cooperation with the CN IGME, CSIC to implement the hydrogeological heritage into sustainable development. Investigations provide a basic hydrogeological characterization of the UGGp by means of the identification of four main hydrogeological units, the inventory of springs and their uses, and the determination of the water chemical facies. The studied main water bodies present a Ca bicarbonated and Ca sulfate hydrochemical facies. The sampled rivers show the same hydrochemical facies either upstream or downstream with the exception of the Quiroga and Castelo das Eiras streams, which will be the objective of further research in the future. On the whole, preliminary investigations enabled the definition of geosites of hydrogeological interest, advised the in-depth development of hydrogeological research and suggested the surveillance of rivers, springs, and caves, reporting the occurrence of solid urban waste that were cleaned in two stream sections and two karst caves.

**Acknowledgements** This work is funded by the Courel Mountains UGGp and the project IGCP-730 Hydrogeological significance of Mediterranean geoparks, and supported by the International Geosciences Program of UNESCO. We thank local people and volunteers for their collaboration in the hydrological research and geoconservation actions. We acknowledge Prof. G. Veni for his constructive review.

## References

Ballesteros D, Caldevilla P, Vila R, Barros XC, Rodríguez-Rodríguez L, García-Ávila M, Sahuquillo E, Llorente M, Diez JB, Fuertes-Fuente M, Timón-Sánchez SM, de Lombera-Hermida A, Álvarez I, Pérez-Cáceres I, Acebo M, Orche Amaré P, García JH, Martín-González F, Alemparte M (2022) A GIS-supported multidisciplinary database for the management of UNESCO Global Geoparks: the Courel Mountains Geopark (Spain). Geoheritage

Ballesteros D, Caldevilla P, Vila R, Carlos X, Martín B (2021) Linking geoheritage and traditional architecture for mitigating depopulation in rural areas: the Palaeozoic Villages Route (Courel Mountains UNESCO Global Geopark, Spain). Geoheritage, pp 1–17

Cunha S, Silva A, Flores Herráez C, Pires V, Chazarra A, Mestre Barceló A, Nunes L, Mendes M, Neto J, Marques J, Mendes L (2011) Atlas Climático Ibérico-Iberian Climate Atlas. Agencia Estatal de Meteorología, Instituto de Meteorologia de Portugal, Madrid

Han J, Wu F, Tian M, Li W (2018) From Geopark to Sustainable Development: Heritage Conservation and Geotourism Promotion in the Huangshan UNESCO Global Geopark (China). Geoheritage 10:79–91

Martínez Catalán JR, Hacar Rodríguez MP, Villar Alonso P, Pérez-Estaún A, González Lodeiro F (1992) Lower Paleozoic extensional tectonics in the limit between the West Asturian-Leonese and Central Iberian Zones of the Variscan Fold-Belt in NW Spain. Geol Rundsch 81:545–560

Pérez Alberti A (2018) Geomorphology of O Courel. Bóveda: Grupo de Grupo Desenvolvemento Rural Ribeira Sacra-Courel

Valente E, Casaburi A, Finizio M, Papaleo L, Sorrentino A, Santangelo N (2021) Defining the geotourism potential of the cilento, vallo di diano and alburni unesco global geopark (Southern italy). Geosciences 11:1–20